DESIGNING COMMUNITIES

Science & Technology Education Library

VOLUME 3

SCOPE

The book series *Science & Technology Education Library* provides a publication forum for scholarship in science education. It aims to publish innovative books which are at the forefront of the field. Monographs as well as collections of papers will be published.

Designing Communities

by

WOLFF-MICHAEL ROTH

Lansdowne Chair, Applied Cognitive Science and Science Education
Faculty of Education, University of Victoria
Victoria, BC, Canada V8W 3N4

KLUWER ACADEMIC PUBLISHERS
DORDRECHT / BOSTON / LONDON

A C.I.P. Catalogue record for this book is available from the Library of Congress.

ISBN: 0-7923-4703-X
ISBN: 0-7923-4704-8

Published by Kluwer Academic Publishers,
P.O. Box 17, 3300 AA Dordrecht, The Netherlands.

Kluwer Academic Publishers incorporates
the publishing programmes of
D. Reidel, Dr W. Junk and MTP Press.

Sold and distributed in the U.S.A. and Canada
by Kluwer Academic Publishers,
101 Philip Drive, Norwell, MA 02061, U.S.A.

In all other countries, sold and distributed
by Kluwer Academic Publishers,
P.O. Box 322, 3300 AA Dordrecht, The Netherlands.

Printed on acid-free paper

Printed in the Netherlands.

To Sylvie and Margot

TABLE OF CONTENTS

Preface   xiii
  Learning in Moussac   xiii
  Overview of this Book   xvii

Acknowledgments   xxiii

PART I   1
FOUNDATIONS   1

1. Theoretical Foundations   3
  1.1. Practices and Resources   3
    1.1.1. Discursive Practices   4
    1.1.2. Language Games Applied   6
      1.1.2.1. Designing Workplaces   6
      1.1.2.2. Designing and Studying Science Classrooms   8
    1.1.3. Implications of an Epistemology of Practice   9
  1.2. Communities of Practice   10
    1.2.1. Learning as Participation   11
    1.2.2. Authenticity of Practices   13
    1.2.3. Actor Networks   14
  1.3. Design and Designing   16
    1.3.1. Designing as Professional Activity   18
    1.3.2. Design as Learning Context   18

2. Empirical Foundations   21
  2.1. Institutional Context   21
    2.1.1. Parent Community   21
    2.1.2. Teacher Community   22
    2.1.3. Problems and Constraints   23
    2.1.4. Improving Science Teaching   24
  2.2. Study Context   25
    2.2.1. Participants   25
      2.2.2.1. Teachers   25
      2.2.1.2. Students   27
    2.2.2. Data Construction   29
      2.2.2.1. Becoming Part of the Culture   30
      2.2.2.2. Assessment of Knowing and Learning   32

2.2.2.3. Interaction of Data Collection and the
            Emerging Curriculum   35
2.2.3. Data Interpretation   36
       2.2.3.1. Constructing the Evidence   36
       2.2.3.2. Credibility of Interpretations   38

3. Engineering for Children Curriculum   41
   3.1. Intended Curriculum   41
        3.1.1. Conceptual Considerations   41
        3.1.2. Teachers Goals   42
        3.1.3. Overview of the Activities   43
   3.2. Engineering Design Activities   43
        3.2.1. Tools and Materials   43
               3.2.1.1. Handyman Tools and Materials   43
               3.2.1.2. Engineering Log Book   44
               3.2.1.3. Engineering Techniques Board   44
        3.2.2. Preparations for Engineering Design   45
               3.2.2.1. Strengthening Structures   45
               3.2.2.2. Stabilizing Structures   46
               3.2.2.3. Designing a Creature   46
        3.2.3. Design Challenges   47
               3.2.3.1. Designing Towers   47
               3.2.3.2. Designing Bridges   50
               3.2.3.3. Designing Domes   50
   3.3. Teaching Strategies   51
        3.3.1. Creating an Engineering Language   51
               3.3.1.1. Building on Existing Language   51
               3.3.1.2. Creating New Language   52
               3.3.1.3. Linking to Canonical Language   52
               3.3.1.4. Reflecting On Action   52
               3.3.1.5. Encouraging Emergent Design   53
        3.3.2. Questioning Techniques   54
               3.3.2.1. Context of Questioning   54
               3.3.2.2. Content of Questions   55
               3.3.2.3. Responses and Reactions to Questions 56
               3.3.2.4. Gittes Questioning of Students in the
                        Context of Their Work   56
        3.3.3. Creating a Community   58
               3.3.3.1. Mediating Trouble in Collaborations 58
               3.3.3.2. "Sharing"   60
               3.3.3.3. Teachers as Learning Members of the
                        Community   61
   3.4. Teacher Learning   62
        3.4.1. Opportunities for Growth   63

3.4.2. Continuing Struggles   64

4. Knowing Engineering Design   66
  4.1. Engineering Design Prior to "Engineering for Children: Structures"   67
    4.1.1. Associations with and Talk about Engineering   68
    4.1.2. Pre-Unit Engineering Challenges   73
  4.2. Post-Unit Assessment of Engineering Design Practices   74
    4.2.1. Classification of Engineering Design Knowledge   76
    4.2.2. Engineering Design Language   80
    4.2.3. Associating Engineering Design   81
    4.2.4. Writing Engineering Design   85
    4.2.5. Talking Engineering Design   90
    4.2.6. Coping with Complexity and Interpretive Flexibility   94
    4.2.7. Knowing to Negotiate Plans and Courses of Action   95

PART II
TRANSFORMATIONS OF A COMMUNITY: THE
EMERGENCE OF SHARED RESOURCES AND PRACTICES   99

5. Circulating Resources   101
  5.1. Case Studies of Resource Networking   104
    5.1.1. Case Study 1: The Canadian Flag   104
    5.1.2. Case Study 2: The Thimble   112
  5.2. Inventors, Copy-Cats, and Everyone Else   116
    5.2.1. Insiders   119
    5.2.2. Outsiders and Marginals   121
      5.2.2.1. Outsiders   121
      5.2.2.2. Marginals   124
    5.2.3. Copying a Resource   126

6. Circulating Material Practices   129
  6.1. Technology, Society, and Knowledge   129
  6.2. Socio-Technical Evolution: The Case of the Glue Gun   130
    6.2.1. Brief History of Events   130
    6.2.2. Limited Resources   131
    6.2.3. Changing Practices   133
    6.2.4. Changing Settings   136
    6.2.5. Circulation of Practices   139
      6.2.5.1. Unsuccessful Circulation of a Practice   145
  6.3. Cultural Production and Reproduction in a Community of Practice   147
    6.3.1. Embodiment   147

6.3.2. Evolving Networks of Practice   149

7. Emergence and Circulation of Discourse Practices   154
7.1. Trajectories of Competence   154
    7.1.2. Snapshot of an Evolving Community of Practice   155
    7.1.3. A Trajectory of Competence in Triangular Bracing 159
        7.1.3.1. A Traditional Lesson about Triangles   159
        7.1.3.2. Significant Teacher Scaffolding   163
        7.1.3.3. Contingent Emergence of Triangles   165
        7.1.3.4. Competent Practice   168
    7.1.4. Actor Network Approach to Changing Discourse
           Practices   170
7.2. Learning to Tell Engineering Design Stories   173
7.3. Engineering Design Conversations   182
    7.3.1. Presenting the Artifact   183
    7.3.2. Extending Language Games   184
    7.3.3. Using Artifacts as Conversational Anchors   186
    7.3.4. Integrating Personal Experiences, Classroom
           Discourse, and Formal Engineering   187
    7.3.5. Sustaining Student-Centered Discussions   189
7.4. Teachers as Network Builders   193

PART III
NETWORKING ACROSS INTERSTICES   197

8. Networking Humans and Non-Humans   199
8.1. Heterogeneous Design Processes and Design Products   200
    8.1.1. Design History of an Earthquake-Proof Tower   201
    8.1.2. Material Basis of Designing and Design   212
        8.1.2.1. Networking Tools   212
        8.1.2.2. Networking Materials   214
        8.1.2.3. Networking the Current Artifact   215
    8.1.3. Social and Psychological Basis of Designing and
           Design   216
        8.1.3.1. Networking Individuals   216
        8.1.3.2. Networking the Embedding Culture   216
        8.1.3.3. Networking Teachers   217
8.2. Ontology of Resources   218
    8.2.1. Interpretive Flexibility of Plans and Artifacts 218
    8.2.2. Ontology of Rules   221
8.3. Artifacts as Structuring Resources in Interaction   225
    8.3.1. Inextricability of Thinking and Acting   226
    8.3.2. How Artifacts Constrain Interpretive Flexibility 228
8.4. Toward a New Conception of Problem Solving   233

      8.4.1. Case Studies of Problem Solving   233
            8.4.1.1. Flexible Constitution of Problems   237
            8.4.1.2. Ontology of Problems and Solutions   238
      8.4.2. Micro-, Meso-, and Macro-Problems   241
      8.4.3. Negotiating Problems and Solutions   243
   8.5. Designing as Context for Learning   246

9. Networking Individuals and Groups   250
   9.1. Networking Within and Across Groups   254
   9.2. Case Studies of Networking   259
      9.2.1. Networking Within Groups   259
      9.2.2. Networking between Groups   266
      9.2.3. Teachers' Contribution to Network Construction 268
            9.2.3.1. Instituting Constraints   268
            9.2.3.2. Scaffolding the Construction of Accounts of
                    Collective Activity   270
   9.3. Networking and the Emergence of Culture, Power, and
   Norms   271

PART IV
CONCLUSIONS   277

10. Designing Knowledge-Building Communities   279
   10.1. Designing for the Circulation of Resources and Practices 281
   10.2. Artifacts and the Networking of Communities   283
   10.3. Designing and Assessing Collective Learning Experiences 285
   10.4. Designing for Authentic Problem Solving   286
   10.5. From Research to Practice: Curriculum on Simple
   Machines   288
      10.5.1. Whole-Class Conversations around Teacher-
            Designed Artifacts   289
      10.5.2. Small-Group Conversations around Teacher-
            Designed Artifacts   291
      10.5.3. Small-Group Conversations around Student-
            Designed Artifacts   291
      10.5.4. Whole-Class Conversations around Student-
            Designed Artifacts   291
      10.5.5. Making it Work   292

11. Epilogue   294
   11.1. Participating is Learning   294
   11.2. Networking Teachers – Learning to Teach Science by
   Participating in the Practice of Science Teaching   297

11.3. Reflexive Coda   302

References   304

Index   312

PREFACE

The study described in this book arose in the context of a three-year collective effort to bring about change in science teaching at Mountain Elementary School.[1] This opportunity emerged after I contacted the school with the idea to help teachers implement student-centered science teaching. At the same time, the teachers collectively had come to realize that their science teaching was not as exciting to children as it could be. They had recognized their own teaching as textbook-based with little use of the "hands-on" approaches prescribed by the provincial curriculum. At this point, the teachers and I decided that a joint project would serve our mutual goals: they wanted assistance in changing from textbook-based approaches to student-centered activities; I wanted to collect data on learning in student-centered knowledge-producing classroom communities.

I brought to this school my new understandings about classroom communities from several earlier studies conducted in a private high school (e.g., Roth & Bowen, 1995; Roth & Roychoudhury, 1992). I wanted to help teachers create science learning environments in which children took charge of their learning, where children learned from more competent others by participating with them in ongoing activities, and teachers were responsible for setting up and maintaining a classroom community rather than for disseminating information. After I had completed the data collection for the present study, I watched a documentary about an elementary school in the small French village of Moussac (Envoyé Spécial, TV5, September 14, 1994). It epitomized some of the central issues in this book. I provide the following description of this broadcast to attune readers to my central concerns for doing the research and writing the book. My description includes translated excerpts from interviews with the teacher in Moussac.

LEARNING IN MOUSSAC

It is morning in the French village of Moussac. Through the viewpoint of a camera, we see children walk towards their one-room elementary school as they explain that they come to school when they want to. In fact, their teacher Bernard does not want all of them to come at the same time. Once at school, the children show the visitors around. There are computers where children publish their own newspaper and write letters to pen pals all over the world. The children explain the function of the music room including home-made

---

[1] Throughout the book, I use pseudonyms for school, students, and teacher (though the teachers' names are provided in the Acknowledgements).

percussion instruments and a tape recorder, the discussion room with its long table, the classroom for the little ones (about K-3) and that for the older ones (about Grades 4–6). Then there is their daily "curriculum" which children establish themselves through collective efforts.

Journalist:	*So, if I understand well, the children in this class do what they want?*
Teacher:	*Not entirely, not entirely. It's not what they want, but what the collective activity of the class, what the events globally bring about. So what I try is to order these events somewhat.*
Journalist:	*Because I get the impression that they, by and large, get by without you?*
Teacher:	*Yes, they do well without me. My own problem is to make this group function as a community.*

Throughout the school day, one can see five-year-olds mounting a puppet show or gathering for one of their collective meetings with Jean, their current, but weekly-changing chair person. In these discussions, Bernard, the teacher, is but one of the members, who waits for his speaking turn as any other member of the community. There is the chess lady who comes every Wednesday. Bernard emphasizes that her presence is more to provide for interactions with the children than for learning the rules of chess. A couple of older kids bring their letters to the post office where the official indicates that the class has become part of normal village life. In the shop, children work on their own projects, Bernard watches the youngest ones, but without interfering. He explains that it is only when people get nervous about failure and accidents that failure and accidents come about. There is also a parent who comes to garden with the children; there are some children in the music room, experimenting with the percussion instruments; a boy who sits in the corner by himself and listens to music with his headphones; in another corner, a boy writes a letter while others read with a group of smaller children.

Journalist:	*Don't you have this fear weighing over you, "I have to teach them to read, teach them to write, and teach them to count"?*
Teacher:	*No, I don't have this fear at all, absolutely not at all. For slowly I learned that when the children are part of a group that really exists as community, when there is a real setting, when the interactions with this setting, with other children and adults, when this context really exists, at that point, all children without exception learn to read. How? Now, this is another thing. But this isn't really my problem. The gardener's problem is not really how plants grow tall, strong, and well. His problem is to put them at their right spot, to plant them somewhere, to recognize that this might not be the right place, and to replant them somewhere else, and so on.*

And I have this gardener's job. And when the garden, the community is like a garden, when it really works, children inevitably learn to read. It's like their teeth, they will inevitably come. But although the children's mothers know about it, they still worry. So here, it's about the same thing. One has the responsibility of not preventing children from learning how to read.

In the end, Bernard talked about failure that students experience in schools other than his own. In his view, it is not the children who fail, but teachers, schools, and the system who fail them. Learning is natural, emerges from participation in the collective activities of a community; learning and participation cannot be separated, they are irremediably bound up with each other.

This documentary about the school, children, teacher, and community of Moussac contains some elements that have become central to my thinking about learning in schools. What we see here is a school where the transition to out-of-school life in the community is much more transparent than in most other schools. School has become part of village life. Even within school, life operates as an open community in which members pursue activities at different levels of competencies, the teacher and other adults are only part of the community. The fact that children of different ages work together in the same open classroom contributes to the distribution of competence similar to the distribution one can observe in the village itself. These adults do not "teach" as such, that is inculcate students with pieces of information that they need to memorize in order to succeed. Rather, the teacher and the entire village of Moussac allow the children to participate in daily activities, that is, to develop into increasingly competent members of the community at large by actively participating from young ages.

In this village, children learn as part of engaging in real activities, activities that are meaningful to them, over which they feel ownership, and that they plan on their own. They publish a newspaper, they communicate with peers in other parts of France and the world, they keep journals to write about the contents of their reveries and reflect on their learning. Younger children learn as they participate in legitimate peripheral ways in the activities of older children. There are meetings of the collectivity, times when older students read to and with younger children; at other times, adults from the community interact and make available their competence as resources to the members of the school community.

An important aspect of the community in the school of Moussac is that children have a sense of self-determination and control over their activities, and with it, over their own learning. But the activities of individual children are not independent of each other, not an odd collection of individualist activities disconnected from those in which other children engage. Rather, the activities of individual children are in part determined by the "sum total" of the collective activity. Individual interests and those of the community influence each other in a reciprocal way: they are mutually constitutive. Coming to school at their own time, going to the post office, or determining their daily curriculum are but a few of the outward signs of the children's ownership over their learning contexts. In such a community, teachers' activities change from those they traditionally performed. Here, teachers have to be centrally concerned with setting up the community, keeping it going, and not interfering with children's propensity for learning. They do not have to disseminate information, but only help order emerging collective activity. Bernard's comment that children learn inevitably when community really exists, resonates with Lave's (1993) assertion that we do not need to force children to learn. Learning occurs inevitably, as part of our being-in-the-world and participating in collective activities with everyone else.

When I talk to teachers about this documentary, I inevitably hear lists of reasons why this is an idyllic situation and cannot be transported elsewhere. However, I have made observations at a private school where I taught for a three-year period that show some strong resemblance with the scenes from the school in Moussac. The physics laboratory became an open place where people engaged in learning and did activities at their time and of their interest. There was a community in which older students helped younger ones in grappling with intractable problems in their physics experiments. Mathematics teachers came to do their work on the available computers. (When all computers were used, teachers did something else first, and came back later rather than enacting their "rights".) Other science teachers came to work with students on experiments, to observe and participate. But in this physics laboratory, it was not just the physics students who learned, and it was not just physics that was learned. Students came to do mathematics projects, write their religion and geography essays, produce graphs for their economy class. Again, there were interactions between students at various levels of competencies, interacting with each other and myself (sitting in an adjoining office with the door slightly open, doing my own work). Like the teacher of Moussac, I saw my job as keeping the community going, creating an environment in which members learn, inevitably and because they want.

What do children learn when they work in such an environment? How do community (teachers and peers) and the physical setting (materials, available tools, etc.) shape the artifacts children design? How does the distribution of knowledge in the community change over time? In other words, how do children learn from each other? What is the function of individual groups in the maintenance of the collectivity? These and similar questions drove the research program from which this book emerged. The classroom I observed and in which I participated over a four-month period had some striking similarities with that of the village of Moussac, notwithstanding the undeniable differences which are described later in this book. The most important similarity was that there existed a sense of community in which collective learning was celebrated over individual prowess. Interacting with others, co-participating in collective activities, and getting along with others were some of the things the teachers had brought about and maintained.

OVERVIEW OF THIS BOOK

This book is centrally concerned with knowing and learning in communities, including the roles of setting, tools, artifacts, materials, and community members (who engage in activities with varying degrees of competence across a variety of practices). My ultimate goal, to which the current book contributes, is to construct learning environments in which students have significant opportunities to take charge of their own learning; to construct learning environments that are fundamentally oriented toward democratic ideals – independent of the age of the learner – rather than the preparation of "obedient bodies" (Foucault, 1975) who become fodder for factories and exploitation.

The research presented in this book falls into a category of research labeled "design experiments" (Brown, 1992). As a design scientist, I construct innovative learning environments and simultaneously conduct research on teaching and learning therein. My emerging understandings are fed back into the same classrooms to bring about, and amplify positive conditions of learning. Because of the closeness to the classroom, design experiments constitute research efforts that are not only suitable to generate theory from practice, but – because of the thick descriptions they can provide – inform practice and practitioners in meaningful ways. That is, because design experiments construct theory in practice, they can lead to interventions that work by recognizable standards and are reliable and repeatable in the same and different settings (Brown, 1992).

My initial observational and theoretical descriptions of student and teacher learning in this knowledge-building classroom took different frames suited to the audiences of the journals in which they appeared (Roth, 1995c, d; 1996a–e; 1997a). For this book, I chose one theoretical frame, embodied in the analogy of network and networking, and predominantly draw on data not previously published. Interested readers are encouraged to consult the journal articles for additional and complementary data, for even a book of this size only allows the presentation of a small set of the original data sources.

This book is divided into four parts. In Part I, I lay the theoretical and empirical foundations for the studies of knowing and learning reported in Parts II and III, and provide descriptions of the curriculum and learning outcomes that I observed. Chapter 1 develops the theoretical discourse that guided my observations and the observational descriptions resulting from the study. Four notions are central to the project:

- Knowledge is conceived in terms of *material* and *discursive practices* and *resources* which constitute the knowledge in communities.
- The locus of knowing is the *community*, a collective characterized by

specific practices and resources that are common to its members, but different from those in other communities.

- *Discursive practices* (or, synonymously, language games) are the most important practices in a community because they produce and mediate communication, embed all other practices, and constitute the members' world.
- Communities and the activities of their members are effectively modeled by actor network theory, a way of understanding the world as a network of human and non-human actors.

Chapter 2 provides a description of the study's setting, the institutional context and teachers' activities for improving science teaching. Vignettes are included that describe a few characteristic children who feature more frequently in my subsequent accounts of knowing and learning. I provide details of the data collection and the – for design or teaching experiments, characteristic – interaction between understandings generated by the research and subsequent teaching episodes. Finally, an account is provided of the emergent character of the understandings that are at the heart of this book. Chapter 3 provides a description of the curriculum, "Engineering for Children: Structures" (APASE, 1991), as it evolved from the experiences of the participants. My descriptions detail teachers' efforts, children's activities, and the general context in which learning occurred. These descriptions of the larger context permit readers to situate knowing and learning as it is described in later chapters. In Chapter 4, I analyze children's learning by contrasting their views and understandings of engineering at the beginning of the unit with those toward the end of "Engineering for Children: Structures". I draw on different forms of assessments and thereby take into account the heterogeneity in which children's knowing expressed itself. Although one might argue that historically, assessments of learning outcomes are made after learning processes have occurred, I report on children's learning in Part I to provide readers with an advance organizer for situating the learning process studies in Parts II and III. Knowing what the children learned provides readers with a sense of where the processes described in Parts II and III of the book ultimately led.

Parts II and III form the core of the book. They feature descriptions of learning processes in a knowledge-building community. Learning here is conceptualized in terms of two mutually-dependent analogies: (a) the networking of human and non-human actors into actor networks and (b) the circulation of resources and practices in extant actor networks. My central argument in this book is that we should not think of cognition as something that occurs exclusively in the mind of people, but as something that arises from the interaction of human actors with their social and material settings. In a series of chapters, I describe the ways in which children's engineering design can be understood as the construction of networks that include various actors, resources, and practices; the results of this construction were the design artifacts and children's competence in a range of engineering design-related practices.

Part II deals with the learning of a community. Chapters 5 through 7 address the central question, What are the processes by means of which resources and practices come to be recognized as shared within a knowledge-producing community. In Chapter 5, I describe the circulation of resources, that is, facts, artifacts, materials, tools, and so on. In the course of the engineering unit, children created many facts (ideas, Canadian flag), artifacts (materials), and stories (bungy jumping) which were appropriated by other students. I use the dispute over the attribution of an idea (using Canadian flags to decorate structures) to illustrate how "shared" knowledge can be detected in communities.[2]

Chapter 6 deals with knowledge-production in a community in terms of its material practices. In the glue gun, I observed a paradigm case for technologies that become sites of transformation in communities; in such transformations, communities do not only change but the technological resources change as well. A second remarkable transformation was related to practices. Such practices changed for individuals, and simultaneously, because of individuals' membership in a larger whole, also changed the practices available to the entire community. Using hot-gluing as an example, I describe the trajectory of students from their status as newcomers to the glue gun practice (they participate legitimately but peripherally) to the point where they are core members of the practice and begin to teach other students. I use this case study to theorize the appropriation of material practices.

The classroom community was transformed by the development of discursive practices and the circulation of stories. In Chapter 7, I describe the appropriation of the "triangle" practice into the classroom discourse in which the two teachers played a crucial part. The teachers' work can be understood in terms of an advertisement analogy. This analogy links the developments of this classroom community to those generally observed in the everyday world of science and engineering. Notions such as *recruitment*, *enrollment*, *attachment*, and *buying into* show the nature of teaching and learning as cultural reproduction, and are more helpful descriptions than the corresponding language in the conceptual change literature. A critical element in the enculturation of children to canonical discourse was interactions with a very competent teacher. Through these interactions, children learned to take new and different, canonical points of views. Classroom conversations – which were to an increasing degree student-centered as the unit advanced – contri-

---

[1] "Common sense" is one of the most intriguing phenomena imaginable. It is so pervasive and yet too difficult to investigate and theorize (Chapman, 1991; Pollner, 1987). How do we come to do everyday things without explicit instruction? Understanding the structures of these activities are at the heart of the ethnomethodological enterprise.

buted in important ways to children's growing discursive competence. In Chapter 7, I provide observational and theoretical descriptions of the roles of artifacts, teachers, and peers in the children's construction of design accounts. That is, through the presentation of their work followed by question and answer sessions, children learned to construct legitimate stories of engineering practices and artifacts.

Whereas Part II focuses on networking of individuals into a community and the concomitant circulation of resources and practices in community-constitutive networks, Part III documents the ways in which students' actions were networked with their material and social settings. That is, the third part of the book illuminates the microprocesses that are constitutive of the material and social basis of cognition. My central claim is that the situatedness of cognition arises from students-acting-in-settings, an organic whole, rather than from the interaction of separate and independently modeled minds and settings.

In Chapter 8, designing is characterized as a *heterogeneous* process in which multiple elements – which I treat as actors within actor networks – including tools, materials, community standards, teacher-set rules, current state of the design artifacts, individual preferences, and past discursive achievements fuse to bring about design artifacts. At any one moment, design artifacts enfold past achievements and provide affordances and constraints for future developments. Although various actors are networked to give rise to children's designs, none obtains an absolute ontology. That is, these actors do not have singular meanings and applications, but are interpretively flexible so that they mean different things to different people and at different moments. What then was the role of the non-human actors in children's discourse and design? Artifacts played a central role in networking discursive and material aspects of children's worlds. Student designers' discursive and material activities were over, about, on and with the emerging artifacts. The students' discourse was about, and practical activities were directed toward, the artifact leading to a product of the design process; the activities were over and with the artifact in that these themselves became tools and media in the design process. The artifacts supported students' sense making, planning, thinking, and testing of ideas. In this, design artifacts became *self-reflexive structuring resources* which, in their development, opened up opportunities and constraints for their own future development.

Chapter 9 is likewise concerned with networking processes across interstices, but this time across phenomena often separated by researchers' disciplines. I document the microprocesses by means of which individuals and groups were networked to form what we understand as a community. Communities consist of individuals who, in interaction with others, produce communities through their interactions. Most interactions are within small groups so that communities (their rules of conduct, patterns of behavior, social phenomena) are produced in face-to-face encounters between individuals. This chapter is devoted to my observations of the production of a classroom community; the site of this production was in small groups where networking occurred. My description of children's collaborative constructions of accounts of their work illustrates how students (and teachers) bring about first accounts which later become collective stories. I also describe how students shifted between different types of activities: individual (parallel), collaborative, and community work. In the shifting of these activity types, individual, group, and community emerged as part of the overall network. The structures of group activities allowed small groups to organize themselves, determine leadership, and negotiate future actions.

In Part IV, I draw implications of the study to the design and enactment of curricula that allow the emergence of knowledge-building communities. In Chapter 10, I discuss conditions that must obtain so that networking and circulation of resources and practices can occur. An example is provided that describes how I used the experiences from this study to design a curriculum on simple machines in a Grade 6–7 classroom. In the epilogue, I propose that we begin to think about networking beyond the immediate classroom and school context and therefore: (a) design learning environments so that students learn by participating in activities of the society at large; (b) provide teachers with opportunities to work at each other's elbows; and (c) conceptualize writing as a networking process.

ACKNOWLEDGMENTS

A research project such as that described in this book is difficult to impossible to complete without a supporting network of people and organizations. In its entirety – support for the teachers' development effort, teaching, research, writing, and editing – the project was made possible by grants 812–93–0006, 410–93–1127, and 410–96–0681 from the Social Sciences and Humanities Research Council of Canada. Besides the financial support, many people contributed in various ways to the project and this book. My most sincere thanks go to my partner Sylvie Boutonné who shared the experience in the classroom, transcription and analysis of the videotapes, and the preparation of many manuscripts in which I reported my initial understandings about learning in a knowledge-building community. We had many conversations over and about the videotapes and her intricate knowledge of the database assisted me in weaving together the network of human and non-human actors that make this book. Without her understanding and support this would not have been the same project.

Much support was provided on site during data collection. Bridget Walshe and Christina Schnetzler were enthusiastic teachers who wanted to, and made, a difference in the lives of the children they taught. They were always prepared to talk about their experience as teachers in this class. Their collaboration helped me to understand how team teaching can provide for learning through co-participation, an experience that is common in other professions but quite uncommon in teaching. I am grateful to the students for their participation, their willingness to talk to me about their work, and their patience with the intrusion that camera work usually constitutes to the lives of people. Loretta Ceraldi and Josette Desquin were gracious administrators who facilitated my work in many ways, especially by making available a portable classroom which was used for meetings, interviews, and storing of research equipment.

Allan MacKinnon and Fiona Crofton provided assistance during the early part of the study, and particularly during the intake interviews with the students. The Association for Promotion and Advancement of Science Education (APASE) assisted by making available the engineering curriculum developed under its auspices. Although I take final responsibility for any errors, Michelle McGinn conducted a thorough final edit of the book. She was helpful as editor and sounding board for papers and articles based on the project during which I initially formulated my ideas about learning in communities.

PART I

FOUNDATIONS

Most of this book is about processes: the design of artifacts in a knowledge-building community; the emergence of networks in which knowledge – as resources and practices – is produced, re-produced, and circulated; the transformation of a community; and the networking of individuals with collectivities and the material world surrounding them. In Part I, I lay the foundations for understanding the claims about these processes. This part also contains important descriptions of how to set up a classroom so that collective knowledge-building becomes a central and characteristic feature. My theoretical understandings, elaborated in Chapter 1, focused my observations and further developed through interactions with the database. My descriptions of the research and development efforts (Chapter 2) contextualize the nature of the data and the teachers' engagement in making a difference. The description of the curriculum, both in its conception and its enactment (Chapter 3), establishes the context within which students learned, and within which the processes I later describe were situated. Finally, Chapter 4 provides readers with a framework for what students ultimately learned from their experiences in this engineering design curriculum: a description of their understandings of engineering before and after this unit.

1. THEORETICAL FOUNDATIONS

The study in this book draws on, and weaves together, theoretical notions that arose in different scholarly communities. Among these are anthropologists interested in everyday cognition within and across cultures; sociologists and anthropologists engaged in interpretive (rather than structural) studies of science and technology; and ethnomethodologists concerned with the methods scientists, engineers, and ordinary people (which Lave (1988) empathetically calls just-plain-folks [*jpfs*]) employed to structure and make sense of their everyday lifeworlds.

1.1. PRACTICES AND RESOURCES

In recent years, an increasing number of researchers have begun to view human knowledge in terms of *practices* (Bourdieu, 1990; Ehn, 1992; Lave, 1988; Lave & Wenger, 1991; Rorty, 1991; Schön, 1987). Practice, as the social and material construction of reality, replaces picture theories of knowledge and views of the mind as a mirror of nature. Practices are patterned activities which people enact to get the things done that matter to them; they are specific to the physical, social, and technological situations in which they occur; and are continuous so that there cannot be distinctions between non-linguistic and linguistic, or material and discursive, practices. Practices are complemented by resources, tools, materials, words, information, and everything else that is necessary for engaging in practice.

Practices structure and produce human experience in and of the world. Participation in practice constitutes knowing and learning; to share practices is to share an understanding of the world with others. Among shared practices are mathematical, scientific, and technological practices that people engage to accomplish practical tasks. Shoppers in supermarkets make best-buys, dairy factory-workers prepare shipments of dairy products, child street-vendors maximize profits as they buy and sell candy, scientists construct factual knowledge about the world by means of various mathematical practices, and engineers design individualized urban transport systems.

Practices are specific to the material, social, technological, historical, political, and economic context in which they occur. Similarity between any two contexts does not guarantee similarity in the enacted practices that can be observed. For example, physicists and engineers use the same mathematical equations, but use them in significantly different ways and for different purposes (Brown *et al.*, 1989); carpenters, cabinet makers, and wood cutters use the "same" chisel but in ways characteristic for their own communities; these ways differ across the different communities. The "same" words used

3

by members of different communities mean different things, and the "same" engineering drawings give rise to characteristically different discourses whether they are used by a firm's accountants, supply managers, workers on the shop floor, or designers. Hence resources such as tools, instruments, artifacts, and words do not embed unique meanings available in the same way to all people but draw their meanings from their use in the everyday practices of particular communities, within the contexts of their use; their structures are a function of the social, material, and instrumental conditions of the context. This has significant implications for science (and math) education, for we can no longer assume that the structure of the world reveals itself in a unique way. One can therefore not expect students to derive Western scientists' use of some tool nor can one expect isolated indigenous people to discover the practices of using a chain saw if they found one on the forest floor.

1.1.1. *Discursive Practices*

Among the most important social practices are those of linguistic nature. Following others who operate in a pragmatic framework, I take an anti-representational view of language. Language is not a representational medium, but a resource we use to get matters done. From this pragmatic perspective, using language means participating in discourse, or, as Wittgenstein said, playing *language games*.[1] Hence there is no more meaning to an expression than the overt use to which the expression is deployed (Wittgenstein, 1995/1958; Quine, 1987). We learn to play language games, as any social practice, by participating in them with others, through mutual observation, emulation (or mimesis), and correction in collectively-observable situations. Language is action, situated action. Two expressions have the same meaning when they can be employed in the place of each other without difference. To have meaning, therefore, is to have a place in a language game. Knowing from this perspective does not mean getting reality right, or having or holding a right re-presentation.[2] Rather, it means engaging competently in actions for coping in and with reality (Greeno, 1991; Rorty, 1991). Are all language games equally good? Rorty (1989) reminds us that

---

[1] I use *language games* and *discourses* synonymously.

[2] Traditional cognitive science uses the notion of representation to conceptualize purported mental entities that are manipulated by mind to bring forth "thinking". However, this notion is inappropriate to denote my own way of understanding cognition as it is portrayed in this book, namely as arising from our being-in-the-world. I therefore refer to re-presentation to emphasize the distinction between things in the world, and ways of presenting them again: as mental images, written documents, spoken words, artifacts, and so on. This distinction was made by the philosopher Heidegger (1959, 1977) who, to describe cognition, used the German word *Vorstellen* which literally means to put before oneself, and thereby evokes the separation of the knowing subject and objects of knowing. *Re-present* captures some of the flavor of Heideggers language.

Great scientists invent descriptions of the world which are useful for purposes of predicting and
controlling what happens, just as poets and political thinkers invent other descriptions of it for
other purposes. But there is no sense in which *any* of these descriptions is an accurate representation of the way the world is in itself. (p. 4, *italics* in the original)

But this does not mean that any language game is as good as any other.
Although there are no objective criteria for deciding between language
games, some are more useful than others. Thus, canonical discourses should
be considered in terms of their usefulness rather than their truth value.
Furthermore, discourses also take time to develop and to become shared
ways of providing observational and theoretical descriptions.[3] New discourses
are not adapted from one day to the next, or even within the span of a few
years. Rather, old language games fall out of use, and new ones take over
with the change of generations.

Looking back at an historic period during which a major shift in scientific
theories occurred, one can observe shifts in the ways particular vocabularies
were/are used. Rorty (1989) describes the period of the Copernican revolu-
tion as one of "inconclusive muddle" (p. 6) which ended with Europeans
finding themselves fully immersed in a Copernican language game. Further
support for the view of learning in science as a shift in language games comes
from microanalyses of individual scientists at work. In a detailed analysis of
the history of the electromagnetic motor, Gooding (1990, 1992) illustrates
how ontologically flexible objects in Faraday's world and his descriptive
language slowly converged to produce a new language and, with it, new
objects in the world. The period of invention and "discovery" can best be
described as one of "inconclusive muddle" because it is not clear what
Faraday's objects are or what his language describes. From interaction with
the objects and the slowly changing descriptions (in words, images, drawings,
numbers, etc.), a new language game emerged. And with this new language
game arose a new world full of magnetic fields, electric fields, rotors, coils,
and electromagnetic motors. Saying that an old language game was inappropriate for describing a specific segment of the world amounts to saying that
now, after having developed or appropriated a new discourse, one is better
able to describe this segment.

What then does it mean to construct a common understanding, or shared
meaning? Initially, two or more individuals talking to each other do so
based on the default assumption that understanding is shared. If, during the
unfolding conversation doubts about this assumption emerge in one or the
other participant, they will engage in a repair sequence intended to re-

---

[3] It has been claimed that all observation is theory-laden (Feyerabend, 1975; Hanson, 1965).
Several studies of emerging discourses in science classes have convinced me that it may be more
appropriate to speak of observational and theoretical descriptions (Roth, 1996f; Roth & Duit,
1998; Roth et al., 1997a). This conceptualization allows me to account for (a) students' descriptions of observations that change into theoretical descriptions and (b) for my observations that
particular observational descriptions appear to facilitate students' development of canonical
descriptions (observational and theoretical).

establish the default state. In this way, understanding is always something like a "passing theory" (Rorty, 1989) or situated "conversational inference" (Lee, 1991). These theories of understanding are passing, because they have to be corrected in the face of mumbles and stumbles, metaphors, malapropisms, gestures, tics, and so on. This talk occurs against background assumptions about context, interactive goals, and interpersonal relations. From these background assumptions, people derive frames within which to interpret the utterances of others. To be able to understand each other, conversation participants then have to be able to converge in their respective passing theories about the situation and its development.

Everyday language games are the default systems in terms of which all other language games are interpreted:

Any individual, in understanding his or her world, is continually involved in activities of interpretation. That interpretation is based on prejudice or (*pre-understanding*), which includes assumptions implicit in the language that the person uses. That language in turn is learned through activities of interpretation. The individual is changed through the use of language, and the language changes through its use by individuals. (Winograd & Flores, 1987, p. 29)

Thus, there is a background of common sense, a sense of the mundaneity of certain social practices which do not need explication. These practices go without saying; they are so evident that formulating them would be considered strange. These implicit beliefs and assumptions, common sense, often cannot be made explicit. The upshot of the impossibility of making all assumptions and beliefs explicit, and the primacy of practice is that much of our knowing does not exist in the form of mental representations (Winograd & Flores, 1987). We can drive a nail into a wall by using a hammer without an explicit representation of the hammer. My competence of hammering comes from familiarity with hammering, not a representation of a hammer. As a consequence, these implicit beliefs, assumptions, and common sense notions have to be learned by participating in the practice itself, a point I take up again in relation to participating in communities of practice.

1.1.2. *Language Games Applied*

The view of language games and their emergence as developed here not only has descriptive value but also has been used in a prescriptive sense for the developing (designing) new workplaces, and in designing and studying classroom learning environments.

1.1.2.1. *Designing Workplaces*

In the traditional design of (computer-based) work environments, designers analyzed the work tasks in terms of some cognitive model, and then developed new environments based on those models. Frequently, these new environments were ill-suited to the workplace for which they were designed.

One of the central problems was that notions that were taken for granted by the designers were not so obvious and self-evident to the users. Now, because language and the world we experience are mutually constitutive, much of what we know does not need to be made explicit and goes without saying. It is so mundane that we do not need to express it (Pollner, 1987).[4] The designers developed work environments on the basis of their own mundane sense, their ways of viewing the world, leaving unsaid what was self-evident to them. But, belonging to a different discourse community, users did not understand. An illustrative example concerns the ways files are deleted in computer environments. In some computer environments, a command such as DEL FILE.TXT removes the file from the disk; in other environments, users move a file icon to a trash can. In the first setting, users have to learn new commands, DEL being one of them. In the other, one could still talk about "putting a file into the trash." In the second setting, the user does not need to know anything new; a familiar way of viewing the workplace, that is, the same language game, was still useful in the new computer environment. As a consequence, relatively short training times were needed in the second environment. It has been argued that the success of the second environment was largely due to the proximity of the old and new language games they afforded (Brown & Duguid, 1992). Similarly, the success of programs such as Lotus 1-2-3 and Quicken was due to their easy integration into the practices that accountants and just-plain-folks enacted with spreadsheets and checkbooks before the advent of computer technology (Winograd, 1996). Lotus and Quicken were designed to allow newcomers to engage in trajectories of increasingly competent practice in the new context.

New theories of designing frame the successes of these two and similar programs in terms of Wittgenstein's notion of familiar language games.[5] By working with the future users in a mock-up environment, designers begin to develop a common language game with the users. This language game and the initial mock-up artifacts are then allowed to develop into more complex forms until the new community has developed the new environment and the associated discourse necessary to find one's way around the environment.[6] The language game is similar to the original workplace discourse – as in my

---

[4] At the same time, this mundaneity is so complex that Winograd (1995) described it as a "'mind-blowing intricacy' [that] becomes an object of wonder, standing in stark contrast to the brutal simplifications that are the stock-in-trade of standard AI methodology" (p. 113).

[5] Descriptions of several design projects conducted in the Wittgensteinian tradition, all situated in Scandinavian countries, are provided by Ehn (1992). A computer environment based on philosophical principles was presented and discussed by Winograd and Flores (1986). A survey of novel, Heideggerian approaches to design can be found in Winograd (1995, 1996).

[6] Greeno (1991) and Roth (1995b) use environmental metaphors to describe how people know. Accordingly, knowing mathematics or science is like knowing a familiar setting with its resources (gas station, drug store, supermarket) that serve to get specific things done (get gas, vitamin tablets, groceries). The computer program "Microsoft® Bob™" is designed along the same lines by providing an interface in the form of a room in a home (Winograd, 1996). The user faces familiar objects such as a calendar, a chequebook, and a notebook scattered around the

earlier example of deleting a file in computer environments – so that other newcomers can readily appropriate the new language game based on the extant family resemblances.

1.1.2.2. *Designing and Studying Science Classrooms*

I have used a language game view of knowing for successfully describing how Grade 11 physics students developed useful, situated language from earlier "muddles" (Roth, 1995a, 1996f; Roth *et al.*, 1996). Here, the objects and language were not ontologically fixed, but flexible and they changed in the course of students' interactions over and about events in a computer-based Newtonian microworld. At first, students converged within their groups on common ways of talking about these events. Eventually, and in part shaped by interactions with me, the teacher, students evolved a Newtonian way of looking at and talking about the computer displays. Subsequently, this allowed students to evolve new observational and theoretical descriptions. Objects and events served as starting point for conversations. These conversations shifted, and with them the way students experienced the microworld. It was in part through my interactions with them that students' talk converged on canonical ways of talking. The microworld itself could not bring about canonical talk, for it only placed some constraints on the emerging language games without limiting the language games to canonical descriptions.

In another series of studies, my graduate students and I illustrated the emergence of new language games as students engaged in a variety of material and discursive practices (McGinn *et al.*, 1995; Roth, 1996g, 1996h; Roth & McGinn, 1996). For example, we documented students' work of building a community in which an appropriate (from a scientific perspective) language game about mechanical advantage came to be shared by a large majority of members. Although the notion of mechanical advantage was introduced by me, the teacher, "advantage" and "disadvantage" already were resources in everyday out-of-school language games. There is an extant family resemblance between the canonical notion of mechanical advantage and the everyday notions of advantage and disadvantage. Students brought these everyday notions as part of their familiar discourse to the classroom. Because of the family resemblance, the shift from "advantage" to "mechanical advantage" is small, so that I observed a rapid appropriation of this new way talking about machines by students. The mathematical practices associated with mechanical advantage require the calculation of a ratio, a comparison between effort and reference loads. Even if students did not calculate mechanical advantage at first, they could easily assess whether their machine provided an "advantage" or a "disadvantage." That is, in the case

---

room. By clicking on these objects, the user starts specialized applications associated with household tasks such as scheduling, bill paying, and letter writing.

of mechanical advantage, the classroom community had the opportunity to develop a "special network of recurrent conversation" (Winograd & Flores, 1987, p. 158). Across situations – small-group and whole-class conversations, design activities, and structured investigations – mechanical advantage became a central aspect of the language game about simple machines.

However, there were situations in which student-generated language games were not readily appropriated by the community and therefore did not significantly change the existing language games. For example, "Lana's law", "Randy's rule", or "Alain's rule" were the labels of a variety of appropriate mathematical practices to solve balance beam problems (McGinn *et al.*, 1995). Although individual students first developed and proposed these language games, there were few opportunities for an emerging network of recurrent conversation equivalent to the mechanical advantage case; when levers were involved in the design of machines, other concerns predominated in the language game which did not require any of the formal, mathematics-based rules as discursive resource. Thus, despite students' development of new language games and despite repeated opportunities to participate in them, these discourses were not appropriated widely in the classroom community.

1.1.3. *Implications of an Epistemology of Practice*

The epistemology of practice outlined above has important implications for teaching, learning, and researching in classroom environments. If we accept on the one hand that language and culture are contingent and fundamental human achievements, and on the other that the world does not tell what language games to play, we have to reevaluate traditional views of experiments and "hands-on" activities in science classrooms. For this means that we have to abandon the belief that students, by interacting with specific artifacts or by doing specific experiments, can "discover" or "induce" scientific ways of talking. Students must do this, in part, by interacting with someone who already has achieved competence in these practices. In this, my interest in practices parallels that of ethnomethodological studies of practical action and practical reasoning concerned with the explication of ways in which the rationality, intelligibility, accountability, or reproducibility of actions and their settings are organized, maintained, and made visible (Jayyusi, 1991). Descriptions of practice are provided when we study the way in which members make their sense available to each other, in their own accounts and descriptions, and how these accounts and descriptions are embedded and interact with other practices. For there is no meaning *out there* and predating our experience, "there are only acts situated within discursive and embodied access to a world that is always and already shot-through with meaning" (Lynch, 1991, p. 82).

If we, as Quine (1987) suggested, abandon the idea "idea" and focus on perception, speech, and action we are led to the same methodological assumptions that underlie the ethnomethodological research program. We then assume that the phenomena of interest are observable, that is, visible and audible rather than, as Mehan (1993) suggests, beneath the skull and between the ears. From an ethnomethodological perspective, embodied activities are self-organizing rather than the product of externally articulated rules; people produce culture and community, rule-following and rule-adherence, and so on, rather than mindlessly follow, as cultural dopes would do, pre-existing rules of social behavior. In the course of conducting their activities, people furnish each other with instructions for discovering the sense and interactive implications of their talk and actions, and make the same understanding of the situation available to the analyst. It is the analyst's task to recover the sense of participants from the visible and audible documentary evidence. Furthermore, if descriptions are always locally organized, situated actions to achieve specific purposes, then there cannot be universally correct descriptions. This also applies to my description of events that occurred in the Grade 4–5 classroom which is the focus of this book. This book constitutes *my* account of the events in the classroom, not *the* universal description.

1.2. COMMUNITIES OF PRACTICE

In recent years, the notion *community of practice* (Lave & Wenger, 1991) has had considerable impact on the theory and practice of educational research. Communities of practice are identified by the common tasks members engage in and the associated practices and resources, unquestioned background assumptions, common sense, and mundane reason they share. Communities of practice form and disband with emerging and disappearing needs, or may be stable over longer periods of time. Communities of practice are not homogeneous such that all members know exactly the same things, or have obtained the same practices and resources.[7] Rather, expertise is usually distributed in social and material terms.

Communities can change in a number of ways. First, by accepting newcomers who, in part, assert their individualities, new competencies are introduced. A graduate student in physics who eventually receives a Ph.D., becomes a post-doctoral fellow, and then obtains a tenured position constitutes an example of a change agent in the physics community. Second, oldtimers may obtain new resources and participate in new practices so that the number of individuals engaging in a specific practice increases. Examples of such changes are physicists who go to another university to learn how to build a functioning TEA laser (Collins, 1982) or biochemists who study a new technique in another laboratory (Jordan & Lynch, 1993). Third, com-

---

[7] Strangely enough, one of the ideologies of education is to have all students go through the same sets of experience and meet the same sets of standards. In my view, this ideology, because everyone is dealt the same standard fare, leads to the situation that many students are not challenged to their maximum while others do not succeed.

petent old-timers may experiment with and develop new practices that are made necessary in the pursuit of their current goals. For example, when physicists theorize and design new detectors, they expand the range of available practices for physicists (Traweek, 1988). All three changes in the existing resources and practices of a community are constitutive of "progress" and learning more generally. All three examples also underscore that learning, in essence, is participation.

The fundamental problem of schools is that the social contexts they offer are different from everyday communities of practice where newcomers enter existing communities; in most school classrooms, there usually is one teacher who is responsible for a common learning curriculum. Furthermore, students are grouped homogeneously in terms of age. Such homogeneity of learning and age is uncharacteristic of everyday out-of-school communities. On the other hand, the teacher of Moussac had created a working community that could produce and re-produce itself by accepting new members and losing old-timers. Participation and learning were not primarily dependent on his presence – he only contributed a regulatory role whenever necessary to maintain the community. In most classrooms, however, the homogeneity in age and competencies is a constraint, because cultural reproduction depends to a large extent on the presence of the teacher. Reliance on one person, however, limits opportunities for co-participation in the production and re-production of culture. To me it appears as a precondition to all learning that the new language game make sense to the learners, and that there are opportunities for them to engage in the practice. However, the design of artifacts provides a context in which a community can develop a shared language game that makes sense to all its participants, including students and teachers. Building a classroom community therefore means finding appropriate activities, artifacts, materials, and tools (technologies) that support the emergence of shared language games, and at the same time allow new language games to emerge that have some family resemblance with those others already play. In this sense, I maintain the usefulness of the notion "community of practice" in school science classrooms.

1.2.1. *Learning as Participation*

In this book, I identify little with traditional understandings of knowledge as things people hold in their heads and teach in lectures.[8] I am fundamentally concerned with a different conception of knowing and learning. Knowing can be recognized in and as competent participation in everyday activities.

---

[8] Historically, lectures are pre-print forms of circulating texts. I have always found it more fruitful to read two or three *different* authors on a topic than to attend the lectures of a single professor. Reading allows me to engage in learning at my own pace and access different pages at will. Lectures continue irrespective of my understanding. In my view, lecturing and associated testing of lecture-specific content knowledge is part of an ideology that leads to memorization and grade grubbing.

Rather than engaging in the reproduction of knowledge, people, including students, are first and foremost engaged in the production of knowledgeability:

Knowledgeability is routinely in a state of change rather than stasis, in the medium of socially, culturally, and historically ongoing systems of activity, involving people who are related in multiple and heterogeneous ways, whose social relations, interests, reasons, and subjective possibilities are different, and who improvise struggles in situated ways with each other over the value of particular definitions of the situation, in both immediate and comprehensive terms, and for whom the production of failure is as much part of the routine collective activity as the production of average, ordinary knowledgeability. (Lave, 1993, p. 17)

Knowledgeability comes from participating in a community's ongoing practices. Through this participation newcomers come to share a community's conventions, standards, behaviors, viewpoints, and so forth; and sharing comes through participation. Thus, one primarily becomes a scientist through apprenticeship with senior researchers (Toulmin, 1982; Traweek, 1988); one becomes a member in a technological community by means of learning by doing in practicum-like settings (Constant, 1984; Schön, 1987); one becomes a Mayan midwife by participating in tasks which increase in importance as one makes a formal commitment to becoming a midwife (Jordan, 1989); and one becomes a high-school jock or burnout by hanging out with the cliques of older siblings, their friends, or others who are already part of these groups (Eckert, 1989). The process of becoming a full member in such communities can thus be described as a trajectory of *legitimate peripheral participation* of increasing intensity in the ongoing practices of a community.

Lave and Wenger (1991) made it quite clear that legitimate peripheral participation is an analytical notion, not a prescription for the practice of designing teaching-learning contexts. This concept therefore needs to be redefined in the context of schooling. Here, the community is bounded by the activities of the students, involving teachers and possibly other adults. Legitimacy provides the possibility for participation in the ongoing activities; peripherality the possibility to choose the level of participation in relevant activities. A second shift needs to occur in our understanding of mastery. Participation always requires negotiating and renegotiating meaning in the world; it implies the constant interaction and mutual constitution of experience and understanding. What the teacher of Moussac intuitively realized was that mastery resides not in the master, but in the organization of the community which led to the, for traditional views, astonishing feat that all children learned to read, write, and do arithmetic despite the lack of direct instruction. If "masters" are present in the community, their key responsibility should be to the organizational structuring of the available resources rather than the dissemination of information and the suppression of newcomers cultural production.

The practice view of cognition changes our view of learning and makes recommendations for the design of learning environments that are quite different from those based on models of individual cognition and traditional conceptions of learning. Productive knowledge is circulated along with the building of a multi-leveled community of interpretation (Brown & Duguid, 1992). Traditional models of learning focus only on the explicit knowledge that can be stated in the form of formal propositions and procedures. The community of practice view sets up learning environments in which students learn explicit and implicit knowledge.

Computer support groups have been used as examples of communities that have bootstrapped themselves from newcomer to old-timer, without formal teachers (experts) that disseminate information, and based on distributed expertise so that each member is in some respect a newcomer while in others an old-timer (Collins *et al.*, 1989). One training system for technicians explicitly designed to model multi-level expertise so that all levels train together has had an enormous success (Brown & Duguid, 1992). Every participant had the opportunity to see others as newcomer and old-timer in various aspects of the subject matter, and hear discussions of others who operated at different levels of competence. Reportedly, 20 hours of training with the system produced the equivalent of eleven months of on-the-job training.

1.2.2. *Authenticity of Practices*

Over the past seven years, my main interest has been with school science classrooms that can be described as communities of practice or knowledge-building communities. In all the classrooms I have researched, five conditions obtain:

- Problems are either ill-defined or defined so loosely that students can impose their own problem frames.
- Students experience a sufficient level of uncertainty and ambiguity in finding solutions.
- The curriculum is predicated on, and learning driven by, students' current state of knowledge.
- Students experience themselves as part of a community in which specific practices and resources are shared and scientific and technological knowledge are socially constructed.
- Students can draw on the expertise of more knowledgeable others, whether they are peers, teachers, or outside experts.

These conditions (which parallel existing conditions in the scientific world) assure that the learning environment bears resemblance with environments in which scientists do their work and where knowledge-building is the central concern that drives all activity (Roth & Bowen, 1995; Scardamalia & Bereiter, 1994). The notions frequently used in such contexts are those of *authentic* learning environments and *authentic* practices. In anticipation of criticisms, I elaborate my understanding of "authenticity".

Authenticity is not determined by an identity relation between some everyday activity of scientists or mathematicians and what students do; this confusion has led critics in the past to ironically refer to my students as "little scientists" or "little mathematicians". At issue here is whether or not the artifacts and activities of the setting allow students to engage in practices that bear resemblance with those in out-of-school communities. Students develop and employ practices to reach their own goals so that they experience ownership over products and evaluations. Authenticity evokes the identification of individuals with the goals, processes, and products of their activity. Therefore, the learning environments I have researched are authentic not because they mirror the real thing – designing towers and bridges, designing experiments – but because they give rise to language games, reflections and interactions that share some fundamental similarities with everyday environments of designers and scientists.

1.2.3. *Actor Networks*

Actor network approaches constitute important theoretical and empirical tools for describing and theorizing communities and how those communities develop common practices and resources. Actor network theories have an advantage over other approaches in that they do not make *a priori* distinctions regarding the ontology of the objects under study. That is, actor network theories do not *a priori* determine which aspects of a community are responsible for new developments: they can be human or non-human, psychological or sociological, social or technical aspects. Actor network theories also give us the vantage point of making the familiar strange by looking at phenomena through the eyes of quite unlikely actors. Here, I develop my view of actor networks as they relate to the studies of knowing and learning in the later parts of the book.

Actor network approaches have been developed by researchers in science and technology studies (Callon, 1986, 1987, 1991; Latour, 1987, 1988a, 1992; Law, 1987, 1994; Law & Callon, 1988; Pickering, 1995). Actor networks constitute analytical tools used to model success and failure of scientific theories and technological inventions in a symmetrical manner. They constitute tools for investigating and understanding the evolution of scientific and technological communities. Actor network approaches view scientific and technological communities as networks of actors; each individual, group, technology, company, belief, finance, raw material, or artifact constitutes an actor in a network. Thus, actor network theories do not distinguish between human and non-human actors (artifacts), between individuals and institutions, between mental and social or material activity. From the perspective of actor network approaches, such distinctions are a consequence of the analyses, not those of presuppositions. In this, actor network approaches constitute "symmetrical anthropologies" (Latour, 1993, p. 101). Hence, actor network approaches allow researchers to eschew traditional reductionist approaches that focused, *a priori*, on the individual or society, society or technology, content or context, human or non-human actors. Actor network approaches integrate all these dichotomies, an integration that is frequently referred to as the "seamless web" (Schot, 1992). This view of social reality is consistent with Davidson's conception of human behavior and Derrida's (e.g., 1986) construction of texts as "centerless networks of relations, networks which can always be redescribed and recontextualized by themselves being placed within some larger network" (Rorty, 1991b, p. 59).

Any aspect of a collectivity, technology, or classroom can be understood by following and/or taking the perspective of one or more actors in a network. Understandings that arise from actor network analyses are likely to be heterogeneous, for they stress the co-evolution of technical and social aspects, content and context, and human and non-human perspectives. The tools of the resulting symmetric anthropology (Latour, 1993) are therefore appropriate for those analysts who view cognition as arising from the interaction of individuals, their tasks, and their material and social settings. Actor network approaches are appropriate for modeling (a) the "heterogeneity" of actions, goals, motives, and activities and (b) the "seamless" way learning arises from situated activity. Actor network approaches also account for the mutually constitutive relationships between individual and culture during enculturation: As individuals move along trajectories from being newcomers to becoming old-timers in a culture, they not only participate increasingly in the practices of the culture, but transform available practices through their own embodied experience; and with it, they change the available resources. In a similar way, technologies such as hydro-electric networks, railways, airplanes, or computers do not merely impact on culture, but co-develop and co-transform with culture.

Actors should not be thought of as well-defined, stable entities, because they can redefine their nature, identity, and relationship to transform the network in unforeseen ways: actor networks are reflexive entities because they are able to redefine and transform what they are made of. In many situations, actors themselves can be seen as actor networks. For example, a class of students consists, for the teacher, of a number of human actors. From an administrative perspective, such as the principal's, the class may be considered as one actor. At yet another level, a number of classes may be viewed as one actor such as when a principal refers to "what happens in the intermediate grades" or when a superintendent talks about "achievement at Mountain Elementary School".

Symmetric anthropology is important to me, because I want to understand school learning without making premature commitments to the contributions of material and social settings, tools and information technologies, distributed and individual aspects of knowing, and nature versus nurture debates. Here, actor networks provide a promising perspective on knowing and learning in complex classroom environments. Actor network approaches handle a variety of issues which are otherwise handled only by combinations of approaches.

Actor network is a core analogy which I use throughout the book. Thus, the emergence of "shared" knowledge is modeled as the networking of actors and the circulation of resources and practices in extant networks. Readers will notice that the actor network view not only describes (as ethnography), but also models learning of the community as a whole; it not only explains the appropriation of practices from more knowledgeable others (cultural reproduction), but also accounts for the transformation of practices (cultural production); it models learning at the classroom level in situations with much teacher effort as well as apparently spontaneous learning; it not only recognizes the social aspects of learning like social psychology or traditional sociology, but also its technologically and materially situated nature; it models successful and unsuccessful teaching by including such heterogeneous actants as teachers' rhetoric practices, the strength of hot-glued joints, or students' interests; and it allows me to understand the transformation of a classroom community in terms of the distribution of physical and social resources.

There are a number of methodological implications of actor network theories. Understanding arises when we study, from the point of view of diverse actors (Latour, 1987):

● how causes and effects are attributed,
● what nodes are in the network,
● what size and strengths are each of the links between nodes,
● which spokespersons are most legitimate, and
● how each of the elements and links is modified and transformed during a controversy.

Because network analysis includes all actors (nodes), it refuses re-presentations of achievements that leave out aspects of individuals' work that are in fundamental ways distributed; it acknowledges the primacy of simultaneous multiple membership in different communities for each actor in a network; it recognizes the existence of multiple meanings; and it addresses the needs of researchers who desire a symmetric approach to the description and explanation of success and failure of sociotechnical phenomena, the success of scientific facts and theories as well as the stability of beliefs incompatible with scientific cannon (often termed misconceptions, alternate frameworks, naive conceptions, or folk beliefs). In the past, this form of analysis allowed me to model phenomena such as the stability of scientific and folk discourse about nature in the face of multiple viewpoints (Roth, 1996f), conceptual change in science and science classrooms as social achievement (Roth, 1995a; Roth & Roychoudhury, 1992), and the affordances of technologies in science classrooms as mediating agents in the construction of social and natural facts (Roth, 1996g; Roth et al., 1996).

1.3. DESIGN AND DESIGNING

According to the British Royal College of Art (1976), design is a broad field of human activity, an area of human experience that reflects a human concern with material culture and human activity. Consequently, human capacity for designing is as fundamental as the capacity for language and other practices. Among the activities characteristic of designing we find: trying to create environments that reflect our aspirations; using tools and materials in purposeful and goal-directed ways; making value judgments about objects and places; responding to visual messages of advertising, products, signs, buildings, films, and television; and creating visual photographic images and making qualitative value judgments about them (Roberts, 1994).

There are four characteristics of designing and design which make them particularly interesting for science and technology education. First, the subject matter of design, and the objects of designers' attentions are ill-defined problems. Design problems are characteristically distinct from the traditional fare of "problems" that students encounter in school textbooks. Thus,

Design problems are described as "ill-defined" because there is no way of arriving at a description of the possible provision merely by the reduction, transformation or optimization of the data in the requirement specification. By the same token, it is rarely possible to determine whether or not the finished design is "the correct," "the only" or "a necessary answer" to the requirements. (Roberts, 1994, p. 174)

Design problems are the epitome of open-ended problems and designing is a paradigm case of problem solving.[9] Most problems in the everyday out-of-school world are ill-defined so that the extent to which students engage in designing in school, creates a parallel structure between everyday out-of-school and school activity. Moreover, problems emerge as snags in the course of on-going activities. These snags are always relative to particular people and their contexts rather than "problems" in some absolute sense (Winograd, 1996).

Second, designing is essentially cognitive modeling. In designing, a person evaluates systems as they are or might be, transforms them to gain insight about how to structure them, and assesses the quality of fit between alternative conceivable structures for the system. The design of artifacts has the advantage that it is a form of cognitive modeling that unloads cognitive activity into the environment so that the relationship between different aspects of the system can be physically tested.

Third, designing is an intentional activity in which the objects of the designer's attention require change. That is, in designing, individuals act on objects and events of ambivalent structure and bring about changes towards the open horizon of an unknown and unknowable future.

Finally, designing results in a distinctive kind of knowing and knowledge. Much as in learning a language, designing results in more than knowing *that* (resources) and knowing *how* (practices): through designing people learn much of the tacit aspects of knowledge that resist formalization but are nevertheless central to any practice. Designing relates acting, making, and

---

[9] For this reason, I discuss "problem solving" in the context of the evidence for the heterogeneous nature of designing and design artifacts (Chapter 8).

being. In this, designing is an activity that makes us aware of the fundamental thrownness of our being. Through designing, humans structure continuous experience into series of overlapping episodes from which, by focusing on designing and ongoing interpretive activity, they are enabled to construct meaning and knowledge.

1.3.1. *Designing as Professional Activity*

When engineers design in teams, they often coordinate their activities by having a single focus of attention. This focus frequently is in the form of a design drawing but also includes the physical or computer models of buildings or transport systems (Bond, 1989; Henderson, 1991; Latour, 1992; Luff & Heath, 1993; Suchman & Trigg, 1993). These material objects, which serve as the focus of attention and often change in the course of joint work, also have important functions to the problem-solving activities of the collectivity. They serve not only as the topic of participants' conversation but also as background against which participants make sense of each others utterances. As such, these material objects are tools for social thinking, negotiating and integrating distributed cognition, and structuring interactions. Together, manipulations of the focal artifacts (drawing, building, taking apart) and talk constitute the "language of designing" (Gal, 1996; Schön, 1983; Schön & Bennett, 1996).

In changing the focal artifact, which is witnessed by participants in the design process, the emerging artifacts embed the history of the interaction so that participants can refer to specific instances in their past conversations by indexing the corresponding parts of the artifact. At the same time, the horizon of future activities is outlined as potential by present states of the artifact. The most important feature of design artifacts in these processes is their equivocal nature (Bijker, 1993; Henderson, 1991; Sørensen & Levold, 1992). Negotiation, progress, and invention are all processes made possible by the different meanings that artifacts have for different participants in the design process.

1.3.2. *Design as Learning Context*

Design is the epitome of a context in which elementary students can learn to solve open-ended problems (e.g., Harel, 1991; Harel & Papert, 1991; Kafai, 1994). In the work of these authors, children's learning through design and professional design are distinct. To avoid any confusion, I make the following distinctions. People frequently use "design" to refer to the planning process where a designer develops an idea on paper and distinguish it from the realization of the design in material form. This seems a limited conception. In my view, the design of artifacts without precedents includes not only the planning phase where students may draw and outline their artifact, but also the development of prototypes. As I discuss in Chapter 8, such a move is necessary because children have, with the emerging artifacts, first opportunities for testing their ideas. As they develop their models, children go through "revolutions" in their understandings which is a process quite distinct from routinely designing some artifact. Faulkner's (1994) notion of "'normal' design" is useful here. Similar to Kuhn's (1970) notion of "normal science", normal design lacks much of the excitement of innovation. Normal design is characterized by its well-established and paradigmatic nature and application of canonical knowledge; it is a more or less linear, logical process of applying well-established knowledge to existing problems as promoted in the following quote:

Engineering requires a large knowledge base to make choices and those choices are not made by chance, but logical reason based on previous experience . . . . A building cannot be designed or constructed in real life without logical reasoning using basic engineering principles. (Aerospace engineer, personal communication, June 12, 1994)

I believe that the engineer in this quote refers to "normal design" rather than design during innovation. Learning through design is therefore not exclusively represented in final products but in the process. While most professional design concentrates on the product as the essential outcome, learning through design activities makes processes the central issue of education. This learning is only partially reflected in the resulting artifacts. Much of the evidence provided in later chapters will support this claim.

Interests and goals are central to meaningful everyday activity; understanding how people pursue goals is critical to understanding cognition (Schank, 1994). Designing is one of the four types of learning environments that make learning-in-practice, as students pursue goals of their interests, a central element (Schank et al., 1994). Designing also fits well with an emerging interest in Science-Technology-Society issues, and in technology as context for learning science: Design is the central activity in technology (Faulkner, 1994). There are also suggestions that design artifacts become "objects-to-think-with" because they allow students to construct and test knowledge by incorporating in the design their ideas and feelings (Kafai, 1994).[10]

Design has also been described as a paradigmatic learning environment for newcomers and old-timers in a domain. Designing is an unfolding conversation in which the designer interacts with the materials of his work; these materials and the emerging artifact continuously talk back to her (Gal, 1996; Schön, 1987; Schön & Bennett, 1996). As she assesses unanticipated problems and potentials, the designer evaluates the moves that have created them, thus linking past with present and future actions. This conversation is not reducible to an application of general rules or theories, although some of its main features are in fact amenable to description. Schön's example of engineering design students who engaged in a design task not for the purpose of learning specific facts, but for learning patterns of activities central to

---

[10] I provide evidence for and theorize the notion of "object-to-think-with" in Chapter 8.

engineering practice (such as designing tests, generating hypotheses, making presentations, etc.), shares key features with the examples of goal-based scenarios provided by Schank and his associates (Schank, 1994; Schank *et al.*, 1994).

2. EMPIRICAL FOUNDATIONS

In this chapter, I describe the contexts in which this study was conducted. Information is provided about teachers and students, as well as about data collection and interpretation. Details of the curriculum including the activities, teaching strategies, and the affordances for learning are outlined in the subsequent chapter.

2.1. INSTITUTIONAL CONTEXT

This study was conducted at Mountain Elementary School, a school in Western Canada that offers in part a regular elementary school program, in part a French Immersion program.[1] The study took place during the fall of the year for which the teaching staff at Mountain Elementary had voted to make science teaching the focus of their staff development efforts. This focus responded at once to the teachers needs to improve their own teaching and to the students' needs. An internal survey of upper-division students (Grades 4–7), conducted with the help of the students, identified a higher frequency of student-centered, hands-on approaches to science as a core need. As part of the overall effort, and in exchange for participating in studies on learning and teaching science, my research team facilitated teacher in-service (as outside facilitators to teacher-organized professional development days) and in-class teaching (by assisting in planning science lessons and through team-teaching). In the following sections, I describe parent community, teacher community, problems and constraints arising from the historical situation, and teacher development efforts that set the institutional context of this study.

2.1.1. *Parent Community*

Mountain Elementary School had a Parent Advisory Committee (as mandated by Ministry), which contributed to the corresponding bodies at district and provincial levels and met once per month with the principal. The Parent Advisory Committee (PAC) was described by both principals as "very active and supportive". Both principals noted that the school had a considerable number of parent volunteers, a sign for the active interests this parent community had in the education of their children. Because French immersion

---

[1] In Canada, where French is one of the two official languages, English-speaking school-aged children can enrol in French immersion programs. Here, their classroom language is French (including textbooks) in all subjects other than English.

is not offered in all schools of the community, some children at Mountain Elementary commute from other parts of the city. Some students even commute over long distances with several transit changes; to allow students to successfully make this commute, the school day ends at 2:45 rather than the usual later end time. The principals estimated that most parents fell into middle income bracket though there was great variation in socio-economic backgrounds within the parent community ranging from very low to "obnoxiously high". (Sometimes the Christmas hampers prepared by the school go to families with children in the school.)

The Parents Advisory Council (PAC) generally was in strong support of the school, and of improving science more specifically. The parent advisory council contributed both in time and money to school affairs. Teachers who needed money to bring about change in their science teaching could apply for small grants for the purchase of supplementary materials. The advisory council worked actively on making connections between the school and scientists and engineers so that teachers could draw on a "visiting scientists and engineers" program whenever they feel the need. In one instance, PAC assisted Mountain Elementary in purchasing microscopes that were sold off by a local biomedical laboratory. In another instance, PAC assisted in the construction of a weather unit that was subsequently used by the teachers.

In terms of school culture, the principals regarded the proximity of parents to teachers an important factor for bringing about a positive overall learning environment – parents were very involved and close relationships existed between teachers and parents. Some parents came to the school often and became part of the school life in academic and leisure terms. For example, every Friday after school, parents and teachers played floor hockey against the students. This activity also offered another example of the strength of the positive climate and inclusivity of the school. Even teachers who were no longer at the school asked to be notified of these activities so they could join.

2.1.2. Teacher Community

Teachers in this school formed a very close community with professional and personal bonds in and out of school. Both principals attributed the culture of the school to a number of things. First, they mentioned informal activities – treat days, monthly lunch days organized by and for staff, frequent staff parties and gatherings, impromptu gatherings (e.g., sometimes a few staff are in the staff room talking and a bottle of wine left over from some previous event will be remembered, pulled out, and shared among them), holiday celebrations, and skiing or skating events – that brought the teachers together and provided a bond between them. Second, the principals offered administrator history as contributing to current culture. One former principal, Werner, had a counseling background and worked hard on establishing a positive school climate. Although retired, he was still invited to gatherings, staff showed great affection toward him, and parents continued to talk about him. The immediate past principal, Kenneth, was also a strong leader who had a positive influence on the established school culture. Stories about Werner and Kenneth still circulated and bound members of this community. Teachers and principals recounted other stories that constituted the community, and preserved the history among those who had not been part of the staff at the time. It has been pointed out that such stories, indeed, constitute a form of institutional or collective memory in which the identity of an institution can be celebrated (cf. Middleton & Edwards, 1990). These stories not only make sense of the teachers world, but at the same time make something of their world, and constitute the faculty as a community.

This strong community identity was thought to be the origin of the low turn-over rates among the teaching staff. Most of the staff had been with the school for a number of years and "people don't easily transfer out of here". This impression was further supported by a teacher who earlier in the year had returned for a visit to Mountain Elementary School (she had left because of a big shift in grade level assignment). This teacher talked about how different it was at the new school in terms of the relations among staff members and the general tone in the school – she wished to be back at Mountain Elementary. It was also evident that many of the teachers had relationships with each other outside school which added to the informality, familiarity and camaraderie evident in the school. The principals also suggested that staff members could voice their needs and had strong commitments to work together which also contributed to the sense of community at the school.

2.1.3. *Problems and Constraints*

Although the teachers at Mountain Elementary were generally enthusiastic about changing their teaching and had an awareness of the weaknesses of their current science program, there existed constraints that did not allow them to construct efficient solutions. When I arrived at the school, teachers suggested things like "there is a need to make science more human", "girls aren't into science as much as boys... I guess it starts with parents at home... girls aren't necessarily less motivated but we need different approaches", "a lot of science is touchy-feely", "still comes back to teaching strategies... some teachers teach the same way regardless", "we don't have the jargon or background knowledge to apply it", "a lot of things are making science more distant than it should be", and "the school expects to have cohesion and blending but there's so much division and difference that it can't be".

Although they had tried in the past to bring about change in science teaching in the school as a whole, all attempts appeared to have failed ("we've tried over and over again to meet with groups but we never have time"). Some of the problems that were identified included: limited access to resources for science teaching; lack of science content knowledge and pedagogical content knowledge for teaching particular units; lack of equipment for teaching science; limited budget to purchase new equipment or supplies; lack of understanding of and approaches to gender-related issues; and lack of time to prepare a class which, because of its short duration, does not warrant the amount of preparation time needed. Above all, budgetary constraints were constructed as a major impediment to reform.

Over the past few years, budget cuts presented continual challenges. The principal indicated that the cutbacks in the past had represented about 5% of the total budget. During the year of the project, the school lost "30% or more" from its budget. This cut was not known at the end of June when budgets had been planned. The outcome was that about 70% of the budget had already been consumed by September. The principals had to meet with the staff to decide about measures to reduce their photocopying because they did not have enough paper supplies till the end of the year. The principal also noted that the new equipment budget was gone, library resources were severely cutback and many budget items were completely eliminated in the new budget. The engineering design curriculum responded to these budgetary constraints because it made use of everyday materials and scraps that were procured from a variety of sources; it also provided teachers with release time for observing other science classes and planning curriculum.

2.1.4. *Improving Science Teaching*

In this context, teachers wanted me (and my research team) to tell them what to do about improving science teaching, and to put on workshops, rather than to begin transforming their own conditions. For the first staff development day, which was to focus on science and science teaching, some teachers initially wanted "experts" (such as myself) to put on presentations. Because of my belief that sustainable change can only come about if the people concerned take charge of their own learning, I expressed my reluctance to put on a "dog-and-pony" show. In the course of three meetings with Sandy, the coordinator for the professional development day (who was later joined by other teachers to form a planning committee), teachers finally decided to take matters in their own hands. With the help of one of my graduate students as a facilitator, they planned a professional development day during which they could brainstorm (a) their ideas about science, (b) their beliefs about science teaching, learning, curriculum, and connections of science to society, and (c) their ideas for a vision statement that would direct their professional development activities for the next three years. An entry from my field notes shows that at that point I was convinced teachers were on their way to changing their own condition:

What we have done so far is to help the Mountain Elementary community get underway towards new dimensions of science teaching and learning. We hope to facilitate this process further as we work in teachers' classrooms. At this point, we won't be able to do more than the three classrooms which I indicated above [where I had committed to facilitate curriculum development and classroom teaching], and we have to see how the teacher community develops out of that.

The professional development day turned out to be a success from the perspective of many teachers and also from the perspective of the two principals. The vice principal noted that for the four years she had been at the school, this was the only professional development day for which the staff as a whole had prepared and in which they showed the highest participation. For the first time, they had been active participants rather than passive recipients of information provided by "experts".

Over the following year, teachers formed a variety of committees that investigated and planned curriculum for the entire school, and ways to improve students' and teachers' attitudes towards science; organized a science supply area; set up and ran a filing system to share teaching ideas, kept a science bulletin board, and published a science newsletter; established a way of surveying and distributing existing resources across the staff; and, with my help, received a grant that allowed them to hire substitute teachers so that regular teachers could pursue in-service related to science teaching at the school. Tammy, the first teacher with whom I worked, and the teacher of the Grade 4–5 classroom featured in this book, volunteered to be the science coordinator and the liaison with me and my research team. She became one of the driving forces of the teaching reform in science teaching at Mountain Elementary School.

2.2. STUDY CONTEXT

2.2.1. *Participants*

2.2.2.1. *Teachers*

The first teachers to be interested in working with me, and hosting a research team were Tammy and Mary. The two shared responsibility for a split Grade 4–5 classroom such that Mary taught Mondays through Wednesdays, and Tammy instructed Thursdays and Fridays. As it turned out, Gitte, who team-taught with Tammy the engineering design unit on which this study is built, had arranged her schedule in such a way that she worked part-time from Monday through Wednesday as a curriculum developer in a local, non-profit agency, and as a graduate research assistant on Thursdays and Fridays. Because of this arrangement, and although very interested in working with us in the class, Mary did not actually teach part of the unit.

Tammy had 12 years experience teaching at the elementary school level, six of which were part-time; she had once taught a related unit on bridge building. She had also attended a conference presentation on building bridges and a three-hour workshop on the engineering design curriculum (described below). Gitte had three years classroom experience in elementary schools at the grade 4 level, and had worked as a curriculum developer and workshop presenter for the past four years. During this time, she co-wrote another curriculum, "INVENTIONS: Ideas in Motion" that provided children with opportunities to learn about simple machines by designing machines. She also wrote the manual for a gender equity workshop, "Gender Inclusiveness in Math, Science, and Technology".

Although the curriculum (better known by Gitte) outlines a series of specific activities, the two teachers planned the activities together to adjust them to the specific needs of the children in Tammy's class. Tammy and Gitte quickly established a very close and personal relationship. They frequently phoned each other to talk for extended periods of time about the approach to be taken during the next lesson. Both felt that their relationship with respect to the curriculum and the class was very symmetric and they felt that they learned tremendously from one another. Tammy learned about questioning children in the context of science and engineering, and learned content knowledge in science and engineering. Gitte learned about maintaining a classroom community in which students' ideas, feelings, and sense-making were validated.

To prepare Tammy and Mary for teaching the unit "Engineering for Children: Structures" ("Engineering for Children: Structures"), I invited them to attend one of the workshops which the curriculum developers offered. In fact, this workshop was a necessary prerequisite for understanding the teachers' guide which included only bare-bones explanations. The workshop was primarily attended by a class of pre-service elementary teachers as part of their science methods course, "Designs for Learning Science". Leena, the workshop leader was, as Tammy and Mary, a Grade 4 teacher. This gave rise to many exchanges between them as to specific details of implementing such a unit with students at that age. Leena led the entire group through a few of the initial activities in the curriculum, such as stabilizing structures, forming joints, and strengthening materials. Later, the workshop participants engaged in the construction of various towers by working in small groups of four to five participants.

Initially, Tammy and Mary did their activities separate from the remainder of the participants. The pre-service teachers, because they knew each other, engaged in lively conversations that excluded Tammy and Mary. As it was, although Tammy and Mary had a lot of experience as teachers, none of the pre-service teachers could learn from their stories of learning in the classroom. Tammy and Mary actively engaged in the activities which they partially knew from a previously attended conference presentation on bridge building. Leena wanted to bring the various groups together in a large setting so that workshop members could share with others their stories of success and failure.

It was interesting to note that Tammy and Mary began developing new ways of describing their experience by creating their own words. For example, they described their activities of adding braces to strengthen a straw tower as "dewobbelizing a tower". In the context of the framework I outlined in the previous chapter, we see here an instance of creating a new language game. This was also interesting, because later in their own classroom, I observed a similar phenomenon: They encouraged children to build their own language games. Children talked about "wiggly" bridges, "spaghetti stretches" (a way to stabilize an elastic by stretching it over a piece of spaghetti), and "unshaky towers" (towers that had been stabilized using "triangles"). Interestingly enough, one of the pre-service teachers pointed out the non-technical nature of such talk, but Tammy suggested that the children in the classroom would talk in this very way.

During the whole-group sessions, the pre-service teachers mostly asked technical questions or about particular strategies for implementing the unit. Tammy and Mary's questions differed. A university colleague and I characterized these questions as "seasoned" as compared to the "naive" questions of the pre-service teachers. Tammy and Mary's questions were related to (a) specific pedagogic issues, (b) the age at which children would be able to do certain activities, (c) the part of the yearly school cycle to teach the unit, and (d) technical aspects that are critical to successfully implementing the engineering design unit. The seasoned teachers' conversation and talk was thoroughly practical, in terms of the everyday life in elementary school classrooms.

2.2.1.2. Students

This investigation was conducted in a mixed-grade French Immersion classroom with 23 Grade 4 (10 boys, 13 girls) and 5 Grade 5 students (3 boys, 2 girls). Ordinarily, French is the language spoken in such classrooms. However, to accommodate Gitte who is unilingual, engineering design was taught in English. Although Tammy felt ill at ease at first, she thought that the advantages for her own and the children's learning afforded by team-teaching the unit would outweigh children's lack of practice in French. However, whenever feasible, she took the opportunity to conduct her conversations with small groups in French. According to Tammy, all students ultimately exceeded her expectations with respect to problem solving, levels of interest and task engagement, and competence for working collaboratively with peers. Some of the students who under normal circumstances and in traditional classroom environments did not achieve well did so in this unit. The following vignettes present students who were representative of the class and who feature more extensively in later sections. The guidelines of this Canadian province mandated at the time anecdotal evaluation rather than grades. The vignettes are based on such anecdotal information provided to me by the teacher.

Arlene and Chris. Arlene and Chris were both 10-year-old Grade 5 girls who were so similar that Tammy talked about them in the same breath. They had been in-school and out-of-school friends for a long time. Arlene and

Chris were very independent, managed their time very well, and always submitted neat and well-organized work. Both were keen and enthusiastic, enjoyed learning and experimenting with new ideas, and very competently related to and communicated with others in the class. Academically, they were above average in all subjects. Throughout the unit, like all other Grade 5 students, Arlene and Chris never changed partners and always worked together. During my initial classroom observations and analyses of the data, I was struck by what appeared to be a rather unbalanced approach to collective tasks. One of my research team members captured their relationship in the metaphor of "the surgeon and the operating nurse". Chris often made one-word utterances ("pin!", "straw!", "tape!") which observers and Arlene ("Yes ma'am") alike heard as commands for materials. However, both girls regarded their relationship as balanced and the stories they constructed of their design activities emphasized joint decision-making.

Andy. Andy was a keen, conscientious, usually well-organized Grade 4 student. He worked diligently and always finished his tasks on time. Andy related well to his peers and worked well in groups. Academically, he was very good in mathematics and demonstrated good problem solving abilities. For example, he was the only student to solve a stability problem that asked students how many straws it would take to stabilize a 15-sided figure given that 1, 2, and 3 straws, respectively, stabilized a square, pentagon, and hexagon. Andy had some problems with spelling, using vowels, and applying rules and patterns in the English language. Although he commented frequently on the difficulty of the tasks, he ended up at Level 1 in my evaluation of the conceptual understanding of the bracing practices (for details see Chapter 7). In group work, he usually followed his team mates. Throughout the unit, he attempted to please teachers; in his reflections, he attributed achievements to the teachers suggestions and he often attempted to give the "right" responses. When he felt unobserved by the teacher, however, Andy enacted silly behavior (even in front of the camera) that attracted the comments of his different partners ("Andy is weird", "he's cracking up", or "Andy is silly").

Clare. Clare, who was nine years old and in Grade 4 at the time of the study, generally achieved at a satisfactory level but had difficulties expressing her ideas. She made many spelling mistakes (her writing was largely phonetic) and the entries in her engineering logbook were very short compared to those of her peers. She appeared somewhat stronger in mathematics. In the past, Clare had problems relating to others. She was generally very concerned how others perceived her. At the same time, she could be very abrupt with peers when they did not agree with her suggestions. According to Tammy, the "Engineering for Children: Structures" unit allowed Clare to mature significantly; Clare learned to solve problems creatively without teacher help and worked well with partners. She frequently expressed pleasure in learning science ("It was fun", "I like science, don't you?", or "I enjoy science because we have almost finished our bridge").

Stan. Early in the school year, Stan, a nine-year-old Grade 4 student, was diagnosed as having an attention deficit hyperactive disorder for which he had to take medication to control his behavior. Tammy suggested that since beginning the treatment, there was a marked difference in his ability to stay on task for long periods of time. When he forgot to take the drug, he appeared restless. Academically, he was very weak in all aspects of the curriculum, although he had slightly improved since beginning the treatment. Tammy thought that this unit tapped Stan's strengths in a particularly good way. The open-ended activities underscored his divergent thinking and spatial skills so that his design artifacts were always very well-made and provided evidence of considerable thought and creative ideas. Stan had problems relating to others. His classmates generally refused to team up with him (even teacher mediation did not help in finding a partner) so that he had to work on his own during the first seven weeks of the unit. However, about seven weeks into the unit, he had an excellent experience collaborating with Tim so that both decided to continue their partnership until the end of the unit.

Tim. At the time of the study, Tim was a nine-year-old Grade 4 student. In most areas of the curriculum, he had minor difficulties with reading, and his writing lacked somewhat in ideas. He usually worked well with others, and Tammy considered him a good problem solver. During the first few lessons of the "Engineering for Children: Structures" unit, Tim was more interested in socializing with others than completing his own tasks. However, he later was very enthusiastic about all tasks, continuously demonstrated problem solving ability, improved his ability to concentrate on tasks, and learned to get along and work with peers other than his normal friends (including Stan). During class discussions, he ventured many ideas and demonstrated interest in, and a good understanding of, engineering issues.

2.2.2. *Data Construction*

Because of its focus on reasoning in complex, ill-defined situations, my investigation was modeled on studies of cognition in everyday activity, studies of work practices in scientific laboratories, and studies in cognitive anthropology (e.g., Hutchins, 1995; Lave, 1988; Lynch, 1985; Suchman, 1987). Direct observations and videotapes of design-related activity; field notes; interviews with students, regular teachers, and visitors; and students engineering logbooks constituted the primary data sources. Data analyses adhered to recommendations by theorists in the domains of interpretive research (Erickson, 1986), discourse and conversational analysis (Edwards & Potter, 1992; Potter & Wetherell, 1987), and interaction analysis (Jordan & Hender-

son, 1995). Any assertion constructed was tested in the entire data corpus
or used to direct further data collection to seek both confirming and discon-
firming evidence.

In this subsection, I describe (a) how I became part of the culture in the
school and the specific classroom in which the research was conducted, (b)
the data I collected, and (c) how my data and data collection changed the
classroom under investigation.

2.2.2.1. *Becoming Part of the Culture*

I began to interact with the two regular teachers of this classroom, Tammy
and Mary, long before I formally conducted my study of children's learning
during the engineering design unit. The actual data collection was intensive
from October to December; less intensive follow ups were conducted in
January and February. In May and June prior to the study, I began to visit
Tammy and Mary's classroom. During this time, I came to know Tammy
and Mary quite well. I found out about their approaches to teaching and
their relationships to the children. By interacting with the students during
their normal classroom activities, I got to know some of the same students
who later participated in the formal classroom study. Toward the end of
June, both participated in the workshop described above.

During the first month of the school year (September), I came to class
once or twice per week to get to know the children who were taking part in
the study, to get to know better the teachers and school, and to become part
of the community myself. The following entry from my field notes describes
the situation in which I frequently found myself.

Mary's classroom. When I got there, four boys approached me and then were all over me. They
wanted to do the drama activities and science with me. Then they fought for the right to sit
next to me. I sat with them in the two activities where students sit in a circle and do, in turn,
some demonstration, the others try to guess what the student was doing or signifying. In the
first activity, we had to use a huge three-hole punch and portray an activity, an object, and the
others had to interpret what we portrayed. As soon as someone found out, the turn went to
the next person. In the second activity, we had to pretend with our bodies to model something
in clay, and the others had to interpret what we modeled. The kids understood this activity
much less, and modeled things that you don't usually model in clay, and some even portrayed
an action; that is, their construction of the task was a completely different one than mine, or
some of the others.

After the drama activities, Mary explained to the students (a) how to correct their spelling
tests, by three times re-writing the word and (b) how to do the worksheet for the night. Mary
and Tammy use quite a number of worksheet activities, and sometimes ask the kids to copy
such things as crossword puzzles (those where you have to find given words) into their notebooks.
Compared to simply giving them a photocopy, this seems quite time consuming (or doing it on
spirit duplicator). I imagine that there are bunches of possible reasons, such as cost of photocopying, a consideration of giving the kids a task which occupies them for a longer time, they can
learn some skill related to the copying itself, like accuracy etc. For my part, I would probably
give the kids the puzzle and have them try finding the words rather than spending their time
on copying the grid.

Frequently, I participated in the drama and sports activities. Tammy or Mary and I sat with children in a circle where we engaged as a collective or with student-partners to our left and right. These activities were always designed to bring about a sense of community; children decided on the nature of the activities and determined the turn-taking. It was a time to build a relationship with the children in this class. At other times, I sat down with groups of children to talk about their writing or mathematics, and talked to the students, in French, about their activities, interests, and other aspects of school life. In one situation, I observed a group of Grade 4 boys completing a worksheet of "problems" at the Grade 5 level. They were to construct equivalent fractions by filling in the missing part in statements such as $\frac{1}{2} = \frac{}{6}$. I saw that, from a mathematics perspective, not a single answer on their worksheet was correct. For example, they had written $\frac{1}{2} = \frac{5}{6}$. They had simply added to the numerator the same number by which the denominator had increased. Rather than "teaching" about equivalent fractions by telling them, I introduced my computer and a mathematical modeling program that allowed me to write fractions in the form to which children were accustomed:

$$\frac{1}{2} =$$

$$\frac{5}{6} =$$

By setting up the two sides independently in equations where their value was expressed as a decimal, children obtained a system that talked back to them. When they "checked" their equations, they found that their "equivalent" fractions had different decimal values:

$$\frac{1}{2} = 0.500$$

$$\frac{5}{6} = 0.833$$

They began to tinker changing the missing number until their fractions were equivalent as indicated by the decimal form. Not five minutes later, the boys got excited. They explained to me that they had figured it all out. They told me that all you had to do to make fractions equivalent was to "times". When the denominator was "timesed by two", the numerator also had to be "timesed by two"; when the denominator was timesed by three, the same had to be done to the numerator; and so on. I realized later that his was significant in terms of having a system that talks back. For here, much like in the engineering design activities in which these boys subsequently engaged, they were able to work with a system that talked back to them about the conjectures they made. Unlike the paper-based worksheet, the mathematical modeling program constituted a medium that talked back. Here, they could express some idea, and test it in a context that provided feedback to their actions. Rather than telling them some correct answer, it allowed them to gain personal experience, to search and construct patterns on their own; here, this resulted in their "discovery" of canonical mathematics.[2]

Such interactions allowed me to become part of the classroom community. Later, when I began filming, children became part of the research community at those moments that I provided them with the opportunities to do some filming on their own. Apart from the difficulties children had with handling the (for children) heavy equipment, these moments allowed me a glimpse at the world through childrens eyes, their interests and intents, which were frequently other than my own.

My interactions with members of the school community at large were not limited to those with Tammy, Mary, and the students. During the early phases (May, June) of the study and during September, I taught or team-taught science lessons in other classrooms and at various age levels. I attended all staff meetings, meetings with the Parent Advisory Committee, professional development days, and parent evenings throughout the year. Spending entire days in the school allowed me to interact with teachers in their classrooms (where I was always invited), in the staff room during recess and lunch, and after school. After completing the study in Tammy's classroom, I continued to work in the school for another two years. I taught and team-taught entire units extending from 10 weeks to four months. This allowed me to continue my dialogue with Tammy, and to seek further information regarding the students' development and Tammy's own views about teaching the engineering design curriculum.

2.2.2.2. *Assessment of Knowing and Learning*

My assessments of children's views about engineering and competencies in engineering design were set up in collaboration with Tammy, for she did not want these activities to resemble formal testing. She preferred my assessments to be integrated and part of the regular classroom activities. I therefore structured data collection around her and the childrens needs.

Pre-unit assessments. The pre-unit assessments consisted of two major activities. For the first activity, children were asked to draw or write a completion of "Engineering is . . .". After the children had drawn and written about their associations with engineering, we invited them to talk about their drawings in one of four focus groups comprising six or seven children and chaired by one researcher. After about 30 minutes, we joined pairs of focus groups into one larger group of thirteen students, jointly chaired by two researchers. Each meeting was videotaped and/or audiotaped. The interviews were semi-structured, but we were interested in eliciting answers to questions such as, What is engineering? and What do you need to be able to do to

---

[2] A critique of "discovery" learning develops from the results and discussions reported in Chapter 8.

become an engineer? The complete interview protocol which we took as a guide is provided in Table I.

The second part of my intake assessment was in the form of challenges which the teachers regarded as motivational preparation for open inquiry, that is, design problems. Together, Tammy and Gitte had decided to offer three such challenges:

- To build a bridge from one sheet of $8\frac{1}{2}$ inch paper by 11 inch that used two inverted paper cups as support, spanned 3 inches, and supported as much weight as possible (pennies in a third paper cup to be placed on the bridge).
- To design different ways to pick up and carry a variety of glass bottles and jars using one straw.
- To support as many nails as possible on top of the head of another nail that securely rested in a piece of board, without using any materials other than the nails themselves.

Post-unit assessments. Children's learning after some instruction was assessed on the basis of the following data. First, near the end of the three-month unit, the children were asked to write one or more sentences following "Engineering is . . .". Then, students were asked to prepare what the local school culture referred to as a mind map[3] that included all the things/activities that they associated with "engineering". Students completed this assignment individually. I also used the glossary prepared by each child toward the end of the unit as an aspect of their engineering-related discourse. Each term in the glossary included a verbal description and a labeled drawing. Mindmaps and glossaries were entries from the children's engineering log books. Finally, I used the records of children's design processes and products during the bridge project (a three-week period beginning Week 10 of the unit) as evidence for the levels of competence in design-related material and discursive practices. This assessment constituted a comparison case for the initial paper-bridge challenge. The design efforts of two groups and their interactions with other groups and the teachers were videotaped.

Process data. Two cameras were used to videotape all lessons from the beginning of October to the Christmas break, but not the dome activity completed in January (circumstances and commitments to other teachers prevented me from continuing to work and do research in Tammy's class). All videotapes were transcribed as soon as possible, most often within 48 hours. In addition to operating one camera, I also collected field notes on index cards which I transcribed and elaborated once I returned home at the end of the day.

---

[3] Teachers also called this technique webbing. Mind maps resemble concept maps but do not have linking words between the concepts and therefore are more webs of associations rather than semantic maps.

TABLE I
Interview protocol for the pretest

What is engineering?

1. Have you heard about engineering?
2. Do you know any engineers?
3. What is engineering?
4. What kinds of things do engineers do?
5. What kind of things do you have to know about to be an engineer? (Problem solving)
6. How difficult is it?
7. What do you know about engineering, engineers?
8. What do you need to be able to do in order to become an engineer?
9. What strikes you about engineers?
10. Can you think of anything at school, home that an engineer has helped with?
11. Do you think you could be an engineer? (Is it for boys and girls, men and women?)

Prior experience

1. Have you built (helped to build) structures?
2. What structures do you like to build?
3. What materials did you use?
4. What structures do you find difficult to build?
5. Do you have LEGO, MECHANO set?

Specific knowledge about structural engineering

A. Strength
1. What materials do we (engineers) use to build with?
2. Which of these are strongest?
3. What materials are strong?
4. Which is stronger, wood or steel? (Why?)
5. Why are many sailboats/houses/??? build from wood rather than steel?

B. Connections
1. What connections between materials do you know?
2. What kinds of materials do people use to connect with? (wood to wood, paper to paper, fabric to wood, rubber to rubber)
3. What other things might there be to connect with?
4. Are there connectors which don't need glue, nails, pins, etc?
5. What other ways are there to connect materials?

C. Shapes
1. What are some solid shapes? (Children might think of solids rather than the shapes; have possible materials)
2. Of these solid shapes, which ones do you think are strongest? weakest? which shapes wobble/wiggle easily?
3. What kinds of shapes have you noticed in the buildings, bridges or other structures around you?

D. Everyday relation
1. What features make bridges/houses strong?
2. If you had to build, which (structural) features would you include?

Attitudes about school, science, learning styles

1. What do you like/hate about school science? why?
2. Do you like to work in groups, with other people? who do you prefer to work with?
3. Do you like working with students from the other sex?

After each lesson, I debriefed Gitte. I debriefed Tammy whenever her schedule allowed. These debriefing meetings were videotaped and, in addition, recorded in the form of personal field notes. During these meetings, the teachers and I talked about what had happened during the previous lesson from our different perspectives as teacher and researcher. As in all design experiments, these conversations constituted not only data for the research, but in important ways influenced events in the classroom. First, Gitte and Tammy, alerted by my own framing of classroom events began to look out for specific events or to direct classroom interactions in such a way as to account for my observations and theoretical considerations. Specific examples are discussed in the following section.

I also debriefed, both formally and informally, any visitor to the classroom. For example, there were two female engineers who talked about aspects of their own work and then interacted with the children about the artifacts they had constructed. There were also teachers from the same and other elementary schools who visited our classroom to learn about teaching a design engineering unit. These visitors' impressions were recorded either on videotape or in field notes.

2.2.2.3. *Interaction of Data Collection and the Emerging Curriculum*

In design experiments, it is typical for the information collected by the various observers to change the setting under observation. In fact, this information can be used to enhance the positive aspects of the learning environment and to suppress other, less desirable aspects. Because I did not conceal information and working hypotheses, there were situations in which Gitte and Tammy changed their classroom behavior. This cycle of data collection, interpretation, revision, and influence on the class is represented in the following vignette.

Tammy and Gitte had set out to teach the unit in a gender-sensitive way. However, during the first week of teaching this unit, there appeared to be an imbalance in the frequencies with which boys and girls were asked to respond to teacher questions in whole-class situations. This working hypothesis was confirmed by a simple frequency count of the interactions in the existing videotapes and the interactions during the next class (4 boys to 1 girl). Both teachers made a conscious effort to shift this ratio to 1:1. The boys reacted by asking, "Why are there so many questions directed towards girls?" whereas the girls asked "Why do we have to give all the answers?"

In a similar way, Gitte and Tammy changed the organization of their classroom as a direct consequence of my observations. In one situation, Andy and Dennis took their materials to the foyer to build imaginative creatures or things from a small set of given materials. There, they worked in relative isolation (apart from two brief exchanges with another student) from the other members of the class. However, as the lesson evolved, some students in the class had changed the rules of the task: they used additional materials or, after negotiating with the teachers, formed groups. Andy, who liked the artifact he had constructed immediately after completing it, changed his attitude once he returned to class. After he had refused to present it in a whole-class "sharing" session, I asked him about it. He said, "after I saw what everyone else had done, I didn't like it anymore". (The sequence of these events is analyzed in some detail in Chapter 5 because it reveals important aspects about learning as an outsider relative to a specific community.) I related this incidence to Tammy who, during the next period made the following announcement.

There were a couple of people that didn't work in the same room as the rest of us, and one thing that was interesting [was] . . . that those were some of the only people who did not use any extra stuff or the ideas of other people. So that just goes to show you that when you are working among other people who are doing a similar kind, you are not really copying; but it is so important to be using the ideas of other people to change your ideas and to improve on them.

In this way, my observations were fed back into the classroom and changed the community I was observing. Similarly, because Gitte and Tammy borrowed videotapes to view them at home, they began to construct certain phenomena or noticed certain patterns in their own behavior. As a consequence, the research influenced the way they were teaching.[4]

2.2.3. *Data Interpretation*

A central aspect of social science research is the construction of an empirical object, that is, some category that is said to "emerge from the data" but in fact is constituted by the embodied practice of an investigator's interactive iterations between construals (tentative understandings) and a yet-to-be structured world (Gooding, 1990, 1992). In series of small rectifications, amendments, sets of practical principles, and so on, construals become constructs that are reified in the world. This cycle of constructions may involve something of an "epistemological rupture" (Bourdieu, 1992, p. 251), a break with viewing the world as we have done before, and a new beginning that includes the bracketing of ordinary pre-constructions and common sense to make them the topic of research. This rupture demands something of a conversion of one's gaze. In this section, I provide an account of my situated interpretive efforts that permitted me to view learning in elementary science classrooms with new eyes.

2.2.3.1. *Constructing the Evidence*

The audiotapes and videotapes were transcribed soon after they were recorded (I did or contributed to most of the transcriptions). The transcripts entered a database which also contained field notes, artifacts, children's

---

[4] I presented extensive observational and theoretical descriptions of Gitte's and Tammy's learning that resulted from their co-participation in teaching this unit (Roth, 1996d; 1998).

engineering log books, and photographs of their projects at various stages of development. In daily meetings with my research team, we generated tentative assertions. These assertions were subsequently tested in the data corpus or during subsequent observations. On the basis of the new information, working hypotheses were discarded, modified, or retained. For example, one day Tammy praised Kathys definition of stability with the following remark:

(Tammy begins to read from Kathy's notebook:) "A thing has to be stable to stay up that's what stability is, the roof is stable. I think that stability is mostly triangles that you put on a shape. I did a dome and something like a cube and two things like a pyramid. I did six things. I worked on Carla's ideas, and she worked on my ideas. When I think of stability, I think of triangles, because when I put a triangle on all my shapes, they were stable. I think it was fun to do stability. One of my shapes is on the page in front. It is a two-dimensional shape. I made a Science World out of tooth picks and marshmallows it was the fun thing to do". Was that an excellent description of what was done? My compliments Katy, I thought that was extremely well written, and it was very clear and interesting, too, how you came up with your ideas.

Following this lesson, I entered the field note, "Tammy values children's discourse, highlights it, and celebrates this as part of the classroom culture". (This episode also shows how a child recognized the reciprocal influence between her own and a peer's work which was an important aspect that contributed to the class as a community.) When the research team looked at Gitte's teaching, we extended and confirmed this assertion. For example, Gitte called on Andy to present his invention, a "spaghetti stretch" (the term was Andy's invention).

Each day we try and look at a couple of techniques of strengthening and joint construction and today I want to point out one that Andy did. Andy, can you come up and show us your "spaghetti stretch" and explain how it came about, how you discovered the spaghetti stretch strengthening technique?

In addition to such confirmatory evidence, we searched the existing data sources and directed our classroom observations toward incidences in which our assertion did not hold. Thus, the following entry in my field notes illustrates that the assertion was not universally valid.

Gitte is looking for specific words. Kids react differently, they are more stifled. First, Gitte attempts to elicit "foundation" then "earthquakes". In both cases, Gitte has kids guessing for a while. As a listener, one has the feeling of a constraining environment. Gitte waits for a specific term, brushes over kids' words very briefly, waits for that specific term to be uttered. When the kids brainstorm, on the other hand, each term is almost celebrated, written on the blackboard, added to the list. In the first instance, classroom talk is narrowed down, guessing is emphasized, getting the right word. In the second, kids look for ideas and divergence.

On the basis of this interpretive process, I arrived at the assertion that in general, teachers valued students' contributions in establishing new, engineering-related classroom discourse. However, on a few occasions, they spent considerable effort "teaching" specific terms – mostly driven by the concern that children should know a few specific engineering concepts.[5]

In this study, I considered the tape-recorded materials and transcripts as natural protocols of students' and teachers' efforts in making sense of events in their material and social settings. Protocols, together with artifacts, provided occasions for construing the discursive and material resources and practices of individuals, groups, and the classroom community. My interpretive work was guided by the ethnomethodological advice that researchers should construct phenomena based on participants' own, documentable understandings (Livingston, 1987; Pollner, 1987).

2.2.3.2. *Credibility of Interpretations*

In interpretive studies such as the one reported here, the classical construct of internal validity is replaced by the notion of *credibility*. Associated with credibility is a set of criteria that ascertain rigor in qualitative inquiry (Guba & Lincoln, 1989). These criteria, widely recognized by anthropologists, sociologists, and others who engage in fieldwork are *prolonged engagement, persistent observation, peer debriefing, negative case analysis, progressive subjectivity*, and *member checking*. After briefly pointing out in which way prolonged engagement, persistent observation, peer debriefing, and member checks have been implicit in my account to this point, I will address in greater length the issue of negative case analysis and progressive subjectivity.

I already described above my prolonged engagement and substantial involvement at Mountain Elementary School which helped me overcome potential effects of misinformation and distortions. My description of the data collection also revealed the variety and density of data collected; that is, my description constitutes an account of extended observations and circumscribes the scope of the study. The daily debriefing meetings with Tammy and Gitte, as previously described, were occasions for testing emerging hypotheses and, in this, constituted member checks. Informal in situ interviews with children also allowed me to test my understandings of the ways in which they constructed their lifeworlds. This allowed me to continually check my own understandings with those of the teachers and students I observed. From the beginning of the study, I had many opportunities to interact with colleagues and graduate students at my institution, and with colleagues all over the world to check my interpretations of various data. Through these interactions, my interpretations evolved. The most striking confirmation of some of my understandings came from an independent analysis of a videoclip done by Jay Lemke (1995). His analysis struck many familiar cords, confirming my own understanding of the events in the classroom.[6]

---

[5] In Chapter 7, I provide a detailed description of one of these attempts.

[6] This does not make my analyses more true. It simply means that two researchers who work from a similar perspective arrived at convergent observational and theoretical descriptions.

The story about Andy and Dennis working in the foyer also became an important part of a negative case analysis. As I report below, students frequently used a tool or idea, or engaged in a practice "because everyone else is doing it" (Chapter 5). While Andy and Dennis worked in the foyer, Jeff began to use additional materials to make his science fiction structure, and Simon and Tim combined their materials to build a giant spaceship. Other students used this as precedence ("everyone else is doing it") to negotiate with Gitte or Tammy about using additional materials or engaging in joint projects. The only ones left out were Andy and Dennis who, working alone in the foyer, did not realize that the rules of the activity had changed. Together with another, similar case recorded one year earlier in a different context (Roth & Bowen, 1995), this negative case was important in establishing my understanding of knowledge-building communities. In particular, this story contributed to my construction of the distinction between outsider and marginal.

My understandings of the process of designing engineering artifacts evolved over time. Consistent recordings of observational and theoretical field notes, and a variety of draft manuscripts and articles written from the beginning of the study to the moment of writing this book, permitted me to track this evolution over time. Consistent with the theoretical framework outlined in Chapter 1, this evolution of understandings is constituted by my changing language game regarding children's designing; and with the changing language game, I came to "see" the videotapes and other data in new ways. Some of my observations in the workshop were, seen from this vantage point of retrospection, important primers for observations to come in the elementary classroom. They were construals, tentative descriptions which I further developed during subsequent observations and writing activities. One of these observations is captured in the following field note.

One group had designed a bridge in the form of an arch. As the bridge emerged, the members of the group found that it was really too steep for vehicles or pedestrians to use. One of the members then suggested to "hang" a platform from their current design, and thus to reinterpret the former platform into an arch that suspended the new platform. Here, they had "invented" a bridge, apparently oblivious to the fact that they had reinvented, in their designing, a widely used engineering principle. Once constructed, they began to liken their artifact to existing bridges.

My description of this event turned out to be significant as it foreshadowed coming observations in Tammy and Mary's classroom. In the workshop, the ill-structured environment allowed students to redefine the meaning of an artifact in such a way that it afforded a new solution, and the reflective construction of an understanding of an engineering principle. It was interesting that in framing the deck as "hanging" from the arch, they had constructed a critical description that supported subsequent theoretical descriptions of the design principle. This became quite clear in the exchange of the group with another student, Larry, to whom the principle was not at all clear.

In the whole-group debriefing, Larry thought that the new bridge designed by the group was not feasible because, as he saw it, the new deck had to carry extra weight. In the discussion with the other students, he created drawings of the two bridges. While it was immediately evident to Larry that in a suspension design, the bridge deck was hanging on cables suspended from towers, he did not "see", the same in the group's design. His description, "bridge deck that carries an arch" interfered with the more appropriate description "bridge deck hanging from arch". His language game did not allow him to "see" in the same way that the designers understood. Only after repeated attempts by his peers, did Larry change his description and observations, and in this, constructed an understanding of the advantage the arch brought to the bridge design.

Here I began to recognize that it was important to arrange for meetings of several groups of designers to bring about exchanges between members. Small group exchanges were not sufficient and were in some sense erratic. Large group meetings, on the other hand, were activity structures in which teachers could increase students' awareness of the distributed nature of expertise. These allowed Tammy and Gitte to help students realize that there were people who had constructed answers to specific design problems, on whose expertise they could draw, and whose solutions could inspire solutions to other problems. I also began to realize that prior tasks such as building arches, using braces to stabilize structures, and reinforcing materials provided resources to the later activities. At the same time, these resources were constraints to children's design activities, because they frequently precluded their consideration of, or search for, alternative design options. I realized the tension that exists between cultural production and reproduction, enculturation and innovation, and repetition and change.

3. ENGINEERING FOR CHILDREN CURRICULUM

In this chapter, readers will find descriptions of the engineering design activities and how they were presented to children. These descriptions provide valuable insights into the assumptions that shaped the teaching of this unit, because Gitte, who was the principal curriculum designer, was also one of the two team-teachers in the class. Readers will also find descriptions of the teaching strategies, with a particular focus on the creation of an engineering language game, questioning techniques, and teachers' actions to facilitate the formation of a knowledge-building community. Tammy and Gitte also contributed to the emergence of a knowledge-building community by portraying themselves as learners. They used their collaboration and the setting as opportunities for learning, and shared their insights with the children.

3.1. INTENDED CURRICULUM

3.1.1. *Conceptual Considerations*

The Engineering for Children: Structures ("Engineering for Children: Structures") curriculum was designed to address the fact that many school children, especially girls, hold negative attitudes and "mistaken" ideas about engineering and science (APASE, 1991). The curriculum developers felt that elementary school children could learn about, and have positive experiences with engineering through this unit. The practical applications of science found in "Engineering for Children: Structures" have the potential for introducing basic science concepts, allowing children to learn about engineering design, and helping children to develop positive attitudes towards science and engineering. Through practical applications, children can articulate and test their ideas about structural engineering. The children's principal "mission" (Schank *et al.*, 1994) in this unit is to design structures that, in an up-scaled form, could have real life applications as towers, strong arms, bridges, and domes. The teachers' guide proposes a "cover story" (Schank *et al.*, 1994) that emphasizes children's roles as engineers who are in charge of "real life problems".

The teachers' guide suggests three important aspects for teaching the unit: "questioning skills", "challenge instead of competition", and "gender sensitive teaching". Basing themselves on Harlen's (1985) work, the curriculum developers de-emphasize single answer questions that encourage visions of engineering as static collections of data to be memorized, and emphasize reflective and analytical questions that promote views of engineering as a dynamic search for answers. The curriculum developers call for cooperation

41

based on the observation that everyday engineering is a collective enterprise.[1] Thus, teachers are encouraged to have children design in collectivities and thereby promote student self-confidence, positive group interactions, and increased student learning. For this reason, the cover story announces tasks as "challenges" rather than competitions. The curriculum guide also calls for special attention to gender issues, engineering as a predominantly male domain could easily be amplified if teachers are not cautious. The following suggestions from the teachers' guide were important referents for Tammy's and Gitte's teaching of "Engineering for Children: Structures":

- Science and engineering should be presented as democratic rather than elitist domains.
- Opportunities should be provided for girls to take leadership roles.
- Male and female pronouns should be used and reference should be made to objects that are familiar to boys and girls.
- Invited engineers should be female.
- Focus should be on problem solving and the values of false starts.

3.1.2. *Teachers Goals*

Even in curricula that are student-centered and open-ended, there are some top-level goals set by the teacher. These goals set the framework for learning. In the "Engineering for Children: Structures" curriculum, there were three referents that guided both teachers' actions:

- In designing structures, engineers must account for forces of gravity, compression, and tension.
- In designing structures, engineers need to manipulate and change properties of raw materials and select from a variety of techniques or develop new ones.
- In designing structures, engineers must consider constraints such as time, budget, space, and material limitations.

On the basis of these referents, Gitte and Tammy attempted to achieve various goals: those stated in the curriculum and those they developed themselves during their extensive conversations before and throughout the unit. In *social* terms, they wanted to increase students' abilities to (a) be socially responsible and cooperative group participants, (b) make reasoned thoughtful choices, and (c) accept the consequences of those choices. In *personal* terms, Gitte and Tammy wanted students to increase their (a) competence as problem solvers, (b) appreciation for personal creative abilities, and (c) sense of self-worth. In *intellectual* terms, the teachers wanted to foster students' (a) creative and critical thinking skills, (b) skills in designing

---

[1] Whalley (1991) showed that "independent inventing" has important social components and therefore does not exist in practice. All human practices are couched in a social and material matrix. Growing up in human society, we therefore always come to a world already shot through with meaning.

ENGINEERING FOR CHILDREN CURRICULUM

structures, and (c) knowledge of engineering as the application of science. Finally, relating to *attitudes* and *appreciations*, they attempted to encourage students' awareness of and appreciation for (a) engineering as a process of discovering rather than a body of facts, (b) processes of exploring, experimenting, and discovering, and (c) problems as opportunities for further learning.

3.1.3. *Overview of the Activities*

Before the actual engineering design unit, we devoted two lessons to a series of starter activities designed to introduce students to open-ended problem solving and provide a first assessment of children's competence working with ill-structured and open-ended tasks. The activities asked students to (a) support as much weight as possible with a single sheet of paper; (b) pick up and carry various glasses, jars, and glass bottles with a single straw; and (c) support as many nails as possible on the head of one upright nail. After these lessons, Gitte and Tammy followed the sequence of activities outlined in the "Engineering for Children: Structures" teachers' guide (APASE, 1991).

The "Engineering for Children: Structures" curriculum can be roughly divided into two major parts. During the introductory part, activities are intended to prepare children for the more complex, open-ended, and ill-defined design activities to come. Among these preparatory activities, children are asked to strengthen a variety of materials, stabilize various two- and three-dimensional structures, and combine various materials to form joints. In part two, children designed three types of structures, towers, bridges, and domes. These activities, the tools and materials used, and a variety of other activities (engineering log book and engineering techniques board) are described in the next section.

3.2. ENGINEERING DESIGN ACTIVITIES

3.2.1. *Tools and Materials*

The ultimate success of the "Engineering for Children: Structures" unit cannot be considered independent from the tools and materials, students' written activities in their engineering log books, and discussions around the Engineering Techniques Board where students' first artifacts were displayed.

3.2.1.1. *Handyman Tools and Materials*

The entire unit was conducted with materials readily available in children's homes or obtained, as donations, from local restaurants (e.g., drinking straws and paper cups). Both teachers and the members of the research team brought scrap wood, cardboard in various forms, and other materials slated to be discarded. The children also brought a variety of tools. In the end, the materials used in this unit included straws, popsicle sticks, tooth picks, wooden skewers, spaghetti, cardboard, cardboard tubes and cones, and newspaper. Materials for the construction of joints included pins, pipe cleaners, paper clips, thumb tacks, invisible and masking tape, elastics, staples, wire, string, and modeling clay. Tools included scissors, glue guns, Exacto knives, and staplers. There was also a bucket with more than 400 wooden blocks to be used for testing bridges.

3.2.1.2. *Engineering Log Book*

All children kept an engineering log book where they (a) wrote about their experiences, (b) entered their design drawings, (c) placed photos taken during the construction of the projects, and (d) formulated critique and self-evaluation. Tammy frequently provided a framework for children's entries not, as she said, to stifle their creativity but rather to scaffold the stories of their experiences in these open-ended "Engineering for Children: Structures" activities. For example, to assist students to reflect on their experience with the initial motivational problem set (bottle pick-up, balancing nails, paper bridge), she had prepared a sheet with four frames for drawings, each associated with a "bubble" in which children wrote about their experience. They summarized their story by completing the sentence beginning "I think . . ". In another situation, Tammy prepared a brief self evaluation in which children rated themselves on a scale from 1 (poor) to 5 (very good) on items such as "I listened to others", "I contributed my own ideas", and "I encouraged others". There were also self evaluation forms that asked children to answer a series of questions in complete sentences. Such questions included, "Name as many engineering techniques as you can that you used in your mega structure!", "What do you think would have made it better?", or "How did you feel building your structure?"

An important part of the engineering log book were children's glossaries. Toward the end of the unit (January), each child produced a glossary from the words that they individually found most significant to their work. Much like the examples in the manual that provided teachers with some background information, children's glossaries included a labeled drawing and a short description for each entry. Children were asked to make sure that each one of the represented items demonstrated their understanding.

3.2.1.3. *Engineering Techniques Board*

The products of children's first set of activities, artifacts that illustrated strengthening and joining, were posted on a special "Engineering Techniques Board". Repeatedly, the class gathered in front of this display to talk about some of the techniques developed by children, and where applicable, to show connections that existed to techniques used by professional engineers.

Throughout the unit, Gitte frequently referred to children's work on the engineering techniques board, to help students generate new solutions for problems dealing with the strength of materials or the stability of structures.

3.2.2. *Preparations for Engineering Design*

Tammy and Gitte began the "Engineering for Children: Structures" unit with a series of activities designed to prepare the children for the design challenges in the second part of the curriculum. In these activities, the two teachers implemented the curriculum that embodied to a large extent Gitte's own understanding of learning. My own 10-year teaching experience in middle- and high-schools had shown that students are not too enamored with decontextualized activities such as "strengthening" and "stabilizing" as emphasized in this part of the curriculum. To allow both teachers to take complete ownership over the events, I did not suggest an alternate route to preparing the children. Strengthening a straw or a piece of clay without having a purpose for doing so is seldom motivating. My observations in this classroom reified these views. The children seemed to be much less excited and enthusiastic during this part of the unit than they were during the later engineering challenges. Gitte, upon watching and analyzing a number of videotapes from the early and later parts of the unit, concurred. At the end of the study, she suggested that the "Engineering for Children: Structures" curriculum should be redesigned so that the preparatory activities would be couched in simpler design challenges. The "creature" project described below was included in this unit in response to our discussions about the contextualization of activities that allow children to design with a purpose.

3.2.2.1. *Strengthening Structures*

Gitte introduced the strengthening activities by providing some examples which she thought children might understand.

Use any materials and your objective is to try to make the material strong. Chris, you already came up with an interesting technique, the "ramboly technique". What I want you to do is to make a demonstration, title it, and put it up on the block. The rest of you will have to think of other techniques that you want to. For example, there might be other techniques. This is a technique that I might call "Gitte's Tube Joint," say, and I would tape it here, put it up on a piece of black paper, and call it Tube Joint.

Here, she provided examples by crumbling newspaper, making a newspaper roll, and demonstrating Chris' "Ramboly Technique". She encouraged children to take ownership by naming their artifacts, and more generally, to engage in naming new techniques, phenomena, and so on. Later, she frequently gathered children around the Engineering Techniques Board and asked specific students to talk about their artifacts and to explain their respective names. If appropriate, Gitte introduced the labels engineers used in their own work for a similar technique. In this way, she encouraged a child-based language game to emerge and provided opportunities to make connections with engineers' language. Here, the two discourses are authentic, one for the children who developed it in response to their experiences; the other, for engineers and the community where they participate.

3.2.2.2. *Stabilizing Structures*

In the stabilizing structures activities, students were asked to stabilize two- and three-dimensional shapes. Tammy introduced the activity by showing a square from four straws and asked how it could be stabilized. Following the instructions one student provided, she fastened a fifth straw to form the diagonal between two opposing corners. Tammy demonstrated that the resulting structure no longer moved at the joints. She then asked students to build pentagons and hexagons from spaghetti and marshmallows (or alternatively, from straws and pins), and, to stabilize these structures by using a minimum number of spaghetti noodles (or straws). Later, Gitte and Tammy extended the activity to three-dimensional objects. The ultimate purpose of this lesson was for children to "discover" braces as a stabilizing component of structures. (A more detailed description of this activity is provided and analyzed in Chapter 7 which is, in part, concerned with the near failed attempt to teach a canonical engineering design practice.)

3.2.2.3. *Designing a Creature*

The designing a creature activity was not part of the curriculum as described in the teachers' guide. It arose in response to my debriefing conversations with Gitte about contextualizing and finding a purpose for "strengthening" and "stabilizing". Tammy and Gitte introduced the activity in the following way.

(Tammy:) Now you have learned quite a few different techniques. You know how to make joints, you know how to make things stronger, you've got quite a few techniques under your belt. We are going on to make towers next week, and bridges after that, and big structures after that. We thought we'd just give you some time not to play, but to use some of the materials in a way that was more creative and that was more suited to just your imagination. We thought we would give you a small amount of materials, five straws, one cone, some pins, and if you'd like to use other materials like cardboard or toothpicks or small amounts of plasticene or something like that, you may. But your basic supplies are going to be five straws, one (cardboard) cone,[2] as many pins as you need. Your job is to create one of two things: it can be a being, but the being would not be a being that would just live on earth, but a being that would live in space, under water, under ground, in a strange environment. The other thing that you may make, it can be something that has a use, so it could be a machine in a way, it could be something that does a certain job, a being would be, like an invention. I am sure that every

---

[2] These cardboard cones were among the discarded materials Gitte had found. Recognizing their potential value in this unit, she brought them to class. In a similar way, others had brought cardboard roles, cardboard of various thickness and strength, and a variety of other materials that were collected from discarded materials at home or in garbage piles.

one of you has a completely different idea in your head of what you would like to do with these things. You can cut the straws as many times as you like, but you will not be issued more straws, so you have to think before you cut. If you would like to use your engineers' logbook to do a drawing before you get going, you may. Or you might be one of those people that love to get their hands on things, then you are gonna have to do it that way. It doesn't matter to me which way you want to do it.

(Gitte:) It might be interesting to develop a story that when you present what you have done to the class so you can say this is my gerbil transmitter cage, and it's useful for these purposes. And you know, create a little story about it, so that when you show it to us that you have a good sense of where your idea came from and how it could be used.

The children spent approximately one lesson (90 minutes) with their designs and another for presenting their designs to the class. Here as for the design challenges, Tammy and Gitte did not simply tell children what to do, but provided a rationale for doing it, provided suggestions rather than instructions, and allowed children to make their own decisions about designing and the nature of their projects.

3.2.3. *Design Challenges*

There were three major challenges in which children were asked to design a tower, a bridge, and a dome large enough to contain several children. From my perspective, the intended curriculum is problematic in its assumption that designing a novel artifact (such as the bridges, towers, domes, and creatures) consists in first creating an idea which is subsequently realized in material. In the course of this study, it became clear to me that designing essentially has to be considered as the entire process from initial conception to artifact construction. In those instances where the teachers actually required children to draw their ideas prior to construction, there existed considerable discontinuities between the drawings and what children produced during construction. Using the tower activity as an example, I provide a description how Tammy and Gitte normally introduced the challenges; the bridge and dome projects are then briefly outlined.

3.2.3.1. *Designing Towers*

Gitte opened the lesson by introducing the mission for the next few lessons: building a tower. By asking, "What are the parts of a tower?", she engaged children in a brainstorming activity to elicit their understanding of the topic.[3]

---

[3] Throughout this book, I use the following transcription conventions:

YES – Words in caps, louder than usual talk.

°Earthquake-proof° – Degree signs to indicate low volume, almost inaudible talk.

(3.2) – Pauses in seconds, one-tenth of a second accuracy.

(.) – Period in parentheses, audible pauses shorter than 0.2 seconds.

= – Equal sign indicates "latching", i.e., the normal period of silence between the end of one speaking turn and the beginning of the next does not exist.

that – italics indicate a greater emphasis on a word or syllable.

(???) – Question marks in parentheses signal unheard words. The number of question marks

01 Ron: Column things, because roofs are over them.

02 Gitte: So you think it is really important for the structure. Jeff, how about you?

03 Jeff: It's not very important, because if there is no roof, then there would be nothing to hold it up. There would be nothing else for the columns to hold up, so there have to be sides and base.

04 Gitte: So you think the roof is just as important as the sides and the base?

05 Jeff: Because if there is no roof, there would be nothing to (be held up?)

06 Gitte: Thats a good point.

07 Tom: Sides.

08 Gitte: Does anyone else have a part of the tower which is most important? (5.3) Ron has said that the columns are, Jeff is disagreeing and saying that base and sides are just as important otherwise the columns wouldn't have anything to support, what do you think Tim?

09 Tim: Beams and columns and the joints to put them together.

10 Gitte: So you are adding another part, the joints involved in the tower.

11 Tim: Because if they never had the beams and the columns, it wouldn't be the tower it would just be something like this ((gestures something flat)) on the ground.

12 Gitte: So you are saying the beams and column are the most important.

13 Tim: It keeps it up, because those is what keeps it up.

The teachers then introduced the children to specific challenges – which Tammy and Gitte had discussed the night before. With the specification of challenges, they did not want to constrain the children but rather provide them with a few starting points to begin their designs. As Table I shows, the children had much freedom for designing, especially if they chose Option 12. (Students' names are listed by the challenge they addressed.)

Before students started, the teachers emphasized that (a) engineering problem solving processes can be frustrating, (b) first design ideas might

---

indicates the approximate number of inaudible or undecipherable words.

(alien?) – Word followed by question mark in parentheses, possible but uncertain hearing: The speaker uttered a word that could have been "alien".

((Points to structure)) Physical actions that are concurrent with utterances are embedded in the text by means of double parentheses.

Grea:::t – Lengthening of phoneme.

[– square bracket to indicate overlap of speakers and activities with respect to ongoing talk. "[" marks the moment where a second speaker or a physical action began with respect to the previous speaker.

?!. – Punctuation marks speech patterns such as questions, exclamations, stops, and full stops, rather than grammatical units.

TABLE I
Engineering design challenge topics for the tower project

1.	Build two towers that can support a . . . above the ground.
2.	Build a tower that is wider at the top than at the bottom. (Brigitta & Melinda)
3.	Build a tower with a turning thing on it (like wind mill, or horizontal on top). (Sandy & Tom)
4.	Build two towers that allow an object to slide on a line between. (Arlene & Chris; Carla & Jane)
5.	Build a tower with a crane-like, mobile arm. Damian, Dennis, & Doug; Jeff & John; Patricia & Anna)
6.	Build a tower that will support the weight of a large book as high as possible above the floor.
7.	Build a tower with an elevator in the center. (Renata & Maggy)
8.	Build a tower that will support a ball as high as possible above the ground. (Kathy & Kitty)
9.	Build a strong arm that will support a weight 45 cm from the wall. (Clare & Shelly)
10.	Build a tower that will withstand a force of a fan's wind 2 m away.
11.	Build a tower that can endure an earthquake. (Andy, Simon, & Tim; Stan)
12.	Build a tower that . . . . (Removable roof: Peter & Ron)

need modification before an artifact reaches its final stage, and (c) engineering means perseverance. Whenever either teacher saw an instance in which students struggled, they reiterated the three points and often asked students in whole-class sessions to reflect on their problems and resolutions. Tammy and Gitte also pointed out that it might be necessary to work through a number of problems (including "catastrophic failure") before students could arrive at their finished product. They asked the children to think about tower building not as a competition, but as an opportunity to work together, get inspiration from others, and to share tools, materials, and ideas.

Gitte made a final announcement intended to increase the challenge. By adding constraints such as those in her announcement, Gitte intended children to develop an appreciation for the essence of "real" engineering.

In today's exercise we are going to make it a little restricted, and the only way it is restricted is that you are not allowed to have a foundation. And I tell you why. The challenge today is to try and build a tower whose design is so sturdy above ground that when a wind, or if you put a fan on it, it wouldn't topple over. What I am saying is, you cannot tape your structure, your tower, or whatever it is on a piece of cardboard to the ground. The reason for that is if you did tape it, it would be a test of how strong the tape is that was taping it to the ground. It would be a test of the foundation, whereas what we want you to concentrate on today is the design of the tower.[4]

As usual, Gitte did not simply constrain children's activities, but provided a rationale. Here, she wanted children to be concerned with stability as it arises from the design of the tower rather than from the material that attached

---

[4] As with many other rules, Jeff found out how to flexibly interpret this one about attaching the artifact to the ground. His tower used staves (made from string) that he tapped to a support (Table I). He successfully argued later in the whole-class discussion that the staves but not the main tower were attached so that he was, in fact, not violating the rule,

it to a solid foundation. While she indicated to students in a different situation that "real engineers" make use of the strength of the foundation, the purpose in their tasks was for children to learn about structural and material strength and stability within the artifacts themselves. However, she also hoped that the children would "discover" on their own some canonical design principles to which she would simply have to attach the canonical label. Altogether, children spent three 90-minute lessons designing their towers and two lessons presenting them to the class.

3.2.3.2. Designing Bridges

Using a range of materials and fastening methods children were asked to design a bridge. Gitte challenged students to make the bridge as strong as possible. The strength was to be tested by using a bucket with wooden blocks that allowed a quantification of the strength. Gitte imposed three conditions on students designing:
- The bridges had to span a 30-cm gap and support weight.
- The bridges had to be free-standing, that is, they could not be fastened for support
- The bridges had to be built from one or two types of materials, and every engineering team was to have a different set of materials.

Contrasting my own understanding of designing, Gitte conceptualized it as the coming up with an idea and producing plans. These plans guided later implementation phases, the construction work. Hence, children were asked to make drawings of their bridges beforehand. However, when it came to construct the bridges, Gitte and Tammy relaxed their requirements that children followed their drawings, but allowed each group to negotiate any design they wanted. The children spent six lessons designing their bridges and two lessons presenting them to the class.

3.2.3.3. Designing Domes

In the dome project, which members of the classroom community also referred to as "the mega structure", children were to build a dome entirely from newspaper. For this, teachers suggested that they make hollow cylinders out of sheets of newspaper by rolling one sheet at a time tightly around wooden dowels. Before removing the dowels, the children placed a piece of tape around the rolled newspaper tube. There were three conditions:
- Use only the materials provided (newspaper).
- Your dome must rest on the floor without attachments.
- Your dome must be strong enough to withstand the stress created by fitting all members of your challenge team inside.

3.3. TEACHING STRATEGIES

Important aspects of any enacted curriculum are teachers' practices that encourage specific types of learning episodes which shape learning environments as the learners' "dwellings" (Heidegger, 1971), and provide opportunities and constraints for particular events. In this study, three facets of Tammy's and Gitte's teaching were pivotal: (a) moves designed to allow the emergence of a child-centered language game of engineering design, (b) questioning techniques that scaffolded students' participation in increasingly competent material and discursive practices, and (c) actions that brought about a knowledge-building community. Each of these aspects of their classroom practices is described in the following subsections.

3.3.1. *Creating an Engineering Language*

Central to Gitte and Tammy's work were strategies designed to bring about an engineering design-related language game. However, the teachers' intent was not, primarily, to impose an unfamiliar language game, that is, canonical engineering design discourse, but to allow child-centered language to emerge. Although the teachers were not entirely consistent with their intents by trying in several instances to encourage the community to adopt specific changes in their language (see Chapter 7), the evolution of the classroom discourse normally began with children's own ways of talking engineering design. In this way, the teachers (a) built on children's existing language, (b) provided children with opportunities for creating their own new language, (c) provided opportunities for children to link their own language to canonical discourse, (d) facilitated the emergence of new descriptions by scaffolding children's reflections on their own actions and understandings, and (e) encouraged children to engage in emergent design.

3.3.1.1. *Building on Existing Language*

The teachers encouraged children to use their prior experiences as resources for inventing new ways of doing things, as a spring board for dealing with problematic junctures in their design work. For example, Chris described at length how she produced her "lashing joint". When Gitte asked her how she got this idea, Chris explained that she had done something similar when she made baskets as part of a Girl Guide activity. Gitte used this as an opportunity to reiterate that the other children should draw on their out-of-school experiences as a resource.

What Chris is doing here is putting some of her experience from *Guides* doing a craft in *Guides* for the joint center. How many of you thought about doing that. How many of you got your ideas from something you made before? What's that, Crafts, or Girl Guides or Boy Scouts? Or something you may have made in art class? How many of you used those ideas in the joints?

Gitte continued to engage children in conversations about the importance of prior experiences to design.

3.3.1.2. *Creating New Language*

For both teachers it was important that children began to describe their experience in terms of their own language games. Although Gitte and Tammy (as well as the two engineers) introduced some of engineering technology, their primary goal was for children to develop their own language game, and to use their everyday discourse as a resource. Furthermore, by attaching students' names to specific descriptors, the teachers wanted to increase the likelihood that children would take ownership of their language games. A few examples were "Ron's flexi-joint", "Andy's spaghetti stretch", "Chris' lashing joint", "Tom's tacky-joint", or "Chris' ramboli". The teachers emphasized that descriptive names were particularly appropriate ("He used a tack to connect the straws, and the name tacky-joint really does describe what the joint is about").

3.3.1.3. *Linking to Canonical Language*

Gitte frequently highlighted children's work that made use of techniques which she knew were used by engineers and which she felt could be helpful in children's later design work. In these cases, she helped students to "see" similarities in the techniques used by various children or techniques that were variations of one theme.

There are another couple of joints that definitely deserve mention. One is, I don't know if you can see it on that, but you can look at it a little later, Renata used it, Shelly used it. I am not sure anyone else used it. A couple ((Points to these joints on the engineering techniques board)), a couple of you used it in a different way, it's a technique where you have a straw and at the end you put in some kind of stoppers so that whatever was on your straw didn't slide off. Renata, do you want to explain to us how you came up with that idea? With those little pins in the end?

After Renata and Shelly had explained how they had come up with the ideas for their technique and how they built the particular artifact, Gitte pointed out that engineers often used techniques that prevented joined materials for moving out of place. In this way, Gitte provided the children with opportunities to link their expressions such as triangle, layering, and overlapping to the equivalent canonical expressions of brace, lamination, and cantilever techniques.

3.3.1.4. *Reflecting On Action*

Both teachers encouraged the children to reflect on their actions as another technique to generate new ways of describing their experience, to increase their awareness of design processes, and to use reflections as resources in dealing with breakdowns.

What I think Gitte is really getting at, is for you to think about what makes you think of things and use it, and actually go into your brain and think, "What do I think of?," "What am I thinking about?" Because part of this science is not just doing your thing and doing it, but is also becoming aware of what kind of things help you think and what you are thinking about when you are learning and when you are experimenting. So *do* think about what kind of things go through your head.

Both teachers created opportunities for children to reflect on their experiences, the problems they encountered, and to share these with others in the community. For example, Tammy began such sessions by asking a specific student, "Maybe you can just start by showing us what you have done, what you think is good about it, and what problem you have". In the following excerpt, Tammy asked Kathy and Kitty to talk about their ongoing project in which they designed a tower that was to carry a ball.

01 Kitty: Well, we started with a base, this base, and then support this
02   part and then put this up and then another one of these
03   ((middle part)) and then the little, (?), like some straws place.
04 Tammy: OK, and what are you having problems with?
05 Kitty: Well, we are having problems with it staying up, because we
06   are pushing it, so we are trying to put more straws on the
          bottom.
07 Tammy: So you have been putting your supports parallel like this, is
08   there a different way of putting supports so that it would be
09   more stable? (4.2) Do you have any ideas? (3.1) You can just
10   add straws like this, is that what you are planning?
11 Kitty: Yeah, we are planing on putting more straws going like this
          ((horizontal)).

After Kitty provided an initial account, Tammy specifically asked them to talk about and reflect on the problems they had constructed (line 04). This provided Kitty with a context to talk about how they had framed their problem ("with it staying up", line 05) and how they planned to resolve the issue (lines 06 and 11). Tammy attempted to encourage more reflection-on-action, but at this early stage in the unit (Week 4), she was not yet very competent in the questioning techniques that would have allowed her to do so. Despite considerable wait times (three and four seconds, lines 08 and 09) after her questions, she was not able to elicit additional ideas from the girls. She then resorted to a more direct approach by seeking the girls' confirmation to what she had constructed as their intent (lines 09–10).

3.3.1.5. *Encouraging Emergent Design*

Contrary to the curriculum guide that emphasized pre-planning, both teachers encouraged students to use current artifacts as resources for generating new ideas. In the following excerpt, Gitte explicitly encouraged students to design in a way that has been described as "bricolage" (Turkle & Papert, 1992) or "conversation" of designer and medium (Schön, 1987).

So what you are saying is that just by building the joint and having materials there, that gave you ideas. So in a sense, your materials were telling you what to do. Did anyone else have that experience where they looked at what they were building like the cup and it gave them ideas of what to do next? Did anyone else have that experience? Simon, a little bit? ((He nods)) What about Jane?

Building on previous designs by modifying materials was one of the possibilities for emphasizing the emergent properties of designing ("So what you have done here is substitute the materials, which is another way of being creative").

3.3.2. *Questioning Techniques*

An important contributor to the success of the engineering design unit was Gitte's questioning. Tammy, who had taught a bridge-building unit before, repeatedly brought up this issue. In our debriefing meetings, she talked about Gitte's competence in asking "really good questions" and frequently remarked to the children in class, "Gitte always asks these wonderful questions". To support her statement, Tammy pointed out that the children in this class learned so much more than during the bridge-building unit she had earlier taught on her own. I focus on Gitte's questioning techniques because she was largely responsible for the content of the unit (Tammy concentrated more on the social aspects. An extended analysis of Gitte's questioning technique that focuses on the context of questioning, the content of questions, and responses/reactions to questions, has appeared elsewhere (Roth, 1996e). Here, I provide an overview and one example from Gitte's interactions with a group of children designers.

3.3.2.1. *Context of Questioning*

Following recent trends in science education, Gitte felt compelled to provide a student-centered and activity-oriented learning environment. However, she knew from past experience that letting children engage freely in "hands-on" activities does not lead to canonical knowledge. She described canonical knowledge, here from the domain of engineering design, as existing out there and being authoritative. Like many other teachers, Gitte had interpreted "constructivism" to mean "anything goes". Questioning children in the context of their own inquiries helped her to resolve a conflict. When she combined child-centered inquiry with appropriate questioning techniques, she did not have to compromise canonical science and engineering content knowledge. Gitte argued that children spontaneously invent or discover many discursive and material practices which are similar, or can easily be related, to their canonical counterparts. However, children will not make these connections on their own or value their inventions as important aspects of adult design activity. Good questions allowed Gitte to combine open-inquiry and the learning of specific engineering discourse.

Through questioning, Gitte attempted to achieve at least six goals:

- To focus children's attention on a canonical concept by "drawing/pulling it out from their experience" ("You know what a plumb line is? Wouldn't it be neat if we saw a kid developing such a technique. It's likely to happen, because I made that connection. If I saw a kid that is trying to make something straight, I might be able to ask questions that would lead that child to thinking about using a string, just by the words that I used".).
- To help children develop a language for talking about their engineering artifact ("those words like 'reinforce', 'brace', 'bundling', those ideas came from the kids, through questioning").
- To "stretch them to improve" their artifacts.
- To make sense of their experience and learning ("questioning helped that conceptual breakthrough to happen").
- To succeed ("I asked a lot of questions that helped them to succeed").
- To cope with temporary failure ("questions make sure that the failure that happens is not that frustrating").

3.3.2.2. *Content of Questions*

To bring about deep understanding through their questioning, teachers need to elicit student explanations, elaborations of previous answers and ideas, and predictions. Research has shown, however, that most teachers ask questions requiring the recall of factual information in a "devised question format" so that teachers mainly ask questions for which they already know the answers (Lemke, 1990). This leads to a common pattern of classroom discourse, IRE, for teacher Initiation-student Response-teacher Evaluation. In Gitte's questioning, I was able to document a virtual absence of IRE-type exchanges (Roth, 1996e). There were only isolated instances of teacher utterances designed to manage conversations ("Can we have one at a time?"); there were a few isolated comments ("Engineers call this a 'raft foundation'"); and, there was about one question for each interaction with a group that addressed emotional issues ("Are you proud of yourself?", "What was the most frustrating part of this whole exercise?"). All other questions related to various aspects of artifact design in an engineering design context.

By analyzing the entire database, I established a typology of the questions

Gitte asked students throughout this unit.[5] Gitte's questioning embedded
five major types of knowledge: knowledge relating to the natural world,
design practice, development and testing of designs, final products, and
knowledge itself. This typology is evidence of the high-level content of Gitte's
questioning, which was more concerned with engineering practices than with
engineering facts. The typology of Gitte's questions matched that established
for engineering design innovation established by sociologists of technology
(Faulkner, 1994).

3.3.2.3. *Responses and Reactions to Questions*

Two aspects of Gitte's questioning, related to children's responses and reac-
tions are noteworthy. First, although Gitte did not evaluate children's answers (as in the IRE sequence), her authority as teacher was undisputed. In
this classroom, teacher authority was asserted and maintained by means
other than the IRE sequence often linked to teacher control. The patterns
of Gitte's questioning and students' answering was symptomatic of authority
rather than a means of control. Second, although initiations requested children's own topics of interest, subsequent questioning sequences clearly elicited discursive accounts that included Gitte's topics: properties of materials,
techniques to change them, and design principles and configurations. Thus,
although question sequences began with students' topical interests, Gitte
ultimately controlled the topic (especially in the initial design phases of a
project) by means of contingent queries that led to the canonical knowledge
domains she had previously identified. (Readers will find more discussion
and evidence related to these issues in Chapter 7.)

3.3.2.4. *Gitte's Questioning of Students in the Context of Their Work*

The following interaction between Gitte and a group of two girls, Clare and
Shelly was typical for this classroom and exhibits a number of important
characteristics of Gitte's questioning.

01 Gitte: What was the biggest problem you had so far? (1.2)
02 Shelly: Getting the glue gun.
03 Gitte: So that's been a real restricting factor for you?
04 Shelly: [Trying to find the
05   glue gun. Well, I think the bottom pillars, trying to get them
               to stand.
06 Gitte: [Yes, now why did you use a glue gun there instead of
07   masking tape or some other connector?

---

[5] This typology and specific questions Gitte asked were developed elsewhere (Roth, 1996e).
Readers will find the typology in Chapter 4, Table I, but without Gittes questions.

08 Shelly: We thought, I think we were supposed to only use a glue gun,
09   and since when you asked me to tape it, you did it all up there,
10   and it could still wobble around, so I decided
11 Gitte: Oh, I see, but what is it about the glue that's really good?
12 Shelly: Well, this is (.)
13 Clare: The glue is strong.
14 Shelly: Yes, it is really strong, and it helped us a lot. That's why we
15   decided to use a glue gun. We only used one drop of glue last
16   time we used it, and it was strong enough.
17 Gitte: That's right, you used different, you used pins and tape the
              last time.
18 Shelly: Pins, tape and clay. Clay doesn't stick as well as pins.
19 Clare: We used the clay so the, hm::, so the, the
20 Shelly: In the place of the.
21 Clare: [Yeah, it was tending to dry out.
22 Shelly: And then I got that idea from the engineering technique board.
23 Gitte: Oh, is that right?
24 Shelly: 'Cause one, you know all the paper clips have something on
25   the back, the one with the paper clip and the spaghetti
26 Gitte: [Yes
27 Shelly: [How it's
              got the clay in the insides?
28   insides
29 Gitte: [Yes, yes, oh thats interesting. Hm:::. Are you happy with
30   your materials?

This interaction commenced with a general question about problems (line
01). Gitte's initiating query had genuine character: she did not know what
course of action the two girls had followed. Before her next initiating query
(following line 30), Gitte asked two contingent queries, the second being a
reiteration of the first (lines 06–07 and 11). Gitte's other contributions to
the conversation can be read as "continuers" (Erickson, 1982).[6] With these
continuers, Gitte relinquished her questioning and invited students to continue their accounts (lines 03, 17, 23 and 26), and confirmed her understanding (lines 03 and 17) or that she was still following their account (lines 23
and 26). It is important to note that none of Gitte's discourse contributions
had an evaluative function. Shelly and Clare's account, facilitated by Gitte's
contingent queries and continuers, covered four of the five domains in the
typology of knowledge in innovative engineering design: in lines 13, 14–16,
18, and 21, they talked about the properties of materials (typology item 1.2);
in lines 18, 19, 24–25, and 27–28, they talked about design criteria and

---

[6] Continuers are contributions to discourse in which one person gives up her/his turn so that
the current speaker may continue. Continuers come in many forms, sometimes as nodding,
sometimes as brief utterances such as "Hm", "Yes", or "I understand" and the like.

design concepts (typology items 2.1 and 2.2); in lines 04–05 and 14–16, they talked about performance of the final product or parts of it (typology item 4.3); and in lines 02, 04–05, and 22, they talked about the location of specific sources of knowledge necessary for their project (typology item 5.1).

Many observers of Gitte's teaching remarked about the competence with which she questioned children. Her questioning techniques are likely to have contributed enormously to the success of the unit. Gitte's contingent queries (which are further analyzed in Chapter 7) were frequent during initial stages of the projects and began to fade as children's stories of designing processes and design artifacts became more complete. Ultimately, the effect of her questioning techniques was that children constructed complex stories of and about engineering. During the initial stages of students' designs, questioning was more frequent, purposefully addressing many issues related to material and discursive practices of engineering (pertaining to both canonical practices and those idiosyncratic to this classroom community).

3.3.3. *Creating a Community*

Tammy and Gitte wanted to create a child-centered classroom community in which students respected each other, and adults interacted with students in a similar respectful way. To bring about a sense of community among the children, the teachers did their best to mediate whenever trouble arose. They also arranged whole-class interactions, "sharing time", so that children could present their work, or engage in understanding and appreciating what their peers had done. By mediating troubles and by providing opportunities for "sharing", the teachers facilitated the emergence of a classroom community.

3.3.3.1. *Mediating Trouble in Collaborations*

An important aspect of this learning environment was the teachers' efforts to bring about a community in which distributed aspects of knowing were emphasized while individualistic and competitive aspects were de-emphasized. Tammy, who took major responsibilities for social organization in the class, frequently engaged the children in brief exchanges designed to make them think about and ultimately to bring about, those interactions with others that fostered the formation of a community. For example, she considered it very important that children encouraged each other to increase participation and overcome frustration ("today we will concentrate on providing encouragement and participation"). Rather than telling children what they should do, she asked them to make suggestions which elicited the desired behavior. Her questions included:

- What does it look like and what does it sound like to encourage participation from those who dont normally contribute?
- How do we get someone who is quiet to participate?

- How do we get someone to give ideas that would help all of us do a better job?
- Tell me in words what you would say to a person to encourage them? What kinds of things can we do or say that encourage others?

She then made a list on the chalkboard consisting of students' suggestions. For example, students generated the following ideas about encouraging others' participation.

- Smile at them. (Jeff)
- Notice ideas, say its a neat idea. (Jon)
- Why don't you help me with? Can I help you? (Damian)
- That's nice. How was it made? (Tim)
- You could tell them how to improve their idea, like "maybe you could do this?" or "that is better than last time". (John)

Throughout their interactions with students, Gitte and Tammy referred students back to their own, "published" suggestions. This was helpful particularly when disagreements arose between two or more students over some aspect of their collective work or when two groups wanted to use the same tools and materials at the same time.

Initially, some of the students wanted to work on their own. However, the teachers discouraged students from doing so. There were only two students who were allowed to work individually. Stan worked without partners for the first seven weeks because, despite Tammy's mediation, other children refused to team up with him. However, this changed when he teamed up with Tim for the bridge and dome projects. Damian was allowed to build his own bridge when it became clear that he would leave the school after the first bridge-building lesson. Otherwise, when there was a request to work individually, Tammy and Gitte encouraged students to view collective work as a positive aspect. Because of their interest in creating a community, Tammy and Gitte encouraged children by emphasizing the collaborative nature of engineering which transcended personal differences.

01 Tom: I didn't want to work with a partner.
02 Gitte: An important skill of engineers is to work with others.
03 Tammy: Look at it as a challenge, How do you get another person involved?
04 Gitte: We will be looking at how well you work with another.
05 Tammy: This is a challenge, to work with others, not to get into a fight.

When there were problems and disagreements between members of a group, Tammy mediated. Her comment in line 03 was designed to bring back the earlier conversation about involving others who, for whatever reason, did not participate in collective decision making and construction. Later, during whole class sessions she referred to such incidences and turned them into teaching episodes for dealing with problems in group work. In the following excerpt, Tammy used several instances in which collaborating students were angry with each other to talk about mediating differences.

I want to compliment these two because of a problem they had in the beginning, and I know they were not alone to have that problem. There were two people who had very good, but very different ideas, on what to do, how many other people found that in their group, that they had really different ideas of what to do. And what they did, they took a little bit of one person's idea, and a little bit of the other's idea, and they really, they have come up with something even better. Congratulations on that you guys. Brigitta and Melinda, another group that had problems at first, but seemed to be working together really well now.

3.3.3.2. "Sharing"

An important aspect in forming the community were whole-class meetings during which children presented and talked about their artifacts and the problems they encountered during the design process. To further the spirit of collaboration, Tammy also asked students to comment on each other's work, especially to "compliment" others on what they had done. Whereas Tammy was most interested in encouraging compliments, another important feature of design cultures, constructive critique, was not fostered in this way. Thus, these meetings brought forth relatively few critical comments from peers that could have assisted children in reflecting on their artifacts, techniques, or material alternatives. The following is a conversation typical for the beginning of the unit. Stan and Sandy presented their "creature" project together because, as it turned out, they had produced complementary items. Stan had made a "coffin" and Sandys "dead body" was just the size to fit inside.

01 Stan:	We are presenting the dead body in the coffin ((laughs))
02 Andy:	The dead body in what?
03 Sandy:	The dead body inside the coffin
04 Stan:	When you open the coffin there is supposed to be a dead body
05	inside, and when you open it, you see the mouth the bum and
06	then the eyes, and then it moves and then the mouth. And
07	Dennis' dogs came to the inside of it, and ate the inside of it. So it has no body in it, no bones and stuff.
08 Sandy:	Starting with, we came up with the idea for making the idea
09	of the dead body in the coffin, then we just we were gonna put
10	pins for the mouth, like these.
11 Tammy:	Thumb tacks.
12 Sandy:	Yeah, no these little pins for the mouth, but they were a little
13	hard to stick in, so we thought of.
14 Stan:	Thumb tacks.
15 Sandy:	Thumb tacks.
16 Stan:	Because it is a lot easier to put in, instead of.
17 Sandy:	And then we wanted to make some pins for here, but it kept

18	on falling of, so we just like did the back.
19 Stan:	So, I actually found out how to make a hinge, in a different
20	way, instead of using this big, heavy metal, and =
21 Tammy:	= Is there any compliments? Oh, sorry Stan, one more thing?
22 Stan:	And we wrote R.I.P, rest in peace on it. And I didn't do it in
23	my blood, though it looks like it.
24 Tammy:	Compliments?
25 Sandy:	Simon.
26 Simon:	I like the way that how you found out to make the hinge.
27 Stan:	John.
28 John:	Its neat how you cut the cones so that can open and close.
29 Tammy:	I like the way Sandy you changed the end of that cone, because
30	it isn't the same, you did something to change the end of it.
31 Sandy:	We, because when we put it in, it was too long, so now it is perfect.
32 Tammy:	You solved the problem. Two more.
33 Sandy:	Doug.
34 Doug:	How did you make the hinge?
35 Stan:	It's a secret.
36 Jeff:	I know how!
37 Tammy:	You have to patent that secret and then sell it.
38 Andy:	I like you put, how you made the man in the coffin, I like how
39	you put the man in the coffin.
40 Sandy:	Thanks.
41 Stan:	I think that worked pretty well too, now I just found a way
42	how to make it work a bit better. Just bent it back. A lot better.
43 Tammy:	That's what engineering is, changing what you've done to make
44	it better that's such an important part of what you are doing, ok, thank you.
45 Stan:	And, if you want to buy it, it costs a dollar 50.

This conversation was characteristic for whole class sessions in several ways. First, students were provided with repeated opportunities to talk about their projects and, in this, to engage in the discursive practices that emerged in the classroom. Second, there were many interactions between children. They selected each others as next speakers so that student-student exchanges were much more frequent than in ordinary classrooms. Third, there were many student questions about how to make a particularly (sometimes spectacular) artifact (line 34). However, Stan's response (line 35) was rather atypical for this classroom; ordinarily, children shared with others their own understandings. Fourth, Tammy (as Gitte in other situations) frequently called for "compliments" (lines 21 and 24), modeled complimenting (lines 29–30) and assisted children in recognizing when they solved a problem (line 32). Fifth, children complied with the request for compliments (lines 26, 28, and 38) which may be an important contribution to the emergence of a sense of community in this classroom. Sixth, girls did not contribute much to whole-class conversations. This episode, as many other whole-class conversations, involved only boys and teachers. Finally, both teachers interacted with children in a very respectful way. Here, Tammy apologized for having interrupted a child (line 21). In other instances, rather than imposing a course of action, they asked children what they thought was to be done.

3.3.3.3. *Teachers as Learning Members of the Community*

An important constitutive aspect of this community was that teachers also presented themselves as learners. Rather than being information distributors, they pointed out, at appropriate moments, when they learned something. Or they asked students to explain uncommon techniques, words, etc. For example, the following excerpt is from a whole-class conversation during which John and Jeff presented their bridge. John pointed to a number of features of the bridge where they had made use of braces to strengthen the bridge and, in this way, increased its carrying capacity (lines 01 and 03). However, when John pointed to a feature that stabilized the piers of this bridge, Jeff disagreed by uttering "deadman!" (line 04).

01 John: Braces, they're right here and inside.
02 Tammy: OK, yes, definitely inside, those are definitely braces.
03 John: They're right here ((points to the bottom of pillar; Figure 4.7)).
04 Jeff: Deadman!
05 Tammy: Deadman, I don't know what a deadman is. Jeff, what is it?
06 Jeff: Well, here is a wall, and see the wall is leaning this way, and
07   it has something coming out of here, cement blocks.

Tammy did not know the term "deadman" Jeff had used, and asked him to explain (line 05). Using his bridge, Jeff then showed how the support for a bridge pier, a deadman, buttresses walls or piers. More crucially, both teachers shared their own learning experience with the children ("Gitte asks you really interesting and good questions, and I really learned a lot, about what kinds of questions to ask, that made it more interesting for me too"), so that the classroom became a learning community (Brown, 1992; Lieberman, 1992). Both Brown and Lieberman point out that for knowledge-building communities to emerge, teachers have to change their roles and become in some fundamental way learners themselves. They need to be active role models of learning, and responsive guides to students' own inquiry processes. In this classroom, this was definitely the case.

3.4. TEACHER LEARNING

Throughout the unit, Tammy and Gitte talked a lot about their intents, their understandings of classroom events, pedagogy, curriculum design, etc. These were opportunities for both to grow professionally and to make changes in their teaching. Gitte described these opportunities to a group of visiting elementary teachers:

My teaching practice has improved so much this Fall, just by watching myself and Tammy interact, by seeing what she does, things that I would never have thought of doing but that are so useful in setting up the unit. And these are more things that have to do with the classroom cooperation and management. Like we had a big session with the children on how to encourage participation from each other. It was wonderful. I think it's really enhanced the unit from when I taught it separately. I know the content, but I am a little rusty on teaching, I haven't been teaching for a couple years. It's just been wonderful, like as a teacher I never had the opportunity to go in, except when I was student teaching and that was completely different and watch a master teacher at work.

As a consequence of Gitte's and Tammy's learning together, the classroom environment changed not only in response to the feedback provided by the research team, the availability of videotaped events, and the daily debriefings, but also because of the continuous professional dialogue that occurred in and near the actual practice. There were two important forms of teacher experiences in this learning environment: opportunities for teacher growth and continuing struggles.

3.4.1. *Opportunities for Growth*

In several instances, I recorded the teachers' conversations over and about children's artifacts, sometimes in the presence of the students' involved in the episodes they discussed. Here, Tammy and Gitte had opportunities to develop and engage in a discourse about pedagogically significant events as they occurred. These instances provided them with opportunities to learn from each other about pedagogy, pedagogical content, or content of "Engineering for Children: Structures". One important issue related to the pedagogical content was that of asking the right question at the right time; it became a continuing topic of discussion between the two teachers. Tammy admired Gitte's questioning for the way it elicited children's thinking, avoided factual answers, probed children's understanding, allowed children's construction of accounts of their artifacts' histories, and scaffolded ("stretched") their learning. Both referred to these questions as "good" or "productive" questions. Gitte on the other hand, admired Tammy for timing her questions so that these actually facilitated children's learning without discouraging them by being too challenging – something Gitte frequently did not realize, especially in the beginning. In the following excerpt, the two talked about one of the incidences where Tammy's questioning had helped students to make a breakthrough in their design.

01 Gitte: Patricia and Anna, and your question actually helped them in
02   this part here. . . . Because I asked them about this, and "why
03   did you put this here?" And they said, "Madame came and

04   said you know, pushed it over, and so we had to be able to
            make it more stable". And they really were proud of it.
05 Tammy: And, they stuck it ((a footing)) out here ((in the back of the
06   tower)), first, and I looked at it and said "Is it working?" And
07   I really had to go, like "Where else could you put it?" because
            they just sort of were stumped.
08 Gitte: Well I think, from my own experience, just experiencing frus-
09   tration with not knowing how to get them to a second level,
            you sort of.
10 Tammy: I think that is kind in the unit too, you know, it is just part so
11   much of this stuff. Thinking about how things move, thinking
12   about how to get them to think about it, you know. I think I
13   have improved so much from when I started.

While still in the classroom, the setting of their teaching, Gitte and Tammy
began to reflect on their teaching. Gitte pointed out a particular instance
with Patricia and Anna where she felt that Tammy had begun to use "good"
questions that helped the girls construct a solution to their tower's stability
problem (lines 01–04). Tammy recognized that her interaction with Patricia
and Anna had scaffolded them and retold the event (lines 05–07). Gitte
refocused the conversation to teacher learning by relating her own experience
of not being able to scaffold children in an appropriate way to achieve the
desired competence (line 09). Tammy then attributed much of this learning
to the curriculum's contribution to the learning about questioning, and reflected on how much she had already learned, just four weeks into the unit
(lines 10–13).

By talking in this way about their joint experience, Tammy and Gitte
began to construct stories with which they communicated what they had
learned. Here, this was a story of success which the two repeatedly retold
to each other, during debriefing, or to any visitor. There were many opportunities for Gitte and Tammy to construct such stories through reflection on
their learning. Over time, and by working side-by-side with Gitte, Tammy
appropriated her colleague's questioning practices. Her questions sounded
more and more like those that Gitte asked. Conversely, Gitte improved her
own competence in asking questions at the appropriate time.[7]

3.4.2. *Continuing Struggles*

Although the team-teaching experience allowed Tammy and Gitte to learn
from each other and solve problematic issues as the course evolved, there
remained issues with which they struggled on a continuing basis. In the case
of providing equal opportunities for boys and girls to participate in whole-

---

[7] Elsewhere, I provide detailed observational and theoretical descriptions of Gitte and Tammy's
learning as it came about through co-participating in the teaching practice (Roth, 1996d; 1998).

class conversations, they could not, despite their positive and ongoing attention to the issue, change the state of affairs to any significant degree. As pointed out above, gender issues in the science classroom were among Gitte's main interests. She was co-author of a set of workshop materials intended to help teachers make their science lessons less threatening to girls. During the initial planning sessions, Gitte enrolled Tammy in these interests, and intended to make this unit a positive learning experience for girls. (Inviting only female engineers was one instructional move coming out of these considerations.)

On her first day in Tammy's class, Gitte noted a gender imbalance. Although both teachers subsequently intended to balance the interactions, large chunks of the whole-class conversation were dominated by boys; girls were rather quiet and hardly ever raised their hands. Both intended to provide equal opportunities for boys and girls during the following periods. However, I was struck by their interactions with the whole class. During the initial lessons, both teachers called on boys about four times as often as they called upon girls. In our debriefing meetings, I alerted the teachers to my observation. Both then attended even more to this issue. They tried to ask as many girls as boys, and to structure classroom activities to allow girls to be more active in conversations and question/answer sessions. They asked me to monitor the frequency of boys to girls called upon and provide them with feedback; they also borrowed videotapes to check the interactional patterns for themselves. However, in the course of the thirteen-week unit, their attempts to bring about gender equity in questioning largely failed. In 14 different samples spread over the 13-week period, at best 34% of the answers were provided by girls (10 girls compared to 19 boys). At worst, there were whole-class interactions during which 17 and 12 questions were asked, but girls had only one or no chance to respond. Despite continued feedback, Gitte and Tammy did not manage to change this ratio in any significant way for whole-class discussions, student-centered presentations, or question-comment sessions. In fact, when several outside observers (elementary teachers from another school) pointed out the gender imbalance of a particular lesson, Tammy and Gitte were astonished. For they were certain they had asked an equal number of girls and boys during the observed lesson. The analysis of the videotapes showed that the ratio was 36 boys to 18 girls for that lesson. Despite these continuing struggles, however, both Tammy and Gitte felt that they had learned enormously in this collaborative teaching experience. Through their team-teaching, they had grown, and enriched their pedagogical competence.

4. KNOWING ENGINEERING DESIGN

In this last chapter of Part I, I provide an account of children's learning during this engineering design unit. Descriptions of children's associations with engineering and their approach to design problems before and toward the end of "Engineering for Children: Structures" allow readers to evaluate where the children started and to what extent they developed. I enhance these descriptions by providing further accounts of students' understandings as evidenced by their glossaries, their conversations toward the end of the unit, and their competencies to negotiate meanings and courses of action in the context of their ongoing work.

In this chapter, I presuppose, in some sense, the findings reported in later chapters. For it is only later that I support my claims of the situated and heterogeneous character of childrens designing. But this procedure is necessary because the study of situated cognition in schools is only in its infancy, and the purpose of this book is to provide first observational and theoretical descriptions that account for children's designing as situated cognition. There are implications of such a treatment of children's designing for the assessment of performance and competence. Human activities are always situated in specific settings, tool-mediated, and object-oriented; tools and objects are structuring resources for knowing and learning such that a change in tools and objects constitutes a change in the task environment with concomitant changes in the actions that can be observed (Callon, 1994; Lave, 1988). Current assumptions make it difficult to compare performance across contexts and time. If human performance is situated, that is, arises from the relation of individuals with tools, material and social setting, and objects and artifacts, competence cannot be assumed to be homogeneous by default. A less stringent assumption is to view competence as heterogeneous, grainy, and always contingent, depending on the relation between individual and setting.[1]

---

[1] Even "setting" may not be assumed to exist in some absolute sense, for the same individual may structure a place in different ways, that is, foreground different aspects at different moments, as a function of different goals s/he is pursuing. McGinn (1995) illustrated this quite clearly in her account of one woman's baking of cookies. The woman, Geena, initially set a goal to make a double batch of cookies. At some point in her cooking, Geena decided that the dough was not fit for making cookies. At that point, a few apples that she had brought home from school, and which had previously not been within the horizon of possibilities, became foregrounded and central to her change of plans. She turned dough and apples into something like an apple cobbler. Thus, how a setting is structured and reveals itself as an agent, has emergent and contingent character (Heidegger, 1977).

A second problem for assessing children's learning resides in the question, "What constitutes expert design-related knowledge?" Pea (1993) suggests:

Expertise is defined dynamically through continuing participation in the discourse of a community, not primarily through the set of problem-solving skills and conceptual structures. Achieving expertise is becoming indistinguishable in your actions and uses of representations in the language games at play from other members of a community of practice. (p. 271)

This definition constitutes expertise no longer as a property of individuals, but as something arising from participation in the diverse activities of specific communities. The level of expertise, however, will vary depending on the specific practice, and on the specific settings in which individuals participate. We observed such variations in a Grade 6–7 class where students designed a variety of machines (McGinn & Roth, 1996; McGinn, Roth et al., 1995; Roth, 1997b). Some of these children analyzed diagrams of pulleys in canonical terms and accurately predicted the tensions in various parts of the apparatus. At the same time, they did not set up appropriate pulley systems when provided with materials. Any level of assessment of knowledge may therefore yield different levels of competence depending on the particular form of assessment and the social and material contexts in which assessment occurs.

4.1. ENGINEERING DESIGN PRIOR TO "ENGINEERING FOR CHILDREN: STRUCTURES"

On the first day of the unit, students were first asked to draw or write about engineering. Subsequently, four researchers (two male, two female) debriefed the students first in groups of about seven, then in two larger groups of about 13 students. Although I had prepared an interview protocol (Table I, Chapter 2), it was designed only as a resource for the other researchers to ascertain that we maximized the quality of the data. During the following two lessons, children were asked to complete three "challenges" that served both as motivating activities for children and for assessment opportunities. Based on my observations of these three lessons, I constructed the following assertion about students' knowing. Support for this assertion is subsequently provided and elaborated.

Assertion: At the beginning of the "Engineering for Children: Structures" unit, students' associations with and understanding of engineering and engineering-related techniques was at best scant. Few students had any experience in designing with engineering-related toys. During the open-ended engineering tasks, students generated few ideas and quickly abandoned what we had construed as challenges.

In the following two subsections, I address each of the two parts of this assertion. Under Associations With and Talk About Engineering, I analyze children's written and sketched associations following the stimulus "Engineering is . . ." and their explanations of these drawings in group debriefings. This subsection also contains examples of children's talk about their prior construction-related experiences and their views regarding material proper-

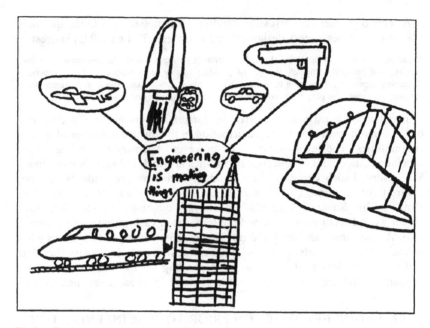

Fig. 1. Tim's pre-unit response to "Engineering is . . .". Although he suggested that engineering relates to making things, he only depicted artifacts. A few students, however, also included people engaged in a variety of activities.

ties. In the second subsection, Pre-Unit Engineering Challenges, I provide an analysis of children's actions during the pre-unit engineering challenges.

4.1.1. *Associations with and Talk about Engineering*

In response to the stimulus "Engineering is . . . ", children generally drew a variety of images that included cars, trucks, trains, and computers. Although children were asked to work on their own, there were no strict rules about "cheating". Consequently, this assessment cannot be taken as an individual test of children's associations. Rather, children later admitted that they had found inspiration from the work of others. Figure 1 illustrates the type of responses children constructed. With a total of seven drawings, however, Tim, together with John, had drawn the largest number of associations. Before the unit, children associated predominantly objects with engineering; there were few figures representing engineers in the children's images. Tim's drawing is typical for this object character of children' initial associations.

A distribution of the children's images in this class is presented in Figure 2. The rather local distributions of some images suggests that one student in a cluster may have originated the image and inspired others. For example, I observed two small clusters of drawings with computers (Anna–Doug,

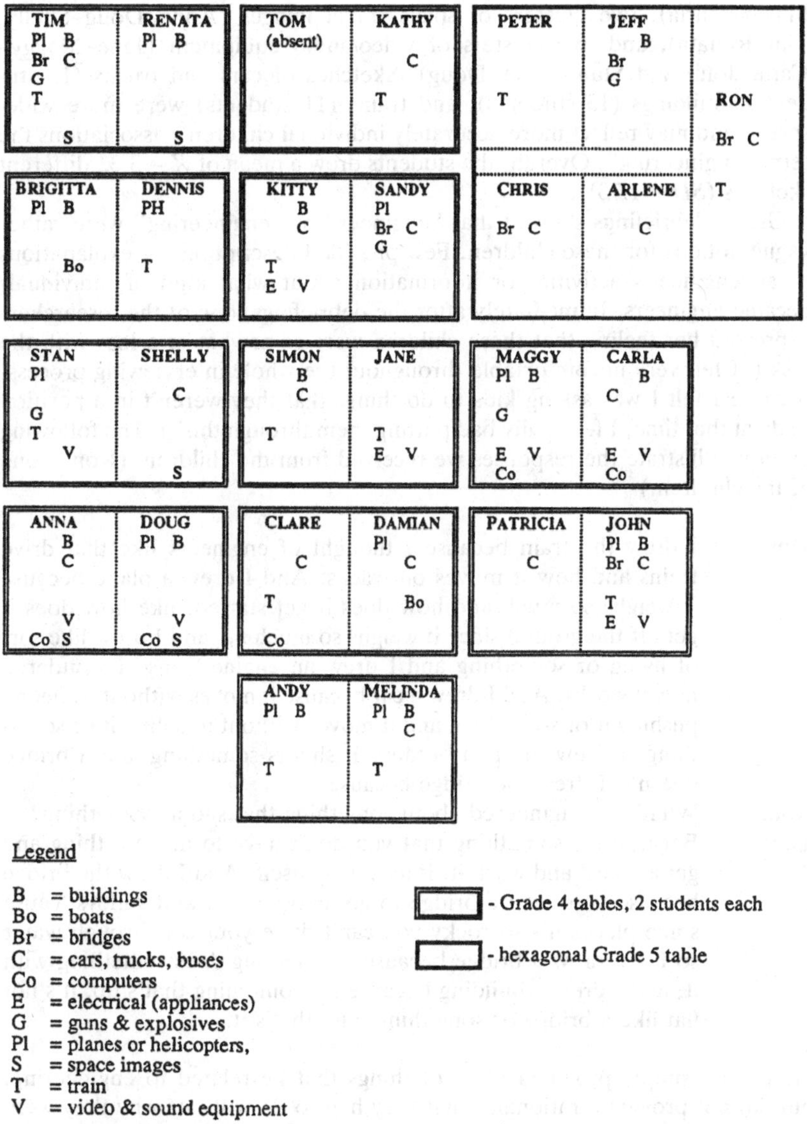

TIM	RENATA		TOM	KATHY		PETER	JEFF	
Pl B	Pl B		(absent)				B	
Br C				C			Br	
G							G	
T				T		T		RON
								Br C
S	S							

BRIGITTA	DENNIS		KITTY	SANDY		CHRIS	ARLENE	T
Pl B	PH		B	Pl				
			C	Br C		Br C	C	
				G				
Bo	T		T					
			E V					

STAN	SHELLY		SIMON	JANE		MAGGY	CARLA
Pl			B			Pl B	B
	C		C				C
G						G	
T			T	T			
V				E V		E V	E V
	S					Co	Co

ANNA	DOUG		CLARE	DAMIAN		PATRICIA	JOHN
B	Pl B			Pl			Pl B
C			C	C		C	Br C
			T				T
V	V			Bo			E V
Co S	Co S		Co				

ANDY	MELINDA
Pl B	Pl B
	C
T	T

Legend

B = buildings
Bo = boats
Br = bridges
C = cars, trucks, buses
Co = computers
E = electrical (appliances)
G = guns & explosives
Pl = planes or helicopters,
S = space images
T = train
V = video & sound equipment

☐ - Grade 4 tables, 2 students each

☐ - hexagonal Grade 5 table

Fig. 2. Distribution of various images drawn by students in response to "Engineering is . . .". Some images were more widespread, such as buildings and vehicles. Other images appeared only in small clusters, e.g., electrical appliances.

Maggy–Carla), two clusters of space travel images (Anna–Doug–Shelly, Tim–Renata), and two clusters of video/audio equipment (Jane–Maggy–Carla–John and Stan–Anna–Doug). Sketches of cars and trucks (15 students), buildings (13 students), and trains (11 students) were more widespread and may reflect more accurately individual children's associations the term "engineering". Overall, the students drew a mean of $X = 3.33$ different sketches $(SD = 1.73)$.

Group debriefings showed that "engineer" or "engineering" were rather vague notions for these children. Few provided descriptions or explanations of an engineer's activities or information about what kind of individuals became engineers. Immediately after the debriefings, one of the researchers expressed her feeling that these children were somewhat at a loss with the task ("I felt very uncomfortable throughout the whole interviewing process, because I felt I was asking kids to do things that they weren't in a position to do at that time, I felt really bad putting them through this"). The following excerpts illustrate the responses we received from the children. (Ron is one of the children.)

Tim: I drew the train because I thought of engineers like that drive trains and how it moves on tracks. And I drew a plane because it weighs so much and how does it get started, like how does it get off the ground since it weighs so much . . . and I drew like sort of flying or something and I drew an engine 'cause I wondered how it works. And I drew a car because it moves without someone pushing it or something and it moves without pedaling it or something. I drew the gun because it shoots something and a bridge doesn't. I drew the bridge because

Ron: What's so engineered about something that shoots something?

Tim: Because it's something that you don't have to make a thing and get a board and wack it, it moves by itself. And I drew the bridge because you need a bridge to go across water and valleys 'cause some places it's so rocky you can't drive your car through water so I just drew a bridge because engineering does something with it, and I drew a building because it's something that's up, it's not flat like a bridge or something, and that's it.

Here, Tim simply provided a list of things that he related to engineering, but did not provide a rationale what they had to do with, or why they were related to, engineering. In this rather typical exchange, he answered Ron's question about the gun by saying that "it moves by itself" and he drew his bridge "because engineering does something with it". Chris, a Grade-5 girl, presented her drawings in this way:

Chris: I drew the bridge because – I don't know. I don't know why I drew the bridge. And I drew the person fixing the car because he's an engineer and likes to take off wheels. And then I drew a scientist having a bad hair day 'cause he's been in too many explosions.

Ron: What does that have to do with engineering?

Chris: Because an engineer came in and he exploded the scientist's work and the scientist got a bad hair day. I got the idea of a mad scientist person blowing up stuff because and the deaths because.

Here Chris associated activities "fixing" and "taking off" with engineering. It was quite typical that the few characterizations of engineers constituted them as male and mad scientists (or engineers), recognizable from their hairdo ("bad hair day"), who blew up bridges and buildings (see also Figure 3, in Chapter 5, which contains engineers that blow up bridges and geodesic domes).

Andy, whose dad is an engineer working for a mining company, was among the four students who knew an engineer personally. He suggested:

An engineer drives a train, drives an airplane, well he fits gas into an airplane, he works in a building. My dad keeps going away, in gold mines and stuff. During air shows he puts gas into airplanes, he works everyday, Saturdays and Sundays, he drives trains. It's like he has a bunch of different jobs.

Even those students who knew an engineer in person answered in stereotypes to the questions "What kinds of people become engineers?", "What attitudes do they have?", and "What do they look like?" Despite personal knowledge of engineers, students associated them with funny (hard) hats, dirt, and smartness. There also existed gender-related stereotypes related to the nature of engineers. Here, Brigitta's response was typical:

Engineers wear funny hats, have oil all over their shirts, covered in dirt, have mustache and stuff. People who become engineers have to be prepared when they come home, for their wives, when they track oil all over the kitchen floor. What else if you live with engineer? Washing cloth, the sink after they wash their hands, oil stains.

Many students did not volunteer ideas at all. There was a particular reluctance on the part of the girls, even when prompted more directly and by female researchers.

Few children had any construction-related experiences. There were five students who indicated that they had previously made structures using Lego or Mechano sets. Chris and Jeff both had built "bridges" in the context of their Scout-related activities.

Jeff: I lashed a bridge together with these long about yeah-high poles. Like from the ground, outside. And I lashed them together with rope. And we made a huge bridge about – from where the cups are to about this table – and we, like, couldn't hold. It looked really weak, and then, and then and all, it was like a bridge and goes like that. And it could, it looked so weak, and you could wobble it, but when everyone got up on it it wouldn't break.

Chris: I made one of those, too. It was up in the forest up by my house. I made a suspension bridge sort of, from one tree to another. I used a piece of really thick rope on the bottom. And I tied it between two trees – like it wasn't over a log or anything. And then I used just normal yellow thick rope, and wrapped it.

Despite differences in prior experience related to constructing, I could not notice differences in the later design activities of the children. Initially, those who had prior experiences with material practices also were those who used the tools. But as described in Chapter 6, others quickly learned to use relevant tools by co-participating with more experienced tool-users.

The children in this class had rather vague notions about the strength of materials. There were very few exceptions. For example, Ron and Tom already had some ideas about material strength and what to do to change it.

Ron: Wood is easier to use though, and you can put it higher. Because with cement you have to have, it has to be on the ground normally otherwise it's really hard and with wood you can like you can lift up and put it high up across here or something, you can do that with wood but with cement you can't? And it takes longer cause it has to dry and everything.

Tom: I've only been to Lego Land. And that's made of plastic, right. And it was just like this, almost, like from the end of the chalkboard to here. It was like this portable and people would climb up on the roof, and like everything was just plastic and it was remarkable how strong it was. You need a lot more Lego to make it that strong . . . 'cause each piece is about that big maybe. Well, Lego is just as strong as wood. Lego's probably stronger than wood if you made it in big blocks, like if you stood on this portable roof it would probably make a little dent in it.

During these interview sessions, it soon became clear that the boys were more ready to speak up than the girls, although all researchers made a special effort to get the girls to talk about their prior experience and ideas. When directly asked to respond, many of the girls gave monosyllabic answers, or simply said, "I don't know". We then addressed the question directly. When we asked why boys appeared to speak more about engineering, girls gave answers such as "Boys are just interested in engineering", "Most building things are boys' things", "Boys are more interested in building", "Because we are shy", "Boys don't let other people talk at times", and "Boys keep talking without putting up their hands". Boys on the other hand explained their higher levels of participation in the conversations about engineering by saying, "We get more involved with the world", "Girls don't love Lego at all", "The girls don't say anything, so we just keep talking", and "Building things are boys' jobs".

4.1.2. *Pre-Unit Engineering Challenges*

Recent investigations have revealed a close relationship between scientists' discourse, instrumentation, experiments, and theories of nature (Gooding, 1992; Rorty, 1989). By investigating scientists' laboratory activities one can find out much about their understanding of nature even if they do not explicitly talk about it. Consistent with this research, I wanted to find out more about childrens' engineering-related practices by analyzing their activities in two of the three challenges that asked them: (a) to find as many ways as possible to pick up and carry a bottle using just one straw, and (b) to construct a bridge from a single sheet of paper that could hold as much weight as possible (measured in number of pennies).[2] The following vignettes from the bridge task are examples of typical approaches.

Carla (Grade 4) uses two cups to support the single sheet, and places the third cup (for holding the load) on top. The paper is not strong enough to even hold the cup. Carla looks around and watches others, then folds her sheet once, just as the boy facing her. She counts pennies into the third cup three times, but the most her bridge can hold are ten. At this point, she appears to lose interest, abandons the task, watches others, and walks about the classroom for the remainder of the time.

Jeff engaged in design practices that were untypical in their sophistication.

Jeff appears to make the paper bridge a personal challenge. He wants to set the class record and tries to fold the paper in a number of ways. He also turns the folded paper bridge to test it in different orientations. He arrives at his best solution when he folds the paper like an accordion and places the paper bridge such that the folds are oriented vertically. These folds therefore function like a series of floor or ceiling joists used in building construction.

Most children, boys and girls alike, reacted like Carla. They tried a single sheet, and, at best, folded the paper once or twice. After several trials to support weight, but without changing the configuration of the bridge, they abandoned the tasks, watched or talked with others about issues unrelated to the "problem". In some cases, students tried to achieve a bridge with a higher carrying capacity by taping the paper to the two supporting cups. This solution, however, was disallowed by the teacher, and soon recognized by other students as "cheating". I made similar observations in the case of the bottle carrier.

Gitte had also set up a third challenge, balancing as many nails as possible on the head of one upright nail hammered part way into a piece of board. Because none of the children could balance more than one or two nails, she

---

[2] Because of the help which teachers provided on the third task (described below), I did not include this challenge in the analysis of students' pre-unit competencies in unstructured design activities.

actually showed them how to construct a solution by interlocking the nails – and in this did something she had previously said she would not do: provide a standard "correct" solution. As a whole, the children in this classroom were overly challenged by these tasks. Both teachers sensed the students' frustration and began to help them in generating solutions, or in providing hints about how the task could be accomplished.

The children's performance and competence can be interpreted in different ways. First, observed performances and competencies could be taken as evidence that children commanded few engineering design-related practices that would have allowed them to generate alternate or satisfactory solutions. Second, the children may have found the task irrelevant so that it could not be considered a challenge. Only a few children such as Jeff persisted and tried a number of different ways to construct a paper bridge or a one-straw bottle carrier. The children's engagement in different activities and in conversations unrelated to the activity makes this interpretation very likely that the task as irrelevant. This interpretation would also be consistent with my observations of older students who showed great engagement with problems that they had framed themselves and which they owned; whereas they exhibited minor interest and inferior competencies in teacher-framed "problems" (Roth & Bowen, 1993). Third, the children may have lacked domain familiarity needed to frame a problem such that appropriate solutions could emerge. This interpretation is especially plausible given the task was imposed on children; there was little opportunity to explore the materials and no cover story that would make the task more meaningful. Finally, children did not have prior opportunities to engage in material and discursive practices related to structural design that would have allowed them to frame the challenges as canonical engineering problems.

It became quite clear that, at this point, students did not command discursive or material practices for dealing with the engineering design tasks, framing problems of stability and strength, or providing solutions to these problems. Children had come to this engineering unit with few images of engineering or engineering-related discursive or material practices that would have allowed them to be more successful. One may be tempted to regard this as an ideal situation for teaching a unit. Unlike in some areas – such as Newtonian mechanics where students' mundane theories interfere with the scientific discourse they are to appropriate in their science courses – students here did not bring a set of practices that could have interfered with their classroom experiences.

4.2. POST-UNIT ASSESSMENT OF ENGINEERING DESIGN PRACTICES

To those who observed this classroom toward the end of the unit, it was immediately evident that the children were learning much about engineering design: The changes in children's material and discursive practices were visible and audible. Visitors to the classroom, too, were quite struck by the remarkable competencies of these Grade 4 and 5 students. For example, two elementary teachers who taught at the same level, talked about what they had observed in the classroom during a debriefing session after a lesson on bridge design. They described their observations in these terms:

Jan: The kids were able to talk about the beams and supports and stays; and I thought I was having a difficult time, when I wanted to speak to them about their bridges, because I didn't have the vocabulary that they did.... No one was overwhelmed by the task and they were all solving their problems and adding more of this and less of that and not seeming the least bit daunted by what was asked of them.

Nora: We also noted they were very articulate, the children were able to express their ideas so well in such a sophisticated manner. It's just that these are so many and all of them that I listened to were able.... There is so much background knowledge.

In these remarks, the visiting teachers expressed that these children commanded an above-average engineering design language; and they remarked about the children's competencies for coping with the complexities of the tasks. These observations confirmed my own and are indicative of the competence children exhibited in designing structures, talking about their engineering projects, and coping with the complexity of the task. In this section, I attempt to provide the reader with a sense of children's competence.

Earlier I pointed out that assessing competencies from a situated cognition perspective is not an easy task because, implicit in this framework, practices are situated and not easily compared across settings and situations. Furthermore, competence is not a uni-dimensional entity, but is, because of its situated nature, quite heterogeneous across situations and settings, even for the same individual. Nonetheless, I provide here my assessment of children's learning during the unit, taking the teachers' top level goals as a starting point. Tammy and Gitte had identified three such top level goals. Accordingly, children were to:

•*Goal* 1. Explore and experience some critical, but not necessarily pre-specified issues involved in designing structures; this includes strengthening individual materials, building and reinforcing joints, and stabilizing structures.

•*Goal* 2. Learn to manage the complexity of open-ended and ill-structured design situations.

•*Goal* 3. Learn to design as part of collectivities, plan and execute group projects, make sense, and negotiate; learning as part of "collectivities" implies that children learn to talk about their emerging products and the processes used to create those products.

On the basis of these goals, I identified five dimensions of children's knowing and learning in engineering design that relate to these top level goals:

- Knowing important aspects about design and constructing structures (Goal 1;
- talking and writing engineering design (Goal 3);
- coping with complex design tasks and exploiting interpretive flexibility for creating innovative designs (Goal 2);
- negotiating individual differences in respect to plans and understandings, courses of actions, and the organization of collective work (Goal 3);
- using a variety of tools and materials, and their properties in new and innovative ways (Goal 1 and 2).

Following a classification scheme for engineering design knowledge, I use these dimensions as a framework to present evidence for childrens knowing toward the end of the "Engineering for Children: Structures" unit.

4.2.1. *Classification of Engineering Design Knowledge*

Much of children's learning in this engineering design unit was not of the type traditionally valued in schools – which is predominantly the "knowing that" type. In this study, children developed many competent practices related to tools and materials, that is, they came to "know how to" do a variety of things. Thick descriptions of these practices and the development of a design culture, where students learned from each other many material engineering design practices, are provided in Chapters 6 and 8. Here, I provide an overview and classification of what children learned in the engineering design unit. I show that children became very skillful in strengthening a variety of materials, stabilizing structures, handling tools such as glue guns, and working with a variety of other materials. Many of the resources and practices deemed relevant by the children were rapidly shared at the classroom level; in contrast, there was a much slower appropriation of discursive practices championed by the teachers.

As a framework against which to assess children's learning in this unit, it is worthwhile to consider the dimensions of knowing that constitute engineering design innovation. Following Faulkner (1994), there are five major types of knowledge that relate to: (a) the natural world, (b) design practices, (c) development and testing of the designs, (d) final products, and (e) knowledge itself. In Table I, I provide a summary of the typology of engineering design knowledge with examples of items from this classroom. Each item is briefly outlined in the paragraphs below. (Italicized items are headings from Table I.)

Knowledge related to the natural world. This category includes *theories* from the domain of science and engineering and *properties* of natural and man-made materials used in engineering design. Properties of the materials constituted a central theme, and Gitte had designed several lessons in which

TABLE I
Typology of engineering design knowledge

Content is related to knowledge of . . .	Description	Students talked about and used
1. *Natural world*		
1.1. Science and engineering theory	Laws of nature (forces), theoretical tools (distribution of forces through structures)	Forces, compression, tension, gravity
1.2. Properties of natural and artificial materials	Strength of building materials and strength, bonding ability, and stability of connecting materials	Glue, invisible tape, masking tape, pin joints, straws, newspaper rolls, skewers, string, toothpicks
2. *Design practice*		
2.1. Design criteria and specifications	Use(r) requirements, specification of design components, "reality bit"	Cars, people, river, earthquake-proof, wind, bridge, railing, elevator, gondola
2.2. Design concepts	Fundamental operating principles, normal configurations	Triangle, (cross) brace, doubling, layering, bundling, stay, cantilever, deadman, foundation, pier, bridge deck, floating pier, pillar, dome, base, suspension, anchor, stability, X-shape, column
2.3. Design instrumentalities and general design competence	Structured procedures such as decomposing problem, ways of doing things, ability to judge between competing demands	All of 1.1, 1.2, 2.1, 2.2
2.4. Design precedence	Prior experience with design artifact in the classroom, or design of real-world structures	*Buildings*: Empire State, Eiffel Tower, West Coast Energy, CN Tower, various sports arenas (Superdome, Skydome, BC Place). *Bridges*: Lion's Gate, Second Narrows, Tacoma Narrows, Lynn Valley, Capilano

TABLE I (Contd.)

Content is related to knowledge of . . .	Description	Students talked about and used
3. Developing and testing designs		
3.1. Developing and testing procedures	Methods of testing, ways of developing the design artifact	*Techniques*: Layering, bundling, doubling, embedding, reinforcing, overlapping, strengthening, cofferdam. *Testing*: Earthquake-proof, shaker table, forces, weakness, stability, catastrophic failure
3.2. Developing and testing instrumentalities	Techniques and artifacts used in research and development of design	Collapse, weakness, stability, strength
3.3. Test data	Predicting test outcomes	
4. Final product		
4.1. New ideas	Entire design, techniques and artifacts to produce it	New, interesting, selling feature
4.2. Operating performance	Knowledge obtained through testing, user experience	Stability, strength, weakness
4.3. Production competence	Pilot production, scale up	
5. Knowledge		
5.1. Sources of knowledge	Location of knowledge, availability of tools and materials, sources of help	*In-class*: techniques board, peers, visiting engineers. *Out-of-class*: Well-known local and global bridges and buildings, (see 2.4.)
5.2. Discourse	Canonical engineering concepts	Teachers & engineers: Brace, cantilever, deck, cofferdam, raft foundation, etc.
	Students' own descriptors	*Students*: Triangle, layering, platform, X-shape, stay, bundling, etc.
5.3. Learning process	Learning from successful resolution of problems; source and process of ideas.	Idea, problem

the children experimented with their materials. Therefore, strengthening materials by laminating, layering, and folding or rolling (or doubling up materials in some other way) were the principle means by which children changed the properties of materials. From a science perspective, a variety of forces – such as tension, compression, and gravity – acting on the structures became regular conversational topics in small group and whole class interactions.

Knowledge related to design practice. Design-related knowledge is a central and vital aspect of innovation in engineering design. Because children began to design without formal knowledge in the domain, what they invented and learned from each other was important in this aspect. Gitte frequently asked students for the purpose of their designs ("I was playing the 'reality bit'"), implying specific needs and *design criteria* and *specifications* that potential users of full-scale artifacts might have. *Design concepts* such as fundamental operating principles and normal configurations of materials to guarantee stability and strength of the artifacts were also themes of the lessons. *Design instrumentalities* encompass procedures and alternatives considered by the children during the process of designing. Finally, *design precedence* categorizes knowing related to existing structures. Local bridges (Lion's Gate, Lynn Valley, Capilano Suspension) and buildings (dome-shaped Science World, earthquake-proof West Coast Energy building, Grouse Mountain Tram) frequently became discursive resources in children's conversations. However, national and international landmarks were also frequent features of the classroom discourse. Among those that we heard were the CN Tower and Sky Dome in Toronto, Empire State Building, leaning Tower of Pisa, London Tower, government buildings in Ottawa, and the Eiffel Tower. Through the small-group and whole-class discussion of existing structures, children frequently learned from each other the relevance of a variety of design practices.

Knowledge related to developing and testing designs. This category relates to the technical aspects of developing the artifacts largely falling into the domain of material practices. *Developing and testing procedures* are accepted ways of developing an artifact and methods of testing including, for example, the use of cofferdams for underwater construction or the use of "shaker tables" (children's term for modeling earthquakes by shaking tables) to test whether structures are earthquake-proof. *Developing and testing instrumentalities* includes knowledge of alternate practices to deal with contingent problems in the development process. Both teachers also asked students to interpret *test data*, to predict the outcome of tests, or to predict the performance of their artifacts in more general terms.

Students' own comments provided indications that they learned to use tools and materials and that their artifacts were non-random. After developing his hinge, Stan said "I actually found out how to make a hinge, in a different way, instead of using the big, heavy metal"; Jeff recalled, "Our difficulties were to strengthen the bridge; we solved that problem by using braces"; and Tim remembered, "Like the shape, we were trying to make the tower stable and try to get the least amount of straws and so we used that idea for this part. Because it moved a lot . . . and now it won't sway that much".

Knowledge related to final products. Faulkners (1994) typology focuses on final, full-scale production of technological artifacts. This dimension of knowing was less useful for assessing children's learning because they never designed more than one bridge, tower, dome, or creature. Furthermore, with the exception of Tom and Sandy's wind mill and Chris and Arlene's gondolas which existed as pilot versions, students never had to consider the problems arising from upscaling a design artifact so that there was little evidence of knowledge relating to *production competence*. However, Gitte frequently encouraged students to talk about design features relevant to them by asking them about the *new ideas* represented in the artifact. These ideas could be about the artifact as a whole or individual materials or techniques relevant to its construction. *Operating performance* pertains to content knowledge obtained through the testing of the final artifact; in general, students demonstrated that they could account for the degree to which an artifact met the original design specifications. In a few instances, children demonstrated that they could deal with problems related to the pilot production and scale up.

Knowledge related to knowledge itself. Gitte and Tammy also asked questions about knowledge itself. These questions pertained to the *source of knowledge*, the knowledge of meanings and names of specific *discourse* items (originating from canonical discourse domains or from students' familiar language), or accounts of difficulties and problems students faced during the process of designing their artifacts (*learning process*).

From a pragmatist perspective, it makes little sense to talk about knowledge apart from the continuity of practices in which people engage. The most central of these practices are those of a discursive nature, that is, the language games people play. Children's use of engineering knowledge will therefore be shown in the context of writing and talking activities.

4.2.2. *Engineering Design Language*

An important aspect of learning to design is growing competence to talk design and talk about designing. In the present study, there is much evidence for children's competence in these two respects. The children had developed a remarkable competence for (a) talking about their artifacts in whole-class meetings, (b) talking about plans, negotiating problems and solutions, and making decisions about courses of actions during the design phase, and (c) designing glossary entries describing important design practices. This discourse was rich and appropriate for the weakly structured design tasks they faced. Concomitant with the language, and consistent with the framework developed in Chapter 1, this elaborate language game allowed children to interact with their social and material worlds in competent and increasingly complex ways. Comments by visiting, experienced elementary teachers helped us to recognize these children's competent engineering design discourse as something special. As the earlier quotation showed, Jan recognized the significance of the well-developed language game students used in the classroom. She pointed out that it was not merely one student who communicated well but the entire community. Being a teacher at the same grade level, Jan was fascinated by the discursive practices in this community where children had populated the discourse with their own intentions. They communicated their design ideas, described specific breakdowns of materials and tools that were relevant during various design phases, and talked about how they resolved the problems *they* had constructed.

In the following, different measures are used as indicators for the competence of students' engineering design language. My analyses proceed in two steps. First, I present a coarse-grained analysis concerned with the occurrence of words as a measure of students' competence related to an engineering discourse ("Associating Engineering Design"). The second level of analysis then focuses on written and spoken language as children deployed it to talk design and talk about designing ("Writing Engineering Design" and "Talking Engineering Design").[3]

4.2.3. *Associating Engineering Design*

Toward the end of the unit, children's engineering discourse had expanded considerably. A rather coarse level indication of children's language games in and about engineering is provided in the mind maps Tammy asked students to construct. For example, Patricia's mind map combined words and images (Figure 3). The words refer to a variety of experiences in this classroom. In the text on top, she listed 15 terms directly related to the in-class projects (buildings, braces, construction, bridges, stability, tension, compression, overlapping, catastrophic failure, X-shape, structures, domes, triangles, elevators, and deadman); and five other terms had to do with the students' trip to the local science museum (robots, Domes, Jacob's ladder, electricity, skyscraper). Her map organizes these terms into clusters, and adds "suspended" to the words related to the in-class activities. The images further elaborate her associations: there are sketches of a variety of objects and processes already incorporated in her list and map (light bulb and lightning for electricity, a building experiencing catastrophic failure, a triangle

---

[3] I chose the titles for the three sections in allusion to Ashmore's (1989) 'Wrighting Sociology' and Lemke's (1990) 'Talking Science' to express the active and child-centered nature of children's construction and appropriation of an engineering language.

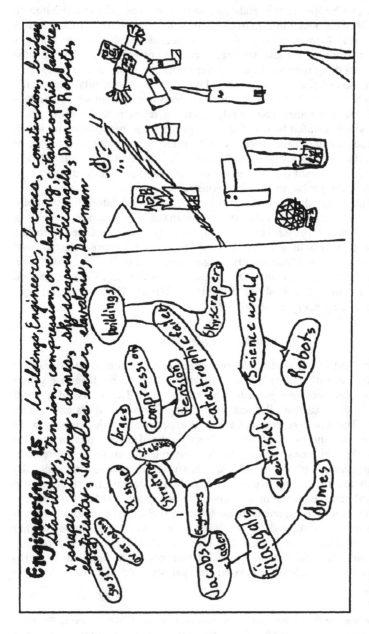

Fig. 3. Patricia's response to "Engineering is . . ." toward the end of the EfCS unit. She included terms related to the design of artifacts and to her visit to the science museum. Some of the terms and images are not related to structural engineering such as the light bulb and lightning.

representing a brace, a deadman, a skyscraper, four stacked rectangles representing overlapping, a robot, and Science World).

Not all children were as elaborate in their responses. Tim's mind map, for example, included marginally more objects than he had drawn prior to the unit (Figure 4a). As before (Figure 1), he associated engineering with making objects and provided a list of things engineers made. None of these items directly related to the unit. As for many children, there was an overlap in those images/words that they used before and after the unit. In Tim's case, airplane, car, building, and gun were part of his answer to both assessments. Kitty's mind map also shows a number of characteristic features (Figure 4b). First, as for most students, some of the associations reappeared (boats, cars, trucks, radio, TV, trains). Second, many children included large clusters of terms related to the in-class engineering design activities. Third, there were references to Science World, the local science museum (Jacob's ladder, light bulb, electricity, and bridges). Finally, she associated the word engineering with the visit of an engineer to class (artificial hip joints, knee joints). The following analysis is designed to provide an indication of the major trends in the entire class.

For this analysis, I distinguished between words directly related to the unit and unrelated to the unit. The words I attributed to the "Engineering for Children: Structures" unit included (a) verbs and nouns related to making stable and strong structures, (b) examples of bridges, towers, and buildings, (c) engineering tools and materials related to structural engineering, and (d) items that could be related to activities in this unit other than designing, including the trip to the science museum, the film presentations featuring bridges, and the visits of two female engineers. Terms that I considered unrelated to the unit were objects related to electricity and gasoline-powered vehicles and trains.[4] A simple comparison of the number of images or words associated with engineering before and after the unit shows a significant increase from a mean of 3.3 ($SD = 1.7$) to a mean of 22.8 ($SD = 7.8$). The analysis revealed that more terms were directly related to the construction activities ($X = 15.5$, $SD = 6.9$) than other engineering items not related to structural engineering ($X = 7.2$, $SD = 5.8$). When each sketch on the students' pretests was counted as one object (as labeled by each student during the group interviews) and compared to the number of unit-related objects indicated on the posttest, there was a significant increase $t(22) = 3.42, p < 01$. A 2×2 MANOVA with gender as the between variable and unit-related/unrelated terms on the posttest as repeated within measure revealed no gender differences or interactions; but there was a significantly larger number of unit-related words, $F(1,21) = 18.21, p < 0.0005$.

---

[4] Similar associations were made by the Grade 5 students in another Canadian province (Bloom, 1993). These associations can therefore be considered part of the common background children bring to their classroom experiences in engineering design.

a.

> **Engineering is** when you build (design) something. Engineers build (design)
> buildings, ships, airplanes, TVs, houses, and make sure they are safe.

gun

boats

TV ——— **Engineers
make**

house

airplanes

buildings

car

electrical

air conditioning

b.

> **Engineering is** building and designing. some times when you build a
> structure and it falls, we call it a catastrophic failure. There are a lot of
> different engineers, some design things and some build things. If there
> weren't any engineerrs, there wouldn't be any structures

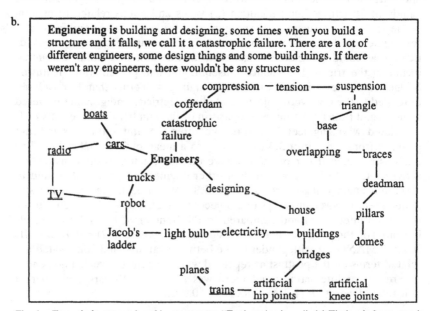

compression — tension —— suspension

cofferdam

triangle

boats

catastrophic
failure

base

radio — cars

overlapping — braces

Engineers

TV

trucks

designing

deadman

robot

house

pillars

Jacob's —— light bulb — electricity —— buildings

domes

ladder

bridges

planes

trains — artificial
hip joints —— artificial
knee joints

Fig. 4.   Two mindmaps produced in response to "Engineering is . . .". (a) Tim's mindmap stood
out because it had so few terms listed. Both mindmaps indicate the overlap with items they had
also listed during the pretest (words underlined).(b) Kitty's mindmap is quite elaborate including
terms that related to designing, the visit of the science museum, and the visit in class by two
women engineers.

4.2.4. *Writing Engineering Design*

Associations with engineering (words or images) are only crude indications of children's competent use of an engineering language game. A much better indication comes from the glossaries (created from Week 11 to 13 of the unit) in which students explained their understandings of terms that had, in the course of the "Engineering for Children: Structures" unit, become part of the common practices in this classroom community. The glossaries were an occasion to study children's language games in-use. Students produced from 6 to 12 entries ($X = 8.6$, $SD = 1.5$). In the 26 glossaries collected, there were a total of 21 different terms, mentioned between 2 and 25 times (frequencies appear in parentheses):

catastrophic failure (25)	compression (22)	tension (21)
brace (18)	triangle (17)	pier (14)
foundation (12)	base (12)	suspension (11)
X-shape (11)	platform (10)	bundle (10)
overlapping (9)	stay (7)	deadman (7)
cofferdam (5)	strengthening (5)	post (4)
anchor (3)	pillar (3)	footing (2)

Asked to explain words relevant to their work in designing structures both in word and sketch, children demonstrated quite complex descriptions that provided insights to their sense making. Figures 5 and 6 are illustrations of the variety of explanations students constructed in this task.[5] These figures are part of the evidence that in this class, students developed their own ways of understanding and using engineering language rather than learning by means of all-too-often standard, prized, and multifariously regurgitated definitions (Poole, 1994).

Tim's glossary items illustrate that he had developed a competent language game to communicate engineering design issues (Figure 5). This competence is particularly noteworthy because he had difficulties with reading and often lacked ideas. (Readers may wish to refer back to Tammy's description of Tim in Chapter 3.) Whereas his mind map constituted a list of objects that engineers design (and build), his glossary items attested to a dynamic view of engineering. Arrows (away or towards some point) and action words (pull, push) express the dynamic nature of *tension* and *compression*. More importantly, he used a variety of different expressions for the same notion, for example, compression. Here, Tim identified compression at some spot of a building; compression as experienced by an individual carrying a heavy load; and (in the text), compression when one person (you) pushes two things together. Thus, he identified compression in inanimate objects, objects

---

[5] Interested readers will find additional illustrations from children's work in Roth (1996a, b).

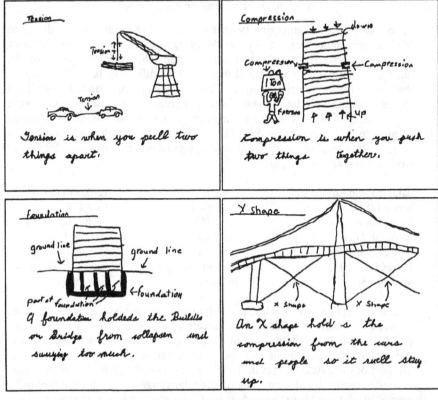

Fig. 5. Four of eight glossary items from Tim's engineering log book. Tim invented the "X-shape" together with Stan while they worked on their collective bridge design. The drawings and texts provide evidence for the deep understandings Tim developed about aspects of structural engineering.

acting on people who experience compression, or people creating compression.

Even more astonishingly for students at this age, Tim illustrated an understanding of the upward forces provided by the ground on a building. Furthermore, in his drawing, tension is distributed through the rope rather than appearing only at the end. In my experience as physics teacher, even Grade 11 and 12 students have difficulties appropriating such aspects of canonical physics discourse. The dynamic character of Tim's understanding is also apparent in his explanations of *foundation* and *X-shape*. Accordingly, a "foundation holds a building or bridge from collapsing and swinging too much". The drawing itself reveals the declarative ("this is") explanation of a foundation. Tim had a special relation to the term X-shape: during his collaboration with Stan on the bridge project, he had "invented" the term which later became part of the common language game in this community. Here, the drawing and the label reveal the declarative "this is" of the X-shape, while the text points out again the function of this technique, "holds the compression from the cars and people so that it will stay up".

Three further aspects of Tim's glossary items are interesting and representative. First, children's explanations frequently expressed their own lived experience of designing structures. Here, Tim had developed the X-shape in response to a structural weakness of his bridge. The videotape shows him repeatedly pushing down on the bridge, turning around to Stan, and noting, "it's still not good enough, see, it's still wobbling". From his pushing actions on the bridge and his subsequent interactions with Stan and the teachers the bracing technique *X-shape* emerged. "Compression from the cars and people" recaptures the compression he had produced in his repeated action of pushing down on the bridge, and in the load test to which the bridge was submitted by hanging a bucket with wooden blocks from the bridge deck. This interpretation receives further support from the fact that his glossary drawing bears an iconic relationship to the bridge he had previously built: posts, ropes from the center pier suspension, and the particular placement of the X-shaped braces.

Second, words did not appear in isolation but, like in a dictionary, they were presented in relation to other words; none could really be understood without understanding the others, creating a self-referential system. Here, Tim used *compression* in the explanation of his term *X-shape*. Frequently, students used similar sketches for a variety of entries: the same bridge could have explained the function of braces, the existence of piers and decks, or the notion of suspension.[6] Third, children frequently used images and descriptions that were and are part of their everyday language games, and thus constituted meaningful ways of expressing engineering design terms. In Tim's drawing which accompanied compression, a person underneath a one-ton weight experiences compression; his drawing made clear that tension exists when one car tows another or when a crane lifts bundles of logs.

The examples from Jeff's and Kathy's glossaries provide further evidence for these phenomena (Figure 6a). Jeff's definition of a brace is in terms of a "this is"; at the same time, it reproduces in an iconic way the deck of the bridge he had designed with John (see Figure 7). In their first test of the bridge's carrying capacity, it soon began to bend in much the same way as his "before" drawing suggests; after repeated cycles of testing and adding braces, Jeff and John designed the strongest bridge in the class. The second example from Jeff's glossary provides an indication of his understanding of compression, especially as it applies to the construction of arches. Here, the downward force at the arch's center is counterbalanced by the ground forces acting upward through the arch. The upward turn of both upper arrows

---

[6] This self-referential nature of language games was exemplified in Quine's (1987) 'Intermittently Philosophical Dictionary' and Ashmore's (1989) 'Reflexive Thesis'.

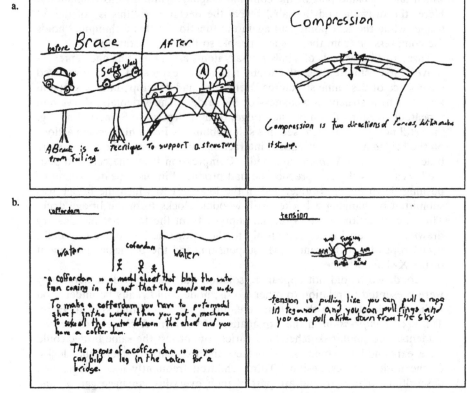

Fig. 6. (a) Two glossary items from Jeff's engineering log book. The uses of braces and his understanding of the forces acting in a structure is remarkably similar to that expressed in the bridge design he presented to the class. (b) Two glossary items from Kathy's engineering log book. Her definition of "cofferdam" includes a proposition, an action, and a statement of purpose. Her definition of "tension" addresses the dynamic aspect associated with forces in action.

to counter the downward arrow shows, together with the inscription "two directions of forces which makes it stand up", that Jeff had appropriated engineering practices (possibly in the science museum) and used them for his own intentions: to account for his "understanding" in rational terms.

Kathy's glossary entries (Figure 6b) equally testify to the sophisticated nature of children's engineering design language. Apparently inspired – as many other girls – by the visiting female engineers, Kathy explained a *cofferdam*.[7] Her explanation next to the drawing included a definition in the

---

[7] It was remarkable that only one boy included in the mind map a reference to the visits of the female engineers, whereas seven girls made such a reference.

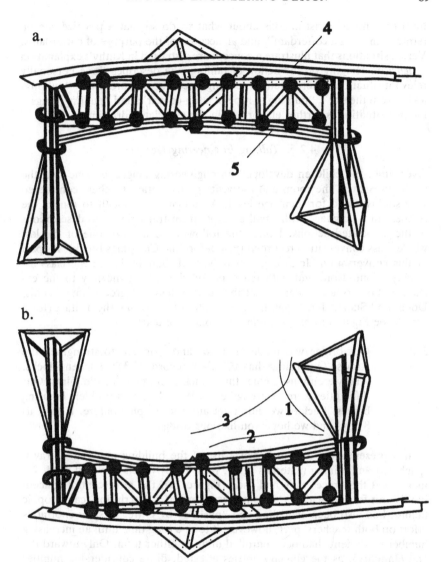

Fig. 7. The bridge Jeff and John designed and which was the topic of a whole-class discussion. The numbers are used to index text and gestures during the presentation. (a) Original bridge design. Testing showed that the weakest part was the piers. (b) Position in which Jeff and John achieved the class record. This position was contested during the presentation.

form of "this is", instructions about what to do so that a practice can be termed "making a cofferdam", and an account of the purpose of cofferdams. Varied situations that illustrate a concept are also used in Kathy's explanation of *tension*. Hinging on the action of pulling, she describes three personally relevant situations where a tension/pulling discourse is appropriate: "pull[ing] a kite from the sky", "pull[ing] a rope in a tug-of-war", and "pull[ing] rings"; the last situation is further elaborated in the associated drawing.

4.2.5. *Talking Engineering Design*

Over time, the children developed an engineering language game that the adults present in the room and the visiting elementary teachers considered very sophisticated for their age level. Although it is difficult to render the richness of a language with small excerpts from transcripts and descriptions of the associated actions, I offer the following analysis of one of the last whole-class conversation recorded (just before the Christmas break). As part of this conversation, John and Jeff presented their bridge to the class assembly. Tom, Ron, and both teachers contributed significantly to the exchange. Four other students contributed to a lesser degree (Tim, Dennis, Doug, and Stan).[8] Jeff began the presentation by showing the artifact (for a sketch see Figure 7a) and speaking the following text:

Jeff: Our bridge was made of straws and spaghetti to strengthen the straws. Our bridge has the class record of 347 wooden blocks. And we only used glue, thumbtacks, and tape for the joints. Our difficulties were to strengthen it. We solved that problem by using braces. Luckily we didn't get any catastrophic failure. It took us 8 days and we hope you like our bridge.

In this presentation, Jeff provided a list of the building materials (straws, spaghetti) and joining materials (glue, pins, tape) which they had used. He mentioned the braces they had used (Figure 7a), the most central concept in this unit for the teachers. Another word in Jeff's description, catastrophic failure, had also been cherished and "pushed" by Gitte. Despite enormous effort on both teachers' part, however, it took a long time until an increasing number of students had been enrolled in using either term. Only toward the end (January), as the glossary entries attested, did a considerable number of children finally use these notions on a regular and unprompted basis. On the day of this presentation, there were still many students who could not even pronounce the word, let alone use it following Gitte's prompts or for

---

[8] The problem of getting girls involved in whole-class interactions, as pointed out in Chapter 3, also existed here. Only boys critiqued Jeff's design, which may in part be a result of the fact that Jeff could make quite snide remarks – although Tammy did her best to quench any attempt toward such behavior.

their own intentions. After this show and tell, the teachers encouraged comments, critique, and questions. Tom was one of the first to be called upon by John:

Tom: I think it would have held a little bit less if it had been the right side up. It would have been better if your bridge hadn't been like between the desks, it would have held a lot more, I mean without legs maybe it might held more. I thought at first it was cheating being upside down, but then I realized it is sort of like mine and Andy's, because it was like between two desks.

Tom voiced a subtle critique of the fact that Jeff and John had tested their bridge upside down, that is, with the piers pointing upward (Figure 7b). He had been one of the spectators during the testing program and had seen that Jeff and John had tested the bridge upside down. Jeff had decided to test the bridge in this way because the piers had shown sign of weakness in a previous test. Tom then related Jeff's bridge to his own, which shared the complex use of spacers and braces that constituted the bridge deck.

Tom's critique shows many features of a competent and sophisticated engineering discourse. First, Tom suggested that the bridge would have held less in the upright configuration. He then proposed that the bridge would have held more, in the upright configuration without the piers. Finally, he related Jeff and John's design of the bridge deck to his own and hinted that it had a very similar symmetry: his deck was identical in both upside down and upright positions. This critique became the starting point for many of the following exchanges. Ron repeatedly questioned Jeff about the symmetry of the deck design, and its relationship to the carrying capacity. Tom's critique also shows signs of reflection on the historical development of his understanding, "I thought first . . . but then I realized . . .". It was characteristic for this classroom, and fully in the teachers' intent, that the children account for learning whether this was because they had wrestled with a breakdown in their design or inspired themselves by watching others. Tammy immediately followed his comments with a praise, "good thinking, I really like the way you changed your thinking on that one". (The transcription conventions are explained in footnote 16 of Chapter 3.)

01 Gitte: And how does that compare to a real bridge that would have
02   those kind of supports? What is different between your bridge
03   there, and say, a bridge you drive over?
04 Jeff: This bridge, most likely the bridge that I drive over doesn't
05   have this
06   [((Points to piers))
07 Tammy: But it has to be held up by something, it does have those.
08 Gitte: Say there is a bridge that has those.

09 Jeff: But see, it is too flimsy holding like this
10   [((holds it as in Figure 7a))
11 Tammy: So in the real world, what would you do about that? If you
12   were a real engineer?

Jeff and John suggested that they could use stays to hold up the bridge or use the piers to attach cables in the way they are used in suspension bridges. Jeff then provided a longer explanation of what he would do if he was a "real" engineer. Rather than viewing his artifact as an image of something real, he constructed the artifact and activities that led to it as an exploration of design principles.[9] Thus, it was not important whether or not "real" bridges could be turned upside down. Holding the bridge in the critiqued "upside down" configuration, he made the following statement. (The numbers in brackets are used to cross-reference hand position or, when in combination with lines, hand gestures made during the talk about the bridge re-presented in Figure 7.)

Jeff: It is free-standing, but we would, if it was an engineer, we, this
             also, this, we could have done it like that because this brings
             also the weight down on the ends [1], it brings the weight down
             on the ends, so it brings the weight down on the ends and it's
             easier for a force to go across here [2] and the tension brings
             it down [3].

What he had done was design a possible configuration and then constructed an even stronger configuration by turning his artifact upside down. From this perspective, if he had to construct a "real" bridge, all he had to do was use the design principles that made his current bridge in its "upside down" configuration as strong as it was. Here, Jeff suggested to make use of the design as he proposed it. The piers weighed down the ends of the deck, additional suspension wires could be used to hold up the bridge deck (increasing its carrying capacity) and he showed how the "piers" created tension in the straws along [2]. When he presented, there was a slight bend in the deck so that the pressure on the piers increased tension in the straws along [2]. Jeff's discursive competence relating to the forces acting within his structure was quite sophisticated as was discussed in relation to his glossary. Jeff did not simply regurgitate teacher or textbook definitions; Tammy could not help him with the forces because she was not competently talking about them herself and they had no textbook to do further reading. Jeff had appropriated this discourse about forces, tension and compression (possibly in part from an earlier visit with his parents to the bridge exhibit at the science museum)

---

[9] Here, Jeff appeared to take the same view of designing as an emergent and heterogeneous process which I describe through this book but particularly in Chapter 8.

and used it here for his own intentions. Other students in this class learned from him.[10]

Not satisfied with the previous explanations, Ron reiterated the questions and critique regarding the strength of the design, but he focused on the symmetry of the bridge deck.

```
01 Ron:     I got a question. If you took the legs off, if you cut them off
02          around the bottom of the bridge, do you think it would hold
03          the same as upside down? Is it build the same?
04 Jeff:                      [Yeah.                    [Well actually,
05          if you cut the legs off, you might try it putting it down like
06          this, without the legs, to see how strong it would be, you might
07          try that. But we don't think it would hold any more.
08 Ron:     Are the bottom and the top built the same?
09 Jeff:    The bottom and the top, yes, 'cause originally this ([4]) was it,
10          and then we built this ([5]) under the bottom, to make it
11          stronger.
```

Implicit in Ron's questioning was the notion of equal carrying capacity if the deck was symmetrical, for he followed the first question "would it hold the same?" with "is it built the same?" Jeff responded that he did not think that there were differences if the bridge deck alone could be tested right side up and upside down. Ron was not satisfied and continued to question the claims about symmetry. In response, Jeff pointed to the top of his double deck (Figure 7, [4]) and explained that they had begun with this part, but that they added the bottom part [5] to increase the carrying capacity of their bridge.

This episode clearly shows the students' competence in talking about their bridge designs that they had developed during the three weeks prior to this conversation. Students' accounts, such as Jeff's explanation for the nature of his bridge deck, also show that they had learned about design through designing, an assessment which agrees with my observation of their design practice. Here, for example, Jeff and John had started the bridge design with little prior experience. With the materials given (straws, spaghetti, glue, tape, pins) and driven to make a bridge with high carrying capacity, they designed the deck in the way shown in Figure 7. Whereas critics may argue that the bridge shares similarities with existing bridges so that one can explain the design by using "copying", Chapter 6 will show that "copying" frequently is not a good discursive resource to account for children's art (practices) and artifacts.

---

[10] How students learned from each, that is, came to participate in the circulation of a discourse is the topic of Chapter 7.

4.2.6. *Coping with Complexity and Interpretive Flexibility*

On the engineering pretest, which asked students to make a strong bridge from just one sheet of paper, they quickly abandoned the task. Some students tried to support the load without manipulating the sheet of paper, others folded it once or twice, and very few attempted to search for stronger solutions. Two months later, the children took the design problem (a bridge spanning at least 30 cm and supporting as much weight as possible) as an extended challenge which they pursued with great persistence over a three-week period. They considered several design alternatives, actively searched for solutions that would enhance strength and stability, and exploited collaboration to achieve better solutions. (Five designs showing different levels of competency are featured in Figure 1, Chapter 7.) Before the unit, students appeared to be at a loss, they walked around the classroom, seemingly not knowing what to do next. Jan's comments, cited at the beginning of this chapter, provide the sense one could get while assisting the lessons. The children evidently had developed great self-confidence that permitted them to deal with the challenges of ill-defined design tasks. Gitte, who had given more than 40 workshops on teaching this unit to elementary teachers, contrasted the children's remarkable self-confidence with the hesitation of many adults working in an engineering design environment. In many cases, adults abandoned the challenges something none of our children designers did during the design challenges of the core unit.

Coping with complexity and exploiting *interpretive flexibility* are interrelated.[11] In this design classroom, students learned to exploit the interpretive flexibility of tools, artifacts, materials, talk, and rules. Rather than accepting limited meanings that might be associated with a specific tool or material, children generated new meanings that allowed them to put things to new use. Take the case of the glue guns. Ordinarily, in what may be interpreted as the canonical use, glue guns are employed to heat and liquefy solid glue sticks by means of electrical energy from AC outlets. In this classroom, glue guns were creatively reinterpreted and successfully used in ways for which they were not designed. For example, Tom used the hot glue gun tip to burn holes into straws into which he placed other straws that he carefully attached with glue. In this way, he produced a new and very stable joint for constructions that employed drinking straws (Figure 1b, Chapter 6). Rather than using the glue gun to exude hot glue, Tom interpreted it as a tool for burning holes into straws. In a similar way, Tim reinterpreted glue guns twice. First, rather than squirting glue from the tip, he heated the sides of two straws simultaneously. He brought the heated areas into contact which formed a solid bond once the temperature decreased and thus created a new type of

---

[11] The concept of *interpretive flexibility* is worked out in some detail in Chapters 6 and 8. Briefly, it indicates that any object or sign does not have a unique and stable meaning, but is interpreted in different ways depending on local contingencies and the horizons agents bring to a situation.

joint leading to a very solid tower (Figure 1c, Chapter 6). His second invention was the "portable" glue gun. He serendipitously discovered that glue guns could be operated even after they were unplugged for a few minutes. He subsequently announced this invention to peers and teachers. Later, I repeatedly observed that other individuals, students and teachers, used glue guns in this way because this allowed them to work on projects at a remove from electrical outlets without carrying the design artifacts around which had resulted in frequent breakdowns.

Designing involves the solution of problems on different levels of complexity (Faulkner, 1994). Because these problems arise with different frequencies, children's learning can be expected to vary with the frequency of design problems they solve. In this unit, children only designed four structures, but they were all different. Thus, little can be said about children"s ability to design towers, bridges, or huts in more general terms. But there is evidence that children learned when they could frame problems such that previously successful solutions could be reapplied. For example, when Andy, Simon and Tim attempted to combine two stable, cubic elements for their future tower, these did not fit in the way they had expected. They found that by only partly completing the second module, it was flexible enough to fit on the other element. When they tried to attach a completed, pyramidal element, they experienced a similar lack of fit. After removing its base, they achieved "a perfect fit" (Andy) of the remaining piece to the existing structure. Later evidence suggests that they learned from these episodes. Rather than completing their next module, they kept its base flexible which afforded them another "perfect fit". Here, children intentionally applied a strategy that evolved from their solution of an earlier problem.

There is ample evidence in the data that students increased their understanding of design and that they intentionally applied this understanding in their subsequent work. This was particularly evident for "design knowledge related to the natural world", one of Faulkner's (1994) five categories of design knowledge. Most children came to strengthen their materials by, for example, laminating sheets of cardboard, bundling straws, spaghetti, and skewers, or producing tight newspaper rolls. They used braces, cantilevers, stays, and arches for stabilizing and strengthening their structures all of which are examples of knowledge related to design practice in Faulkner's scheme.

4.2.7. *Knowing to Negotiate Plans and Courses of Action*

In this classroom, students had many opportunities for participating in collective activity. Initially, a considerable number of students had difficulties working together with others and many students cared more for their own than for collective achievements. Thus, being recognized as the inventor of an idea or practice was very important to students as was designating others as "those who copied". However, as outlined above, Gitte and Tammy continuously encouraged children to work together, solicit participation in collective activity, and value joint learning over individual prowess. Through their collective activity, the children participated in practices that allowed them to increase their competencies in negotiating alternative understandings, plans, and future actions. I illustrate this competence in the following "worst-case" scenario, the collective activity of Stan and Tim. Both have had prior problems relating to others. Tim had been a leader in previous projects such that his peers thought he always wanted things to go his way. Stan, as described earlier, had difficulties working on any task in a consistent way so that his peers did not want to work with him. The project with Tim was Stan's first collective project. The following episode illustrates both students ability to negotiate.

Tammy and Gitte had just announced clean-up and suggested that students seek safe storage places for their projects to prevent any damage (as had happened with some of the towers earlier in the unit). Tim and Stan separately searched for a storage solution. When they came back together, each had identified his own, different solutions. Stan wanted to show Tim a shelf in the foyer of the classroom, where he had already stored the bridge. But Tim intended to shelter the bridge in a box which he was in the process of building. When Stan began to talk in the following episode, Tim continued to work on his cardboard box, barely looking up.

01 Stan:	Tim, Tim, *no*! *no*, Tim (.) I found a good place (.) if you put
02	it in a box it won't work that well (.) I'll show you. Just leave
03	it alone, just (.) come see.
04 Tim:	NO!
05 Stan:	*Ti:::im=*
06 Tim:	=We're gonna put it in a box
07 Stan:	It's better than *any*thing
08 Tim:	Oh, yeah, yeah, yeah (.)
09 Stan:	Tim, ahhh (3.2)
10 Tim:	Let's just see.
11	((About 1 minute passes during which Tim completes his box.
12	Stan gives him a hand.))
13 Stan:	Tim, I wanna show you this *exact* good place, where every.
14	(14.0) ((Tim continues to work on the box.))
15	Come see my place.
16 Tim:	OK, let's just turn this around (1.0) Oh we can still use the
17	box to put it (1.3)
18 Stan:	Put *this* in *that* box (place?)?
19 Tim:	Where is the (?) thing? Where is the bridge?
20	(11) ((Stan and Tim walk to the foyer together.))
21	Yeah take it down to put it in this.
22 Stan:	°How can we put it?° Oh
23	[((They try to put the bridge into box.))
24 Tim:	Jus::t put it (2.1) You got a shield around it now. ((Bridge

25   does not fit into the box.))
26 Stan: NO, I think it is better up there. (1.2)
27 Tim: Free box for sale!

At the beginning of this episode, Stan attempted to convince Tim that the place (in the foyer) he had found was ideal and much better than Tim's idea of putting the bridge into a box (lines 01–03). Tim, however, categorically refused (line 04). Although Stan pleaded for his case emphasizing that it was "better than *any*thing" (lines 05, 07, and 09), Tim firmly stated that they would implement his solution (line 06) and questioned Stan's claim (line 08). But he left open the possibility for other approaches to the problem should his own solution not work (line 10). Stan then helped Tim to make the box. A little over a minute later, they had just completed the box, Stan reiterated his suggestion.

At first, Tim did not seem to listen and finished his box (lines 13–17), despite Stan's urgent plea to follow him. But then he suggested that they might combine the two solutions, put the bridge into the box, and then the box onto the shelf (lines 16–17). They walked to the foyer, took the bridge off the shelf (lines 19–21), and attempted to put it into the box (lines 22–25). Realizing that the bridge was a little too large, Stan proposed to interpret his as the "better" solution to the storage problem (line 26), and Tim accepted by offering his box to other student groups (line 27).

Here, neither Tim nor Stan proved to be intransigent or followed the other without reflection. Tim insisted that they consider his solution. Whereas Stan's solution was already available, it took some time to build Tim's box before they could compare the two options. Stan demonstrated not only flexibility allowing for a comparison between the two ideas, but he actually helped Tim to prepare the second solution. They then based their decision on the relative merits of the two solutions. A video analysis of Stan and Tim's collective bridge project showed that in the course of its nine-hour duration, the two negotiated all their plans and actions. I made similar observations in the case of other students with problems in group situations prior to, or early in the unit. There were only two separate instances early on in the unit when Brigitta/Melinda and Clare/Shelly asked the teachers to mediate their decision about which further steps to take. Tammy confirmed my assessments and was very pleased about the children's progress working together. For example, Tammy thought that "It was neat to see Tim work not with his normal friends but with someone different and do a really good job" or "Tim and Stan worked extremely well together on the last part of the unit which was building a bridge". In the same way, Tammy talked about other children who have had problems working with others such as Clare ("Clare has had some peer problems . . . and she seems to be getting better in that way").

PART II

TRANSFORMATIONS OF A COMMUNITY: THE EMERGENCE OF SHARED RESOURCES AND PRACTICES

In this second part of the book, I provide observational and theoretical descriptions of changes in knowledge at the level of the classroom community. Here, knowledge is conceived in terms of two types, resources and practices. Resources include facts, artifacts, and tools available to the members of a community; practices are the patterned actions in which members characteristically engage. Practices differ from community to community, although the resources associated with a practice may be the same. In Chapter 1, I pointed out that carpenters, wood cutters, and cabinet makers all use chisels but do so in different ways; in each of these three communities, the practices related to the same chisels differ in order to make them most appropriate for the materials and purposes at hand. Conceptual tools also differ between communities although they may be considered "the same". Physicists and engineers share many of the same equations. However, they differ in the way they put these equations to use. The two examples express a distinction between two kinds of practices: material practices, which relate to tools and how they are used, and discursive practices.[1] Part II is concerned with transformations of knowledge available to a community as resources and practices are increasingly circulated among its members. That is, it is concerned with the social aspects of knowing and doing engineering design in a Grade 4–5 classroom community.

The core issues of Part II are how members come to produce new resources and practices and, through processes to be investigated, how this knowledge comes to be accessible to, and appropriated by, other members. I show that building a community is like building networks for circulating resources and practices. The adoption and development of material and discursive practices is inextricably confounded with development of the community. They are mutually constitutive, for no community can exist apart from the network of its practices. Resources produced and new practices developed are universal,

---

[1] These divisions are heuristically useful, although I show below that the distinction has to be abandoned in some situations.

but only within the network they generate. Each of the three chapters in
Part II focus on the circulation of one type of knowledge: resources, material
practices, and discursive practices, respectively. Much of current science
teaching is concerned with the circulation of resources (facts). These
resources, however, seldom come to be used in authentic practices. In knowledge-building communities, on the other hand, the circulation of material
and discursive practices is as important as the prized facts in traditional
classrooms. It has to be reiterated here that the division of knowledge into
resources, material practices, and discursive practices is only a heuristic.

My example of the school of Moussac already showed that the circulation
of practices entails the circulation of resources. Thus, to participate in garden-
ing practices, one needs tools (e.g., rake, hoe), materials (e.g., fertilizers,
water), and facts ("Don't water in the heat of the day!") as resources. In
the case of the children in the school of Moussac, therefore, the circulation
of resources and practices was thoroughly connected.

5. CIRCULATING RESOURCES

An important question for a socio-cultural approach to knowing in school science is how information and artifacts which are known and in the hands of a few individuals, become available to the classroom community as a whole. Every community of practice evolves through the dialectic of its own cultural reproduction – which allows the appropriation of resources and practices by newcomers within and across generations – and cultural production which allows the emergence of new practices – often in response to evolving technologies. To understand a community, one needs to get its own perspective on innovation and reproduction. From an investigator's perspective this means that one needs to follow community members' claims that they invented something, that someone else copied an artifact or idea, or that they used a resource because others were using it too.

Scientists and engineers frequently deal with questions such as Who invented what?, Who constructed a new (arti)fact?, or Who copied information? Copyrights, patents, and reception dates on scientific articles are all designed to determine those who have the right to claim an innovation. Sometimes, such claims lead to fortunes (through marketing of a patented product) or fame (Nobel Prize).[1] However, despite copyrights, patents, and dated submissions, the question about who "really" was the innovator is not self-evident. In such cases, members may ask an independent body (judge, jury) to decide between two claims.

In school science, such debates about claims to innovations will ordinarily not exist – when we think of traditional science classes where teachers tell facts and students memorize, there is no room for innovation. However, in science classes where students pursue projects of their own choice or where they pursue a common research or design question but with their own tools and materials, such debates allow researchers to investigate, and understand the mechanisms of knowledge [sharing] and knowledge [diffusion].[2] That is,

---

[1] An interesting study of the relationship between academic science and patenting culture, and the resulting constrictions on the circulation of information was conducted by Packer and Webster (1996). An interesting case of a discovery dispute existed in the case of the human immunodeficiency virus (Rawlings, 1994).

[2] Garfinkel and Sacks (1986) introduced the convention of enclosing in square brackets any notion that needs to be investigated in its own right. I engaged in the present investigation in part because I did not know what colleagues meant when they said or wrote "in collaborative groupings, students can share knowledge". Thus, I use [share] to indicate that the meaning of this term has to be elaborated through empirical work. In the present situation, this means that I account for how this classroom community came to share certain facts, artifacts, materials, and tools.

debates involving priority claims allow researchers to document learning at the community level and the sites of various aspects of knowledge.

Collecting the data necessary to answer a questions such as, "Who made an innovation?" is not without problems, for as one observes a community pursue its activities, one cannot predict whether resources (facts stories, tools, artifacts, materials) in use somewhere will eventually be accepted by a larger group within the community. Most frequently, researchers realize only too late that some resource has been appropriated by several members and, because of its greater visibility, has increasingly become available to other members.[3] In the classroom community I researched, identifying, following, and backtracking the emergence of ideas, artifacts, and facts that came to be recognized resources in the community sometimes took the work of a detective. This involved going through the existing records, interviewing students to get more information, alerting research assistants to watch for particular resources in use, and so forth.

Resources such as facts, artifacts, and tools are easily circulated in a community-constitutive network. For example, facts, such as the state of a chemical compound as a function of its temperature-pressure characteristics (i.e., its phase diagram), can be looked up in appropriate encyclopedias. Objects, such as specific bacterial strains, plans for an innovative apparatus, or new analytical instrumentation can be copied, bought, stolen, or received as gifts. In the Grade 4–5 community under study, there existed many objects, ideas, tools, and materials used by a larger fraction of the members. The Canadian flag, which six student groups used as an adornment for their towers, became a paradigm case for the emergence of a network in which an artifact was circulated and therefore came to be "shared". Other [shared] objects in this classroom included the use of masking tape thimbles and Scotch tape thimbles that helped prevent pain and injury when students pushed pins into various materials; the ideas of cranes, elevators, and gondolas as part of towers were circulated throughout the classroom as members interacted with each other. Stories, or aspects of stories, that contextualized and accompanied the designs also became part of the resources more widely used by members in this community.

Although the stories differed, some core ideas often remained so that they could be recognized by others. For example, during the "creatures" project, Arlene told her partner Chris a story about bungy jumping. This core narrative of bungy jumping was quickly adopted by other students at the table as a topic of talk and as an explanation for various features of their own

---

[3] Most studies of innovation in technology are therefore historical (e.g., Bijker *et al.*, 1987; Latour, 1992). The success or failure of a technology is analyzed and discussed *a posteriori* and with hindsight. The problem is that the notions of success or failure of a technology can only be applied *a priori*, because the technology itself does not have a stable ontology. Whether or not a program of work leads to a technology is still up in the air. In Chapter 8, the question is raised in the context of one group's design of an earthquake-proof tower which achieved a stable ontology only through the external marker "end of project".

projects. Ron, one of these students, soon related a story about a bungy jumping site he had seen; his conversation with Peter was sustained for quite some time around the topic of bungy jumping so that they made it into a predominant aspect of their project. When Gitte approached the Grade 5 table, Jeff described bungy jumping as a key feature of his own project. The five Grade 5 students, although they worked in three different groups sitting around an hexagonal table, used story lines from other conversations or engaged in conversations as a larger group. Throughout the remainder of the "Engineering for Children: Structures" unit, bungy jumping stories became part of the stories of structures and their designing efforts. Jane's tower, which she presented to the whole class about one month later afforded bungy jumping. In the same way, Renata and Melinda presented a bridge which allowed pedestrians to engage in bungy jumping. Thus, in the course of the "Engineering for Children: Structures" unit, a network formed in which bungy jumping stories were circulated and sustained.

There were also similarities in the projects of groups who worked in close vicinity. For example, Kitty/Kathy developed their initial tower construction near Jeff/John using the same bundling technique for their central construction. After a "catastrophic failure" of their tower, and after moving near Andy/Simon/Tim, Kitty/Kathy's tower began to resemble that of their new neighbors.[4] Based on these observations I began to ask the following (and related) questions:

- How can we understand the Canadian flag as a cultural resource that was used at different sites?
- How did the Canadian flag as a cultural resource become shared?
- Is there an initial invention/conception of the idea of this resource?
- How can there be multiple independent inventors of the same idea?

The search for answers led me to an interesting case for studying claims of priority or ownership of ideas – expressed by the emic/etic descriptor pairs of "I did it first"/inventor and "You (he) copied it"/copycat – and the onset of the sense that the same thing was used by different groups within the community – expressed by the emic/etic descriptor pair of "everyone is doing it"/common practice. Ownership was important to students in the context of material and discursive resources and practices invented and developed by someone in their community, but not in the context of discursive practices introduced by teachers or engineers. The debates about the ownership of ideas and practices were important to me because they were strong indicators that *students* recognized an idea, fact, or artifact as shared; that is, these

---

[4] Gitte wanted students to test their structures for strength and stability, that is, to engage in a genuine engineering practice designed to find the structural limits and critical points of engineering artifacts. "Catastrophic failure" was the term she liked to use in those situations where a test led to the collapse of a structure as a consequence of testing. Although she repeatedly used this term, the students did not adopt it as readily as she would have liked it. See Chapter 7 for details of the appropriation of discursive practices.

debates often foreshadowed the emergence of a (by students recognized) shared practice.[5]

5.1. CASE STUDIES OF RESOURCE NETWORKING

5.1.1. *Case Study* 1: *The Canadian Flag*

The Canadian flag was an important aspect of children's tower design. Debates about who invented the idea of using these flags to decorate structures, who copied these ideas, and who followed what everyone else did permitted me to study the emergence of a network of flag users. Students designed[6] towers in a project which lasted two days for a total of about 5 hours. The project concluded with students presenting their towers to the classroom community during the following two lessons. (Table I in Chapter 3 lists the different projects and their designers.) These presentations also allowed for questions, critique, and comments by the students and teachers in the audience. Near the end of the second day of designing, six towers were associated with one or more Canadian flags – five towers actually flew the Canadian flag; a sixth flag lay on the table next to the tower. At this point in time, three individuals from three different groups claimed to be the originator of the idea (Jeff, Tim, and Chris); a fourth individual had made first public mention of the flag during a whole class meeting (Brigitta).

The map of the classroom during the tower project indicates where the different projects were located, and some of the similarities in these projects (Figure 1). However, although the projects were designed at the particular locations indicated, students freely walked about in the classroom to get materials, tools, advice, and help from others or simply to watch other students during their design work. On the second day of the project, we recorded the following conversation about flags around the hexagonal table (Figure 1) that was used by the Grade 5 students Ron, Peter, Jeff, Chris, and Arlene. Peter and Ron were working together on a baseball stadium with a retractable roof; Chris and Arlene were collaborating on a twin tower linked by a gondola; Jeff and his partner John, a Grade 4 student, had just completed the structure of a tower for a crane; and Brigitta and Melinda,

---

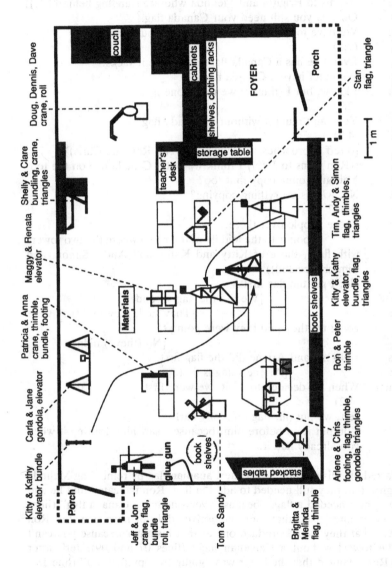

Fig. 1. Distribution of tower projects in the classroom. The material table and the work table (with Jeff and John and Tom's glue gun) were sites of much traffic and many interactions. Several objects that were circulated in the classroom community are also indicated. Canada flag (black rectangle) appears to be concentrated in clusters.

after completing a major aspect of their structure, had presented it to the Grade 5 students.

```
01 Chris:    ((turns to Brigitta and Melinda who are standing behind her))
02           Oh yes, you still need your Canada flag.⁷
03 Ron:      We have to have a Canada flag too, you know.
04           (1.5)
05           Everyone has a Canada flag, otherwise it sucks.
06 Peter:    We don't have a Canada flag.
07 Ron:      I know, but I (think?) we make one now.
08           (2.7)
09           You can't build it without a Canada flag.
10           (4.7)
11           [((Jeff approaches, sits down between Ron and Chris))
12           ((Ron turns to Jeff)) I think we put a Canada flag on the top.
13 Jeff:     You are gonna copy that too?
14 Ron:      No, who is everything copying?
15 Jeff:     Who?
16 Ron:      Those people over there.
17                 [((points to the middle of the room where the two towers
18           with flags, those of Kitty and Kathy and Andy, Simon, and
             Tim are located))
19 Jeff:     No, I get the flag first
20 Arlene:            [We did the flag second.
21 Ron:                        [They said, they said they were
22           going to, they said they were going to.
23 Jeff:                                [No (they?) didnt=
24 Chris:    =No me and Arlene did the flag first.
25 Ron:               [Arlene's flag is first.
26 Arlene:   When we decided to do it, we were first.
27           (4.2)
28           Jeff, we put our flag on really early.
29 Ron:      You put it on before him, because I saw him drawing it when
30           you already had it.
```

Chris started off this episode about the Canadian flag by reminding Melinda and Brigitta that they still needed to add the flag. Ron indicated that he and his partner also needed a flag, "because everyone has a Canada flag" (lines 04–05). Peter interjected that they themselves did not have one, but Ron elaborated that they would produce one at that instant, "because you can't build [the tower] without a Canadian flag" (lines 07 and 09). Jeff, after hearing Ron, claimed that the latter was "going to copy that *too*" (line 13). The stress on the word "too" can be heard as an indication that he had

---

⁷ The transcription conventions are explained in footnote 3 of Chapter 3, pp. 47–48.

already seen others copying the idea or else that Ron had copied other things as well. Ron rejected the supposition that he was copying and, by gesturing toward two groups of Grade 4 students in the center of the room (see Figure 1), attributed such behavior to others. The conversation then turned to the attribution of the original idea. Jeff claimed to have used the flag first (lines 19 and 23), and Arlene seemed to agree by stating that she and her partner Chris were the second group to use it (line 20). Ron appeared to suggest that the girls had started to make a flag before Jeff (lines 25 and 29) and, in this, supported Chris who questioned Jeff's claim to first use (line 24). After a delay, Jeff began to walk away from the table. As he passed Arlene, she modified her claim to "we put our flag on real early". But Jeff did not react (line 27). Nor did he react when Ron provided the additional evidence that Chris and Arlene in fact already had mounted a flag when Jeff only began drawing one (line 29). Did Jeff lie?

On first sight, the exchange may read like a childish conversation about who originated the idea and who copied it. Whereas this explanation of the exchange cannot be excluded, I want to focus on it as a methodological opportunity to learn about the circulation of resources in a community. When I showed the videotaped episode to my research assistants, several of them immediately suggested that Jeff lied about the origin of the flag idea because, so they argued, he wanted to be credited for its invention. Thus, one explanation for the question of who first thought of the flag is that several students knowingly laid false claims. In the present situation, the research assistants thought that Jeff was mainly interested in being in the lime light and sought his teachers' and peers' attention. He may have knowingly claimed false ownership to ideas and inventions to attract further attention. In this and other classes, Jeff was frequently and publicly praised for his creativity, the uniqueness of his projects, and his innovations that, in Tammy's words, "took existing projects into new dimensions". His teachers and peers provided him with constant praise about his uniqueness and outstanding creativity by remarking that he had a creative touch, calling on him more frequently than on any other student during whole-class discussions, or asking him to talk for an entire group about joint projects. Tim could have had similar reasons for claiming ownership. As a Grade 4 student, he automatically stood in the shadow of the Grade 5 students, especially Jeff. Recognition by a larger fraction of the community as the inventor would have allowed him to step out of Jeffs shadow, at least temporarily.

The data I collected in this study do not assist me in supporting or rejecting such an explanation. In most circumstances, the explanation will rest on circumstantial evidence who has laid "false" claim to an idea. Observations in this and other classrooms (e.g., McGinn, et al., 1995; Roth & Bowen, 1995) led me to the conclusion that attributions such as "knowingly laying false claims" are counterproductive for understanding learning in knowledge-building communities. In several ethnographic studies of learning in open-inquiry and open-design classrooms, I found that ideas appeared to spring up almost simultaneously in several groups at once. In other situations, there was a time lag between the moments when different groups began to use some artifact or idea. But in these situations, students often had decidedly different reasons for using the artifact or idea. Can we therefore still speak of copying? There is empirical evidence from several disciplines – e.g. discursive psychology (Edwards & Potter, 1992) and ethnomethodology (Lynch & Bogen, 1996) – which should caution researchers to make attributions of ideas, truth, claims, and so forth. Rather than focusing on untestable attributions of underlying motives and beliefs, researchers are asked to focus on how claims to ideas and truth are established in and as public events. It is therefore appropriate to abandon attributions such as "he lied" and to arrange for classroom climates in which students are encouraged to build on each other's ideas and acknowledge when they do so, rather than focusing on the origins and attributions of ideas.

To find out whether or not Jeff had lied and to develop a better understanding of the emergence of an object common to many, I engaged in the detective work of locating evidence in the site, interviewing students, and going through the already existing records in my database.[8] The following pieces of evidence may serve as exhibits on the basis of which readers may construct their own understandings of the events. Similar to the stance taken by Lynch and Bogen (1996) to their analyses of the Iran-contra hearings in the US Senate, readers should not be concerned with what "really" happened, but with the way the invention was rhetorically attributed and claimed. (See also Figure 2, the timeline for situating the historical development of the case.)

Exhibit 1: At the end of the first day of the tower construction, the class met to show what students had accomplished, talk about the problems individuals had encountered, and to provide helping suggestions. During this meeting, Brigitta presented what she and Melinda had accomplished on that day. She finished her account by noting that they wanted to add a Canadian

---

[8] When I began intensive data collection, I had just completed reading *Aramis or the Love of Technology* (Latour, 1992). Latour provided a case study that integrated fact and fiction. On the surface, Latour's account was constructed as a novel in which a senior sociology professor and a graduate student of technology investigate why an 18-year multi-billion dollar project supported by industrial giants, several government offices, and a city was ultimately canceled. The relationship between the two protagonists is similar to that between the protagonists in *The Name of the Rose* (Eco, 1984) or *Sherlock Holmes and Dr. Watson* (Doyle, 1930). Latour's protagonists guide the reader through a display of documents. Readers quickly realize that the "novel" in fact contains original documents of various sources including excerpts from interviews, government briefs, industry advertisements, excerpts from Mary Shelleys *Frankenstein*, engineering drawings, transcripts of congressional hearings in the US, and so forth. These different textual forms are distinct in their presentation (italics, font type, font size, bars and double bars to the side). In this way, Latour deploys various textual and graphical sources so that readers can form their own conclusions about the "causes" of the technologys failure; and readers do so as if they were piecing together their own detective story.

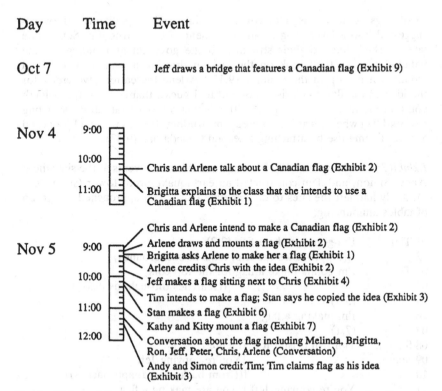

Fig. 2. History of the events surrounding the invention and use of the Canadian flag as a decoration for students' structures. Each event is cross-referenced to the document in the main text.

flag to their tower. Tammy praised Brigitta, suggested that this was a neat idea, and gave the two girls permission to use paper for the construction of the flag. On the morning of the second day (9:16), Brigitta asked Arlene to make them a Canadian flag but Arlene refused. Later on the same day (11:20), Ron told Brigitta that she "still needed a Canada flag". About 10 minutes later, Melinda asked Kitty to make her a flag. Kitty, agreed, got her pens, and made a flag for Brigitta. During the whole-class presentations one week later, Melinda credited Chris for the idea of flying a Canadian flag on her tower.

Exhibit 2: Chris and Arlene talked about making a Canadian flag on the first day of tower constructions (10:18). Chris proposed to use a flag so that a piece of straw sticking out from the top of their building would receive a *raison d'être*. On the next day (9:06), Arlene brought up the issue of the flag while she discussed with Chris a solution for covering the ends of pins that held together the different straw beams in their structure. She thought that small flags could be used to cover the pins, and immediately afterwards suggested also a larger flag as an adornment of their structure. Some time later, Arlene likened their structure to the government buildings in the national capital, then took on the task of drawing a Canadian flag, and mounted it on top of the building (9:13). Arlene thereafter gave credit for the idea of the flag to Chris, "You did it, I guess; thanks to you, I helped; you told me to color the flag" (9:19). They also thought about constructing a second flag when Chris interpreted a protruding straw as ugly and proposed to give it some use by attaching a second Canada flag (9:47).

Exhibit 3:   On Day 2 (10:10), a camera recorded the events at the site where Andy, Simon, and Tim constructed an earthquake-proof tower (see Figure 1). Andy had left the sites to do something else. Tim approached the group of tables announcing:

```
01 Tim:      I'm gonna make a flag (2.2)
02           [((Approaches the site, talks in background))
03 Tim:      I'm gonna make a flag for the top cone
04                               [((searches through his desk))
05           (5.9)
06           I'm making a flag
07           (7.1)
08 Simon:    Tim, isn't, isn't it de[mented
09 Stan:                            [You're copying Jeff
10                                  [((from background, approaches))
11           You're copying Jeff if you are making a flag
12           [((Stops next to Simon; addresses Simon))
13           (1.8)
14 Tim:      No we are not. ((without looking up from his work))
15 Stan:     Look at Jeff
16                       [((Stan and Simon turn heads to look across the room
17           towards Jeff, Tim does not look up from his work))
18           (2.5)
19 Tim:      ((to himself)) Is that supposed to go like this?
20                         [((works on his drawing))
21           (2.3)
22 Simon:    We can do whatever we want to.
```

After returning from a trip around the classroom, Tim repeatedly announced that he would make a flag (lines 01–06). Stan approached Tim and told him that if he made a flag, he was copying Jeff who already had a flag (lines 09 and 11). Tim insisted that he did not copy although Stan pointed across the room where Jeff already flew a flag on top of his tower (lines 15–16). Twenty minutes after this conversation, I asked Andy and Simon about the origin of the idea for the Canadian flag. They referred me to Tim and said that it was his idea. When I asked Tim, he said the flag was his idea. He invented the flag because he needed something to decorate the top of their tower which sort of looked empty.

Exhibit 4: On the second day, at about 9:48, Jeff left his construction site. He sat down next to the two Grade 5 groups, between Chris and Ron. He got paper and pencils from his desk and drew a Canadian flag. Later that day, Jeff repeatedly claimed that the Canadian flag was his idea, as for example, during the above conversation. He also told Kathy and Kitty that they had copied *his* idea.

Exhibit 5: A number of students also provided reasons for not using a Canadian flag. Tom and Sandy explained, "We first decided to make one, but then didn't feel like doing one". Peter and Ron appeared eager to use a flag during the recorded conversation. They had also noted Jeff drawing his flag as a significant event (Day 2, 9:48) and Ron commented on Chris and Arlene's structure, "The only good thing in your tower is the flags" (Day 2, 9:42). When asked about the flag they suggested, "We wanted to make one, but then did not have the time to complete the tower or the flag". Jane and Carla "never thought of making one".

Exhibit 6: Immediately after his exchange with Tim and Simon, Stan also began to prepare a Canadian flag. However, although the flag was completed and lying on his table, he never mounted it on his structure.

Exhibit 7: On the second day of the project (11:00), Kathy and Kitty mounted a Canadian flag on top of their tower. At that time, they were working immediately next to Andy, Simon, and Tim (Figure 1) who already had drawn and mounted a flag on their own tower. Kathy told me that they used the flag because "everyone else is doing one; a lot of people use them, a flag".

Exhibit 8: While Kathy and Kitty were completing their tower towards the end of the second day (11:19), Jane passed by and suggested:

01 Jane: It's like you should have the Canada flag.
02 Kathy: Just like we did.

Exhibit 9: On the first day of class, about four weeks earlier, we had asked students to write and draw what engineering meant to them. During this activity, Jeff drew a flag on the bridge he had drawn (Figure 3).

The conversation and the exhibits provide evidence that Chris and Arlene were the first to consider using a Canadian flag to beautify their tower. Brigitta (who had been working within two meters of Chris/Arlene; see Figure 1) indicated that she wanted to do the same about 10 minutes later during a whole class discussion. There is no evidence in the data sources to suggest whether Brigitta and Melinda had overheard their classmates or that anyone else noticed her comment. Melindas comment had even eschewed all observers. I only recovered this piece of information two days later while transcribing the videotape. There is unequivocal evidence that Jeff, who had built his tower away from his usual seat, constructed the flag after the classroom discussion and after Chris and Arlene on the Grade 5 table had already mounted a flag (see Figure 2). On the other hand, Jeff had used the symbol about four weeks earlier to top a bridge that he had drawn on the first day of the "Engineering for Children: Structures" unit (Figure 3). Tim, who had constructed and mounted his flag still later, claimed to be the originator of this idea. He suggested that he had needed something to decorate the top of the tower. At this point, he had the sudden insight of a flag. He claimed to be the inventor, although Stan told Tim that Jeff already had a flag. Kathy and Kitty, who had started their tower next to Jeff and John and completed it next to Andy, Simon, and Tim, had drawn and mounted a flag "because everyone else [was] doing one". Ron and Peter, who worked next to Chris and Arlene and sat next to Jeff when he made his flag, used the same argument for making and mounting a flag; and Jane's comment, "it's like you should have the Canada flag" can be heard as a statement that Canadian flags *should* be included in the construction. Brigitta and Melinda, although they indicated their intention to make a flag during the whole class session at the end of Day 1, first asked Arlene (who refused), then Kitty (who agreed) to help them in making a flag. Before providing my theoretical description of the circulation of resources, I offer another case study that shows how an idea was circulated in this community.

5.1.2. Case Study 2: The Thimble

The thimble became a tool many students used when they worked with pins. Although the process of sharing this tool was not as dramatic as the Canadian flag, nor was it associated with claims to being its inventor, the thimble is another case that illustrates how objects come to be common in a community.[9] During the first day of the tower project, while Gitte talked with

---

[9] In the absence of ready-made thimbles, students made them on their own. Because of this "making", the case is not pure and involves an element of the circulation of material practices dealt with in Chapter 6.

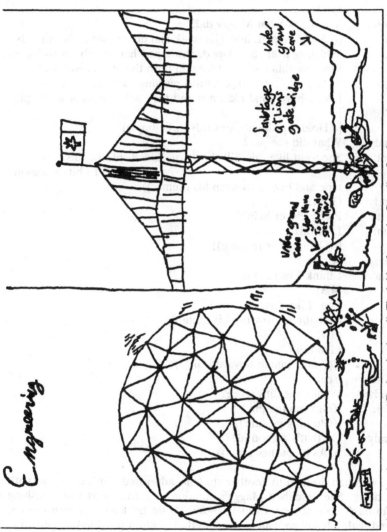

Fig. 3. Jeff's preunit drawings in response to "Engineering is . . .". The bridge featured a Canadian flag four weeks prior to the discussion about whose idea it was.

Andy, Simon, and Tim about their progress, problems, and future design activities, Tim pricked his finger and called out, evidently in pain.

01 Tim: Pins are pains, they give you pain.
02   (3.8)
03 Gitte: You know what Maggy did?
04   [((Andy, Simon, and Tim work and do not notice her thumb))
05   Which is neat. She taped, she taped her thumb up and used
06   that as a thimble. So she could push the pins in with it.
07   [((Gestures pushing motion))
08   Like I have band aid on it and they help me to push my pins
                in
09   [((Holds up thumb, towards three boys))
10 Andy: What did she put?
11 Tim: She went like this. I'll show you. She went like this.
12   [((Tears off a bit of masking
13   tape and begins to wrap his thumb.))
14 Andy: O:::h.
15 Gitte: Pretty clever hein?
16 Tim: Like this.
17   [((Continues wrapping))
18 Andy: A:::h.
19 Simon: I think I've seen it.
20   (4.3)
21   Tim: Like this.
22   [((continues to wrap his thumb))
23   (3.7)
24   Like this?
25   [((Points thumb towards Gitte))
26 Gitte: Yeah.
27 Andy: What's the difference?
28 Tim: You push the pin.
29   [((Pushing gesture))
30 Andy: Oh that's a good idea.
31   [((As he pushes a pin))

Gitte, who had listened to another student, addressed Tim's comment with some delay. She described Maggy's solution to getting hurt while pushing pins through several straws (lines 03–09). Gitte held out her own thumb, covered with a bandage, and underscored the help it provided by gesturing a pushing motion. In answer to Andy's question (line 10), Tim, who appeared to know already what Maggy had done, wrapped his thumb with masking tape while explaining how to do it (lines 11–25). Gitte affirmed that what Tim had just shown was what she meant. Simon also appeared to know about the thimble (line 19). But Andy questioned the use of a thimble (line

27). Tim responded by pushing the pin, but did not provide a rationale. However, when Andy later pushed his first pin, he commented, "Oh, that's a good idea" (lines 30–31).

At this point, Tim remained the only one to work with a wrapped thumb. About 15 minutes later, Simon pricked his thumb while pushing a pin through several straws to form a joint. Tim immediately suggested the use of a thimble ("Put tape on your thumb"), but Simon responded that he was almost done and, by implication, did not need to use this tool. The use of an improvised thimble also became a conversational topic during the whole-class meeting at the end of this first day of designing towers. After Maggy and Renata presented their tower, Gitte asked the former to talk about her sore thumb and how she had dealt with it.

```
01 Gitte:    I just want you to share the problem you had with the pins and
02           how you solved it. When you came up to me and it hurt.
03 Maggy:    Well, I was pushing down and it didn't went through it, 'cause
04           it hurt.
05           So, I couldn't take a thimble. She suggested I get a thimble for
06           my thumb so I could push it down easier or just to wrap around
07           some tape.
```

Gitte followed up by asking how many other students had found thimbles useful. This provided Tim with an opportunity to show his own wrapped thumb to his peers.

Early on the second day of the project, Tim, Andy, and Simon were gathered around the artifacts they have designed so far. Tim offered Andy and Simon pieces of masking tape while suggesting, "Hey guys, put this around your thumb, so you don't cut it. And it won't slip". Simon accepted the first piece and wrapped his thumb. A little later, Andy also accepted masking tape and wrapped his thumb. Around the Grade 5 table, I observed a similar situation. After remarking that her finger really hurt on the previous day, Chris suggested using some cellotape to protect their thumbs. Arlene immediately followed this advice, and Chris wrapped her own thumb. Later, Ron who had watched and listened to the two girls, also wrapped his thumb with cellotape to form a thimble. In this way, an increasing number of students who used pin joints started to use make-shift thimbles on their thumbs.

Here, it is important to note that although the use of a thimble emerged after a conversation between Gitte and Maggy, the evidence as to who had the idea first is equivocal. While I had not recorded this conversation, Maggy's account provides some evidence that Gitte had made the original suggestion – despite Gitte's attempt to shift ownership of the idea to the student ("The problem you had with the pins and how you solved it"). On the second day, the use of thimbles was circulated rapidly, originating at two sites: the tables where Tim and Chris worked. Independently, Tim and Chris suggested using this tool. It is interesting to note that by that time, Tim already had found another reason for using a thimble from masking tape: It decreased slippage. That is, the thimble was not merely used to address the problem of pain, but also a second problem. This is one of the interesting phenomena one can observe with tools: When they are used more widely, they are associated with new interpretations and new reasons for being used. But this is the topic of Chapter 6.

5.2. INVENTORS, COPY-CATS, AND EVERYONE ELSE

The above conversations were about resources – the Canadian flag and the thimble – used by several members of the classroom community. In this, the community had begun to transform as its members adopted new artifacts as resources in their design activities. Three emic descriptors pointed to the existence of resources that were common to a section of the community. According to Kathy/Kitty, Ron, and Jane, the Canadian flag was used by "everyone else"; Jeff, Stan, and Ron recognized that others were "copying"; and Arlene, Melinda, Simon, and Andy credited others for "having done the flag first". Thus, engaging in a practice because a considerable number of others in the community were doing it became an important aspect of networking members and circulating resources. "Doing it first" and "copying it" (or my corresponding etic descriptors, inventor and copy-cat) are also important notions if we want to understand where and how the origin of resources and practices were attributed and began to circulate. These notions – "doing it first", "copying it", and "everyone else is doing it" – are of considerable theoretical importance because they allow us to unpack the notion of [shared] practices in communities of practice[10]. An inventor's designation of someone else as a copycat is an indication that a resource or practice has been recognized as essentially "the same" – the fact that owner-ship of ideas was an important dimension only for student-generated resources and practices still needs special consideration. By the same token, when someone begins a practice or acquires a resource because "everyone else is doing it", it can be viewed as an indication for a member's recognition that others have or do something which is considered to be "the same". [Sharing] therefore could mean that members recognize resources used by different people as essentially the same, although they do not have to be identical. Whether or not an outside researcher considers the claim to same-ness as valid is irrelevant to the community and to understanding what

---

[10] Researchers (social constructivists) frequently use the notions of [shared] knowledge, practice, and understandings without ever unpacking the notion of "shared". In previous studies I rendered this situation problematic. I showed (a) how students normally operate on the default assumption that they are working on the "same" problem and (b) how they renegotiate problem definitions if, in the course of their work, they come to understand that their problem definitions are not exactly the same (Roth & Bowen, 1993).

members of the community do.[11] Sameness is an assessment that members achieve locally and which they may rescind later. At present, because of my interest in understanding communities from the inside, I am less concerned with descriptors such as peer pressure or conformity that could be used to model students' need for following everyone else.

Chris, Jeff, and Tim clearly claimed the Canadian flag as their invention and therefore as their intellectual property. Each believed that s/he was the first to have had the idea. Arlene credited Chris with the invention and recognized that they had mounted their flag before Jeff. Ron, Jeff, and Stan all made remarks about others who had copied the flag. Brigitta wanted to make a flag before Jeff or Tim drew theirs; but she never claimed to have been the first person to have had the idea. Rather, she and her partner succeeded in acquiring a flag from someone else. Others, such as Ron and Kitty, considered adorning their towers with the artifact because "everyone else has a Canada flag". In this way, doing-it-first, copying-it, and everyone-else-has-one became important indicators for identifying the transformation of a community in terms of available resources. Those claiming to have had the idea first, laid claims to having started the circulation and therefore to having started the flag network. "Everyone else is doing it" describes members' recognition that there already existed a network within which resources were circulated.

Especially in the beginning of the "Engineering for Children: Structures" unit, children often designated others as copy-cats, pointing out that they themselves were originators of an idea. On the other hand, both teachers attempted to institute a culture in which some ideas became seeds for developing other ideas. Thus, "sharing" and "building on other's ideas" became important ways for the teachers to talk about community and to help students draw benefit from this community. In this way, they declared that they valued the circulation of resources and practices and de-emphasized the negative connotations of learning from others ("cheating"). Over time, a considerable number of students acknowledged that they had learned from others, copied a technique, or built on some idea represented in someone else's construction. However, a sense for having invented something or for having a new record remained. Jeff, for example, continued stating claims to being first or having the best ("We had the class record of . . .). This study shows that inventors cannot be clearly identified and that several members of the community may claim an invention as their own. One may then ask, how can it be that several people claim ownership? This problem was made more intractable by the observation that facts and objects were never copied

---

[11] The question of sameness has led those researchers astray who are interested in the transfer of problem solving. Because they had constructed two situations as structurally equivalent, these researchers assumed that the problem solving should be similar. They overlooked that the resources and practices in which people engage are considerably different in the two situations. For an extended analysis of the transfer problem consistent with the frame I adopt here, readers may want to consult Lave (1988).

exactly, even if they were copied by a person who already had the object (Kitty made the flag for Brigitta). Although the idea of the flag spread, specific details changed. Nevertheless, "ideas-in-the-air" are a possible answer to the attribution dilemma (Schoenfeld, 1989).

The notion of "everyone-else-is-doing-it" implies that students are part of a community that recognizes certain practices and resources common to their activities. That is, this notion allows researchers from outside the classroom to identify resources and practices that students themselves consider as shared. Kitty/Kathy made and used a flag and Ron intended to make one "because everyone else had one". Andy, while attempting to stabilize various two- and three-dimensional structures also committed to a particular approach after seeing several other members of the community:

I am writing about what I did. Sort of what I started out with. I started out with a square and then *I saw all the others*, and I made a thing *that others, the other people, how they made it*. So I just tried doing that and it worked. I started with a square and then *I saw all the cubes they had made*.

Thus, after some critical number of projects are made using a resource or a practice, new members may join "because everyone is using/doing it". However, it has yet to determined what this critical number of users has to be before newcomers claim that "everyone else is using it".

Here, Kitty/Kathy used this argument after 3 of the 13 projects in the room had mounted a flag; when Ron remarked the same thing, there existed 4 flags. In a study among Grade 8 students, the same argument was used once 5 of 9 projects had begun using a specific technique (Roth & Bowen, 1995). Given that the fraction of projects using the Canadian flag was rather small when Kitty/Kathy and Ron/Peter noticed that everyone else was having one, there is a strong possibility that "everyone else" refers to the immediate setting rather than the entire community. Kitty/Kathy first had worked next to Jeff/John, then between Stan and Andy/Simon/Tim, all of whom had flags (though Stans was lying on the table). Ron had worked on the same hexagonal table where both Arlene and Jeff had previously drawn their flags. He could also see the flags on the towers constructed by Kitty/Kathy and Andy/Simon/Tim. In terms of the networking analogy, this means that neighboring nodes (actors) see surrounding actors. Even if this network is small, or maybe because it is small, everyone else appears to participate in the circulation of a resource. In this case, the actor network also has geographical connotations.

This explanation receives further support from observations during the pre-unit assessment which asked students to draw and write a response to "Engineering is . . ." (see Chapter 4). During subsequent small and large group debriefing sessions, many children expressed that they were aware of the commonality of some images ("others are doing it"). While the images of flying objects (planes and helicopters) were widely distributed throughout the classroom, those of bridges, video equipment, computers, and electrical appliances were rather localized (see Figure 2, Chapter 4). There existed small, local networks within which computers or video/audio equipment were common associations with engineering.

These data illustrate that when students knew what everyone else was doing, they could adjust their actions, redefine their problems, extend the original conditions of their tasks, utilize new materials, or build on explanations and stories they learned from their peers. Locations with high student density were ideal sites for such learning to occur. Whole-class sessions were other settings in which information could be exchanged easily and where students could find out about what others were doing. On the other hand, those who were physically isolated, literally outsiders, did not participate and worked on entirely different problems. This finding sheds light on and complements in interesting ways observations that I made in a different setting (Roth & Bowen, 1995). In this other setting, powerful strategies (graphs) to solve scientific problems did not belong to the domain of common practices in a class where its inventors were located in a marginal position; that is, the inventors were working behind a system of storage shelves that isolated them from other students. In another Grade 8 class, where students worked in close vicinity to each interacted frequently, the same graphing strategies rapidly became common practice.

5.2.1. *Insiders*

There are insiders and outsiders in a community. Insiders see and know what others are doing, follow trends, and observe changes in the conditions of problems and social behavior. Outsiders (physically or socially) experience conditions opposite to those of insiders, which may lead to more experiences that aggravate the position of marginality. When students work in close vicinity to each other or when their movement is not only permitted but actively encouraged, they are insiders. In our case, students easily knew what "everyone else is doing", ideas were quickly taken up by others as a topic of talk, materials were handed around (and thus circulated), and changes to the rules for projects were widely adopted. Insiders knew about these changes and adjusted their own actions accordingly. Ron and Peter, for example, knew that everyone else was using a flag. Even Stan, who could not find a partner, knew that Jeff already had a flag so that Tim was the one who copied. Tim, too, was an insider, for he immediately knew that Maggy had used masking tape to make a thimble. That is, insiders are part of networks; they are literally connected, and participate in and contribute to the circulation of resources and practices.[12]

---

[12] An interesting study compared the circulation of information in four academic domains: mathematics, physics, biology, and chemistry (Walsh & Bayma, 1996). There exist distinct levels of network formation and therefore circulation of resources and practices. Mathematicians and physicists are highly connected across institutions by means of electronic networks. Many researchers in a subdomain know each other, work together, and form a geographically

In this classroom, the circulation of a significant number of resources and practices could be observed among the Grade 5 students who worked around one large hexagonal table (see Figure 1). During the "creatures" project, Arlene's idea of the bungy jumper was quickly taken up as a topic of talk by others around the table who used it as an explanation for various features of their own projects. Arlene's mention of bungy jumping reminded Ron of a bungy jumping site he had seen; his conversation with Peter was sustained for quite some time on the topic and they made bungy jumping into a predominant aspect of their project. Later, when Tammy approached the table, Jeff claimed bungy jumping as the key feature of his project. Small groups used topics from others' conversations, or the five students engaged in one conversation. Around this table, I also observed the circulation of materials. Chris/Arlene gave away part of their limited supplies that they did not need, but which Ron/Peter could use in their construction; and Jeff provided Chris/Arlene with thumbtacks which he no longer needed. The exchanges around this table were not limited to conversations, the circulation of materials, or the transformation of stories that accompanied the projects. A further effect of the proximity of others resided in the transformation of rules for the projects.[13] Here, Jeff obtained a second cone although the rule had said, "You can only work with the materials you received". Arlene reported what she thought to be an infraction of a teacher-set rule. Tammy did not react and thus implicitly legitimized Jeff's actions. Others around the table regarded this as establishing a precedence that allowed them also to use additional materials. Soon, cardboard, various kinds of string, and thumbtacks were additional materials used in these groups. That is, where students sat close, often facing each other, I observed a high degree of networking and the circulation of resources and practices.

Insiders may ask who the originator of an idea is, especially when circulation involves transformation and when many people work on similar projects. In the case of the Canadian flag, the very nature of the projects may

---

distributed community of insiders. Biologists and chemists, who frequently participate in commercial networks, are much less connected, more secretive about their work, and circulate very few resources and practices. When they do circulate something, they often do it in a way that makes it extremely difficult for others to reproduce their procedures and therefore to engage in the practices that allowed them to conduct successful experiments.

[13] I explore the relationship between rules and actions in Chapter 8. The following comment may suffice for the moment. The relationship between rules and actions they are to control is not straight forward, but always an achievement in which the fit is determined retroactively (cf. Barnes et al., 1996; Suchman, 1987; Wittgenstein, 1994/1958). Thus, whether a cook follows a recipe cannot be determined on the spot, but only later when some subsequent action or appreciation is impossible. At this point, the cook may decide that a previous action was not according to the recipe. In learning physics from traditional laboratory activities, this provides students with a dilemma: to know that what they do is what they are supposed to do, students need to know what they see is what they are supposed to see (Roth et al., 1997a). But coming to see what they are supposed to see is the purpose of the laboratory activities so that students are caught in a vicious circle.

have set up conditions for certain ideas to appear in several different sites. Children already have experiences with many public buildings, towers, and bridges, that fly provincial and national flags; they are also dominant features during TV presentations of sports events and are often used on cranes. It appears likely that the children in this classroom drew on past experiences when faced with the question of how to adorn their own constructions. There is also a possibility that they saw another flag in the room without actively attending to it. Ideas-in-the-air are distributed phenomena, and can thus surface simultaneously at different sites (members). For ideas to be "in the air", certain conditions have to be fulfilled. A high level of activity around the same or related problems encourages members to have the same or similar ideas. Such conditions exist within groups that have formed to work on a joint task. In such groups, a second condition is also fulfilled: there must be high levels of interaction between members who associatively connect their respective knowledge. In the present community, such situations of highly concentrated activity existed around the electrical outlets necessary for operating the glue guns, around the tables that served as "warehouses" of materials, and around conglomerates of tables such as the Grade 5 hexagon (see Figure 1).

5.2.2. *Outsiders and Marginals*

Among those who are not insiders, it is helpful to distinguish two types of individuals who have a special relation to a community, outsiders and marginals. Being an outsider means not participating in the circulation of resources and practices of a community. Outsiders see and know the world differently than those within the community. Outsiders are separated from the community in a physical and social sense. They do not keep up with developments within the community. Marginals, on the other hand, know what is happening in the community, engage in its characteristic practices, learn by observing others, and so on. However, they keep themselves apart from others by doing things differently than everyone else or are kept apart because few, if anyone, agree to work or interact with them. Both situations may occur, sometimes temporarily in a classroom community, with possible negative or detrimental effects to children's learning and construction of their Selfs. The following examples illustrate the notions of outsider and marginal, and the constraints these statuses provide for learning.

5.2.2.1. *Outsiders*

In this class, when students worked outside the room they became temporarily outsiders who had severed contacts with the remaining community and therefore did not know about developments within that community. That is, these students no longer took part in the circulation of resources and practices. In one such situation, Andy and Dennis had decided to take their materials and work into the foyer (for its location see Figure 1, p. 105) – one of two regular places students used to work away from others. Their task was to design any structure or "creature", real or imaginary, but from a limited set of building and joining materials. (For the teachers introduction to this activity, see pp. 46–47) Gitte also suggested to students that they should develop their own stories to accompany the structures they designed.

In the foyer, Andy and Dennis generated many associations and ideas for designing objects with the given materials and for constructing stories to go with these ideas. After thinking aloud about jet skies, ski boats, buildings, chop sticks, rockets, sea shells, diggers, and sail boats, Andy settled on designing a boat. Dennis – after he had thought about a range of things to construct including a ski boat, an alien, a fence, a pilot, and a shield – finally settled on making a rocket. While they constructed their artifacts, they were briefly visited by two other students. In the end, both Andy and Dennis were very proud of their constructions. Andy presented it to Dennis and several other students:

See my boat. This is how I made it, I got all my 4 straws, and then put a pin, two pins through it. And then put the cone on top. And then put a pin through the cone and the straw and on the other side. And then it made a boat. See I made, you can really only make one thing, you can't make tons of stuff. This is all you can make the whole day.

After Andy had finished the structure, he wrote about his struggle to build something from the few materials he had received and how proud he was of his construction. However, 30 minutes later and back in the classroom, Andy refused to present his boat in front of the whole-class assembly. When I asked him later to explain, Andy responded:

It wasn't very good. I was very upset with my creation. I didn't like it. It didn't look too good. When I saw everybody else's, then it was different.

When I recorded these events, and interviewed Andy about his later refusal to present his boat, I wondered what may have changed Andy's appreciation of his artifact. Although he was proud of his work at first, he later rejected it. While watching the presentations and reviewing the videos, I noticed that Andy and Dennis were the only two students who had not changed the initial specifications of the task. Within the classroom several students had negotiated with Tammy and Gitte to form groups. Working in groups made twice as much material available for the design of one artifact. As well, a number of individual students and groups had negotiated for and received additional materials. As a result, several groups built large artifacts including a space station, air transporter, and coffin with moveable hinges and a person inside.

As he re-entered the classroom, Andy saw the large projects and noticed that these included more materials than initially provided. His comment, "Hey, Tim got two cones" contrasted Tammys earlier "your basic supplies are going to be five straws, one cone, as many pins as you need". However, to his peers who had remained in the classroom, two cones was not something to be worried about. Once the first student had been able to negotiate for additional materials, and once the first group had formed, the task changed and continued to evolve within the classroom. These changes in the task definition circulated quickly among the insiders (here the network constitutes an analogy in a physical and social sense). As these new task definitions circulated, insiders were afforded new resources and opportunities for designs and their design processes. The outsiders could not participate in this circulation; they were cut off from the new network. Consequently, Andy and Dennis continued on the basis of their earlier task definitions. They lacked access to the resources afforded to insiders.

When Andy re-entered the classroom, he recognized that others' projects shared something that his project did not include. He recognized his momentary outsider status and constructed the results of his work as inferior. Here, it was to Andy's detriment because, after all the effort and pride associated with his project, he developed a negative relation to it. He destroyed his artifact and threw away the story he had written: when I asked him the next day to see his story, he told me that he had lost it.

Situations such as this did not arise frequently, presumably in part because I communicated, without delay, anything I learned or any hypotheses I generated to Tammy and Gitte. As described in Chapter 2 (pp. 35–36), the two teachers used my feedback and acted upon it. In the situation with Andy, I had told Tammy and Gitte about the situation on the very day it had occurred. Tammy used my account to further encourage students to work together and not to view learning from others as an instance of copying:

So that just goes to show you that when you are working among other people who are doing a similar kind, you are not really copying. But it is so important to be using the ideas of other people to change your ideas and to improve on them.

In an open community such as this one, resources and practices are continuously produced and enacted. The interpretation of available resources changes, as do task and working conditions. With it, activities and learning opportunities continuously evolve. By their very condition, outsiders cannot notice these changes and remain unable to adapt to them. It is for this reason that the traditional teacher practice of sending students to an administrative office or a detention center or suspending students altogether is counterproductive. It deprives students from the very learning that the community can offer. In other instances, teachers make arrangements so that students can work outside the classroom community – such as when Tammy, in subjects other than science, permitted individuals and small groups to work in the foyer or on one of the two porches of the portable classroom. In both situations, teachers and school administrations thereby contribute to the construction of individual students as outsiders and take away learning opportunities that others have, usually those that are more compliant with traditional notions of discipline. In my 10-year career as a science teacher, I always found that by creating open and student-centered learning environments, where students are responsible for their own learning (which is necessarily self-paced), one can eschew such problems of control.[14] My goal has always been to create communities in which all students are insiders, although some of them may construct themselves as marginals.

5.2.2.2. *Marginals*

Marginals are networked into the community, but do not participate in networking to the same extent as insiders. It may be possible that an individual actively seeks marginality status and refuses to do something *because* everyone else is doing it. Students may want to distinguish themselves through their projects and work on their own, without severing their ties with other members of the classroom community. In this way, they know what others are doing, which provides them with resources and opportunities for distinguishing their own work from that of others. In this sense, Jeff was a marginal. Although he was not shunned by others and engaged in conversations with them, he worked and interacted almost exclusively with his younger partner John. He was constantly interested in innovation and setting class records. Thus, being an inventor who knows what everyone else is doing (such as copying) and taking a marginal position may in fact be two non-exclusive categories.

Another type of marginal is constructed by their peers. While permitting these marginal individuals to be within the community, other members or even the entire community may shun the person who, in this way, is marginalized. In this study, Stan was the paradigm case of such a marginal. During the first seven weeks of "Engineering for Children: Structures", Stan worked on his own because he could not find a partner, although all other students worked in groups of two or three. His classmates refused to team up with him. Tammy, who knew that Stan was diagnosed with an attention deficit hyperactivity disorder did not try to bring him into the network because she felt that other students were bothered by Stan's attention problems ("He has such trouble staying on task" and "He seems to miss a lot of what goes on"). Even during the projects, many students committed little effort to engage in exchanges with Stan. The following episode was typical for many situations observed and recorded. Andy and Simon worked on the

---

[14] In my 10-year career as a middle and high school teacher, I have had many experiences with students where the negotiation of individual arrangements with "problem" students led to outstanding achievements on their part. These individual arrangements included that a student (who after school earned money by selling drugs) worked on his own and wherever he wanted, under the condition that the negotiated work would be handed in by 3:30 pm on Friday. In another situation, a student read a novel, during five consecutive lessons of mathematics class. Not only did he catch up with his peers afterward, but he ultimately developed a positive attitude toward the subject. Both students thanked me at the end of the year and credited the personal arrangements for these positive experiences they had and the incredible amount that they learned.

same component of a tower, their partner Tim one another component. Stan approached the group (Figure 3).

	Andy	Simon	Tim	Stan
01		This won't work.		
02				What kind of thing are you trying to do?
03		This won't work!	°Earthquake-proof°	
04				Ve::ry in::teresting.
05		Andy, this won't work.		
06	Why?			
07				Earthquake-proof?
08	Why?			
09				Sure, it should be earthquake-proof or it will collapse.

During this entire episode, neither Andy and Simon nor Tim took their eyes of the components they worked on (Figure 3). In response to Stan's question about what they were working on, Tim uttered an almost inaudible "earthquake-proof" but seemed little interested in elaborating his answer. As he watched the three boys, Stan continued to comment (lines 04, 07, 09) but nobody reacted. The transcript quite clearly illustrates that Stan's utterances constituted little more than a monologue. Thus, for example, he provided an answer (line 09) to his own question (line 07). After his final utterance (line 09), Stan left and returned to his own construction site. Several observers of this videotaped episode commented that the group of three evidently "ignored" Stan. Andy and Simon continued working on their problem without ever lifting their heads or reacting to Stan's comments. As readers will see in Chapter 9, an important aspect of networking students into a community is the bodily orientation of individuals. Here, Andy, Simon, and Tim were orientated in a way that did not allow Stan to become part of the network and therefore to directly take part in the circulation of ideas, design plans, stories, and so forth.

Although he interacted little with others, Stan knew what everyone was doing. He walked about in the classroom and visited various groups, observing them in their activities and listening to their stories. He therefore knew that Jeff already had a Canadian flag when Tim decided to make one. Also, although he started out his project without exactly knowing what he wanted to build, he defined his structure as an earthquake-proof building after the "conversation" with Tim. Thus, Stan had the inside track and was part of the network in the sense that he shared particular information and ideas. On

Fig. 4. Andy, Simon, and Tim continue their activities as if Stan (second from the right) was not even present. Stan was a marginal who knew what everyone else did, without being able to interact with other members of the community.

the other hand, he certainly did not get to work with others on problems and had few interactions for much of the initial half of the unit. He was constructed as marginal and contributed to being so. Astonishingly, when Stan actually joined efforts with Tim to work on the second major design project, the bridges, a very harmonious working relationship resulted in an acknowledged design (see for example, the negotiation with Tim analyzed in Chapter 4, pp. 96–97). Furthermore, as I show in Chapter 9, Stan helped out a lot in other projects and became a central figure in networking individuals and groups, and therefore in building a community of glue gun users.

5.2.3. *Copying a Resource*

[Copying] is an interesting notion because it does not simply describe passing on an artifact from one person to the next. Rather, it may involve the enactment and embodiment of a skill. In some instances, [copying] appears to be easy; for example, when students reproduce letters, numbers, and words from the notebooks of their friends into their own notebooks to support claims that they did their homework. Here, I am more interested in situations where copying itself involves a creative act. [Copying] as a process is then much more complex and should not be associated with negative values. Underlying here is the very possibility of learning practices in communities, or, the circulation of skill in communities of practice. [Copying] involves both cultural production, the process by means of which an individual produces a never-before completed sequence of actions, and cultural re-production, the process by means of which a network of practitioners has been enlarged. [Copying], in this sense, involves creative acts in which individuals construct and extend their own competencies. That is, copying a resources is tied to reproducing a practice. This reproduction makes "copying" creative acts in which agents "discover" a social practice in their own actions. This dialectical relationship between re-production and production is evidenced in the following case study.

Gitte moderated a whole-class meeting that focused on joining and strengthening techniques. With the entire class sitting around the engineering design board where more than 100 joints and strengthened materials were displayed, Gitte invited Andy to talk about his "spaghetti stretch".

Andy: Well first Dennis did it, and it worked. And then I tried it and took about 20 spaghetti. And they kept on breaking, because I kept on getting the wrong size. And they broke.

Here, Andy admitted to copying an idea from another student ("First Dennis did it . . . . Then I did it"). But that which worked for the other student, did not work for him, although he had used about 20 spaghettis (trials). The circulation of the "spaghetti stretch" from Dennis to Andy took more than merely seeing the original artifact. The process also included the practice of making the artifact. For the artifact to be circulated and for the community to have another member, Andy had to find a way, through his own experience, to appropriate a technique and with it, as successful outcome, the artifact itself. This did not happen by just observing someone else doing it, but through his repeated trials in solving the implementation on his own. Through these trials, the vicarious experience of making a "spaghetti stretch" became embodied in and as his own practice, for Andy had not really appropriated the technique until that moment when his own spaghetti stretch corresponded to that which he wanted to approximate or copy. Thus, the "transfer" of the artifact was really a circulation of Dennis' skill, now embodied in Andy's actions. Now, he too was able to produce something from a plan (in image form) of the spaghetti stretch. Thus, the spaghetti stretch was not simply lifted off or transferred as one would transfer a flag from one person to another. Andy's work was the embodiment, the conversion of the plan to copy Dennis' "spaghetti stretch" to its successful reproduction. This "transfer" also enlarged the network of "spaghetti stretch" makers and owners. We can see that in this instance, the process of enlarging the network was coextensive with the circulation of a resource (idea of the spaghetti stretch) and a practice (enacting the construction).

6. CIRCULATING MATERIAL PRACTICES

6.1. TECHNOLOGY, SOCIETY, AND KNOWLEDGE

Tools are important components of a community; or, from an actor network perspective, they are important actors that lend their own affordances and constraints to individual human agents and collectivities. In the past, tools – and, technology more generally – have been treated as something independent of society by analysts of technology. Thus, technology was said to influence individuals and entire social entities. On the other hand, individual engineers and scientists usually take credit for having invented a technology. In recent years, such a view of technology has shown to be insufficient in explaining most of socio-technical developments. Sociologists of technology (e.g., collection in Bijker *et al.*, 1987) and historians of technology (e.g., Constant, 1989) now emphasize that it makes little sense to treat the social and engineering aspects of technology as separate entities. Adding social and technological aspects as independent components or taking either a social or technological approach and using the other for dealing with any residual problems also leads to unsatisfactory accounts of socio-technical development. It appears that social and technological aspects are mutually constitutive. In other words, phenomena involving society and technology are best treated in a holistic way, as irreducible socio-technical phenomena (Latour, 1992).[1]

The relationship between technology and learning has become a question of increasing interest with the prevalence of computers and multimedia technology in classrooms. However, scant research exists on the transformation of a classroom community when a tool or technology is introduced. The present study, serendipitously, provided me with an opportunity to study the evolution of a knowledge-building community when an increasing number of glue guns became available to its members. Now, glue guns are relatively simple tools which, nevertheless, require certain competencies to be used. In this Grade 4–5 community, the complexity of the tool was such that it allowed me to study the transformation of associated practices. Furthermore, there were many members who had never used this tool before, which therefore afforded the study of learning and teaching processes relative to glue gun practices.

Some readers may think that material practices related to tools are much

---

[1] The 'Pasteurization of France' (Latour, 1988b) includes a piece called Irreductions, which poignantly makes reference to the author's methodological and theoretical approach as embodied in actor networks.

less important than the conceptually-oriented discursive practices in a community and therefore, may find the study of the emergence of such practices much less interesting than the emergence of discursive practices. There are a number of studies from historical (Gooding, 1990), sociological (Pickering, 1995), and educational (Roth, 1996f; Roth *et al.* 1997a) perspectives which showed that observational and theoretical descriptions of objects and events, and physical manipulations of apparatus cannot be treated as independent because they emerge in a mutually-constitutive fashion. The present study also provides evidence for the integration of the tools students had at their hands and the understandings of various design techniques they developed. The glue gun therefore constitutes an interesting case for studying the emergence of a successful and widespread material practice. The glue gun afforded me the opportunity to study how:

- A limited resource constrained the designing activities.
- Fastening techniques changed as more students began using a tool.
- Physical and social settings changed in response to increasing availability and use of a tool.
- Newcomers to a practice were enculturated through legitimate peripheral participation to become old-timers and core members in a community.

These issues are treated in the following case study of the glue gun.[2]

6.2. SOCIO-TECHNICAL EVOLUTION: THE CASE OF THE GLUE GUN

6.2.1. *Brief History of Events*

During the first month of the "Engineering for Children: Structures" unit and despite Tammy's encouragement to bring tools such as glue guns from home, only Tom had done so. The joints he produced turned out to be extremely sturdy compared to those that others made. The other students developed and used joining and strengthening techniques with different tools and fastening materials (glue, masking tape, adhesive tape, pins). However, the fact that Tom could make strong joints was circulated quickly within this community. During the tower project, and in response to problems they encountered with the stability and strength of their emerging artifacts, an increasing number of students wanted their joints strengthened by means of the glue gun. Students began to ask Tom for permission to use his glue gun. However, despite an increasing frequency of requests, Tom did not allow other students to use his glue gun. Either he offered to help and made the requested joint or he refused and suggested alternative joining and strengthening techniques. Thus, during this initial phase of the "Engineering for Children: Structures" unit, Tom held a monopoly over the glue gun. At the end of the tower project, and one week after Tom had refused to make a

---

[2] A shorter and somewhat different analysis of these events was published in Roth (1996c).

series of joints (Week 6), Tim brought his own glue gun but refused to lend it to others. During the following weeks, more and more students brought glue guns, and even Tammy contributed one for general use: the number of glue guns increased rapidly, until, during Week 9, there were 7 glue guns in the classroom.

With the presence of an increasing number of glue guns, and inundated by a large number of requests to help others, both Tom and Tim abandoned their policy of refusing to lend their glue guns. Old joining and strengthening practices were abandoned and new practices were adopted and developed. With the development of new joining and strengthening techniques, the resources available to designing also changed, affording structures that were not "doable" before.[3] Glue guns, like some other technologies, are associated with geographical constraints in that they require electricity. As a result, the geographical (students' work places are centered around the available outlets) and social orientations (high population density around the outlets) in this classroom changed as a consequence of the glue guns' introduction. Because other students also lent theirs, opportunities arose for students without prior experience to use the tool and thereby engage in the associated practices. With the increasing number of glue guns, the classroom changed in several ways. Adopting a new technology therefore brings forth new affordances and constraints that are inherently unpredictable (Constant, 1989; Latour, 1992; Pinch & Bijker, 1987). Society and technology constantly co-evolve, new interpretations and meanings are attributed by humans to technology, and new affordances and constraints to using technology emerge.[4] The changes that occurred in this classroom with the introduction of the glue gun tool were also not predictable. Each of the following sections considers the different affordances and constraints introduced to the community by the glue gun.

6.2.2. *Limited Resources*

Science educators frequently lament when a particular resource, for example computers, is available only in limited numbers. However, there is little research to indicate how the availability of a resource changes individual and collective cognition in a classroom. In the past, wholesale claims have been frequently made about resources and access. For example, in the 1970s and 1980s, many advocates of computers suggested that, ideally, there should be one computer per student. However, such claims had to be moderated in light of research evidence showing how computers can mediate the construction and negotiation of canonical science discourse (Roschelle, 1992; Roth,

---

[3] For the notion of "doable problems", see Fujimura (1987).

[4] Heidegger (1959) suggested that the meaning pervading technology hides itself but continuously touches us everywhere. We therefore have to remain continuously open to mystery, the continuously evolving unpredictable meanings of technology.

1996f). Computers with small monitors afford different kinds of interactions (depending on the number of participants) and afford different cognitions than do semantic networking activities conducted on table tops (Roth & Roychoudhury, 1992; Roth *et al.*, 1996). The question of the optimal student per tool ratio is therefore still open. A generalized answer to this question is probably not possible. Rather, it is an empirical issue that has to be decided on a case by case basis.

In the case of the glue gun, Tom initially was the only student who had brought a glue gun. He rejected other students' requests to use the glue gun by invoking his father ("My father said, only the teacher or I can use it"). His sister in Grade 7 suggested several months later that it had been her glue gun and neither she nor her father had made requests to limit users. For our purposes here, it does not matter which version of the events we want to believe, for Tom's argument was successful in the community. He did not lend the glue gun, and this set up constraints of various sorts. It is evident that, because other students had no access to the tool, they did not learn how to use the tool, nor did they develop new joining and strengthening techniques. If other students wanted two materials joined or two materials laminated by means of hot glue, they had to make a request. Tom initially tended to grant these requests but, in some situations, suggested alternative ways of solving the given problem. For example, a particular joint requested may have been too difficult or impossible to make. As the number of requests increased and Tom's own progress slowed down, he began to refuse the requests and suggested alternative solutions.

On the second day of the tower project, Tim was one of the students whose request for a hot-glued joint was denied. During the following lesson, he brought his own glue gun to class. As Tom before him, Tim did not lend his glue gun, also drawing on his father to construct an authoritative rationale ("My dad said, I am the only one to use it").[5] Helping others abated his own progress. However, the number of requests for Tim's assistance increased in the same way as it did for Tom. Although he initially maintained his policy of not lending his tool – he completed several joints for Jeff/John and Clare/Shelly – Tim later abandoned his policy. He first permitted his then-partner Stan to use the tool and later extended permission to Clare and Shelly who worked next to him on the same work table.

Once there were more glue guns in the class and once Tim and Tom relinquished control over their tools, the monopoly on the glue gun technology was broken. As more students began using the tool, new networks

---

[5] Again, what Tim's father really said is not relevant here. His father may have said something to this effect; or his father may have uttered something that Tim interpreted in the way he presented in class; or Tim may have simply made up the story. There is much research in sociology and social psychology suggesting that the search for *the* truth, what "really" happened, or what people "really" think is futile (Edwards, 1993; Edwards & Potter, 1992). It is only important how knowledge, truth, and constraints are constituted in social arenas. Using his father as a discursive resource, Tim was able to avert requests without further discussion.

evolved in which the glue gun and glue gun-related practices were circulated. If Tim or Tom rejected a request there were other students who would help. Students no longer had to wait in line for services, or depend on those in control of a tool to judge whether some joint or reinforcement was feasible: Given the availability of glue guns, students could try out on their own whether a joint was feasible. In the process of observing others and experimenting on their own, newcomers to the gluing practice began to traverse trajectories of increasing competence. Notwithstanding these developments, there remained occasional quarrels surrounding ownership or temporary claims on particular glue guns when too many students needed them simultaneously.

A limited number of resources does not by itself have to be a negative constraint to learning. As I pointed out above, to advocates of a one-student-one computer policy, two or three students per computer appear to be a constraint. On the other hand, two or three students per computer afford new conditions and opportunities to learn that are not available to individuals in front of computer screens. As I show below, the limited number of glue guns and electrical outlets brought students physically together and thereby increased opportunities to learn from each other. The effects of a technology on learning are difficult to predict because of the complex nature of socio-technical relations. If it was simply an issue of putting an ontologically stable technology into an equally stable community, it would be relatively easy to model the interactions. This is not the case with technology, as many investigations in science and technology studies have shown. There are often quite surprising developments because technology, people, and communities continuously change. Thus, even when there was only one glue gun, a rejected request to make a joint could bring about new and valuable learning. Tim had decided to construct a lower section to his tower from extra long straw beams. However, Tom refused Tim's request to produce all the long beams he needed. Tom suggested, "Tim, I can't glue that, it's way too hard. Use tape". Tim returned to his construction site where he developed a new and very stable joint (Fig. 1a). Rather than simply inserting one straw into another, he began to tape them. However, the tape did not fasten the beams enough so that, on the next day, Tim added pins to several joints which kept the inserted beams from sliding in and out of other beams. Here, the absence of a resource led to the emergence of another joint. That is, in response to the unprocurable tool, Tim had not only developed a new way of joining straw beams (a practice) but also enriched the community with another type of joint (a resource).

6.2.3. Changing Practices

Tom's monopoly on the glue gun also hindered potential developments in joint construction. Only he could discover new ways of using the tool, such as when he constructed a new type of joint with straws. Because of the glue

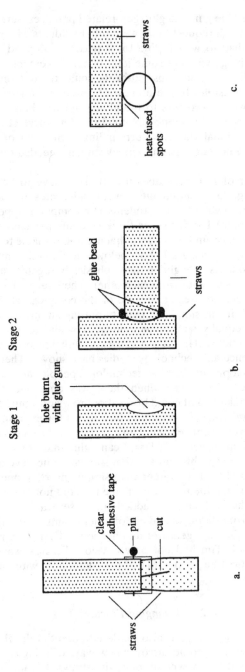

Fig.1. Series of joints developed with and without access to a glue gun. (a) Tim developed this sleeve joint for a long-beam construction after Tom had refused to produce a hot-glued one. (b) Tom's joint achieved by burning a whole in one straw during Stage 1 and recessing and gluing the second straw during Stage 2. (c) Tim's welded joint achieved by carefully heating both straws and joining the heated spots.

gun tip's temperature, he burnt a hole in one straw (Fig. 1b, Stage 1). At first, he was disappointed that he could not glue two straws in the way he had planned. He then realized that the hole afforded new possibilities. He could create a new joint by first inserting a second straw into the hole and then carefully applying some glue (Fig. 1b, Stage 2). In this way, he discovered in his own activities a new practice for making very strong joints.

As long as glue guns were not at hand for other members, such developments and new inventions rested with Tom. However, I saw an increase in such developments as soon as the technology was more widely available. Changes in gluing came about as new users framed new problems and in reaction to specific contingencies; the glue gun practice was transformed, and the range of gluing techniques increased. For example, the accepted practice for joining toothpicks, skewers, and popsicle sticks was to spot-glue them. However, students realized that this practice led to considerable time requirements with larger constructions (relative to those for pin or masking tape joints). After constructing their progress as a problem (too slow), Andy and Tom developed a new gluing practice. They discovered in their trials that they could run strings of glue, and then, in a joint effort, quickly place multiple pieces before the glue had hardened. This new practice permitted them to rapidly complete their project. Andy and Tom did not keep this practice a secret but walked around the classroom to show other students (and the teachers) how to enact their new practice. In a similar way, Tim developed his welded joint (Fig. 1c). This joint required him to heat two straws at once and then hold them together to fuse the hot plastic. This joint helped him to repair his tower so well that it outlasted all other tower projects. The tower still stood when, a month later, a geotechnical engineer visited the class and used it with the children as a conversational topic.

Another change in the practice of gluing was also invented by Tim. At the end of one lesson, he had already unplugged his glue gun. Several minutes later, he realized that it was still hot enough to make a hot-glued joint for a peer who worked at a considerable distance from the outlet. He then proudly announced his "portable glue gun". Later, other students (and the teacher) used the glue gun in this fashion carrying glue guns to the construction sites rather than moving the artifacts to one of the plugged-in glue guns.

The increasing use of glue guns also brought forth a range of problems. Thus, especially when glue guns became widely available, some students abandoned their old practices and applied glue gun practices indiscriminately to their problems. Their construction problems then became circumscribed by the glue gun. In one example, Sandy and Dennis attempted to build a bridge from spaghetti based on the glue gun practice. They really wanted to use the glue gun which led to major, intractable problems for them. They could complete their construction only after they relinquished the glue gun and used masking tape for their initial work to create spaghetti bundles. Then they finished their bridge design by hot-gluing the resulting bundles.

The glue gun practice also had undesirable side effects from the teachers' perspectives. It allowed students to make stable structures while avoiding triangular configurations of building materials to achieve structural stability (braces), the engineering design practice which the teachers had set as an objective to be achieved in these lessons (a detailed case study of the circulation of a discursive practice is provided in Chapter 7). To the great disappointment of both teachers, Peter/Ron and Arlene/Chris built very stable bridges without the "triangles" Gitte and Tammy had promoted so much. Because of students' alternative practices, the task failed to engender the cognitive response in two groups. As they explained, they did not need to use triangles. They had based their projects on practices and resources that afforded them solutions other than those envisioned by the teacher. Traditionally, problems in children's acquisition of new concepts are viewed only from cognitive perspectives. However, the present case shows that there are other aspects which prevent teachers from detecting whether children have learned a new concept. Here, some students did not use triangular configurations which were the teachers' major curricular objective. The children-acting-in-settings gave rise to solutions that circumvented the teachers' objectives.

6.2.4. Changing Settings

The introduction of a larger number of glue guns to the community changed the setting in ways nobody could foresee. The glue guns used by the children designers in this classroom were of the type that plug into a receptacle. This means that without an extension cord, Tom needed to work near one of the electrical outlets (Fig. 2). Because an increasing number of students came to seek help from Tom, the traffic around his construction site increased tremendously and, at times, became chaotic. When Tim brought his glue gun, he initially operated it from the outlet on the other side of the classroom. As the number of glue guns increased, new resources not normally available in a classroom were necessary to allow all glue guns to be plugged in: power bars. However, the two receptacles limited where in the classroom students could work despite additional extension cords and power bars. That is, with the mushrooming number of glue guns and their increasing use, students' work sites were progressively concentrated geographically – although they had been widely dispersed around the entire classroom (including the foyer). As in the case of the table with material resources, this gave rise to areas with high student concentrations and large amounts of traffic. When students wanted something glued, they brought their projects near one of the outlets where there were always several glue guns in operation.

With escalating glue gun use, students began to cluster around the electrical sockets, and this encouraged many student-student interactions. The community began to gravitate towards specific locations rather than the technology diffusing to different sites. As a result, students exchanged many ideas,

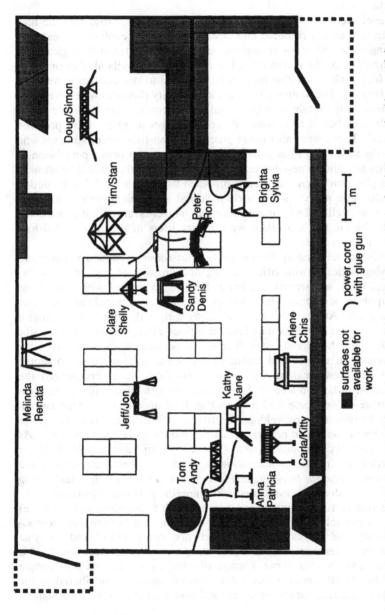

Fig.2. Distribution of the projects during the fifth day of designing bridges. High concentrations of projects are noticeable around the two receptacles in which the glue guns were connected.

taught their peers new material practices, supported others in dealing with construction-related problems, and taught important engineering concepts to each other.

However, new problems, unforeseen by the teachers, arose with the high concentration of the resource and new sets of norms evolved to deal with the changing conditions – changes which cannot be attributed singularly to the technology or the community. For example, the availability of only two outlets forced all seven glue guns to be operated in two clusters around the outlets (Fig. 2). The limited electrical load capacity (fuses blew) evolved new constraints: only certain glue guns could be operated from the same outlet (a smaller number of high-powered versus a larger number of low-powered glue guns). There were also fewer glue guns than there were students who needed them, leading to negotiations in order to get a peer's permission to use a glue gun and to negotiations of the question, "Who should be allowed to use a glue gun given that glue is provided by the school?" The evolution of the glue gun aspect of the socio-technical ensemble gave rise to other problems as well. High concentrations of students around the outlets – enhanced by the fact that many were novice users of glue guns – led to a number of minor burns.

With such developments, the Grade 4–5 classroom resembled a microcosm of evolving societies with cities forming and growing at high traffic nodes (shipping, train, air, street). Industries have to go to sites with easy access and a supply of personnel, which, in turn, entails greater population increases in these areas. As the inhabitants of areas with high population increases know, these developments may lead to serious effects on rush-hour traffic flow. On the other hand, such changes in population density may drive developments in science and technology. For example, attempts to supply Chicago and New England with natural gas from Texas encouraged the development of large diameter, high-tensile-strength, thin-walled, and high-temperature-welded pipe which encouraged advancements in high carbon steel production and arc welding (Constant, 1989). Some such changes in this classroom concerned material practices and the associated resources. As with any complex ecological system (where "system" does not imply linear causal relations in the traditional sense used by systems engineers), the simultaneous presence of agents and forces of comparable weight interacting in non-linear fashions, is associated with unpredictable transformations. The characteristic features of integrated systems (such as individuals in a culture) depend on mutual constitution, and are not reducible to individual components. In my glue gun case, the entire network, comprised of students, glue guns, artifacts, teachers, and accepted design practices, transformed. This transformation brought about a range of changes: where and how people worked and collaborated, which interpretations members attributed to the tool, which practices the tool afforded, and what and how members designed.

6.2.5. *Circulation of Practices*

In this classroom, material practices such as glue gun gluing were circulated as students learned from and taught each other. The associated teaching and learning processes were not confined to isolated interactions, but occurred, in some cases, over the course of several weeks. Thus, although one can speak of the circulation of a practice, this did not occur instantaneously. The following case study shows how Clare and Shelly learned to use the glue gun by working in close proximity to more experienced and competent class mates. The observed interactions included aspects formerly identified with cognitive apprenticeship, modeling, scaffolding, and fading. Learning itself could be understood as a trajectory of legitimate peripheral participation (Lave & Wenger, 1991).

The two girls, Clare and Shelly, had not used a glue gun before. In the following conversation, Tim modeled the use of the glue gun, and provided instructions about how to hold the two pieces to joint.

01 Shelly: We want to 'em attach on like this, as soon as I get it ready, I can let you use it.
02 (1.6)
03 Stan: This is a really weird glue gun.
04 Tim: Yeah, see it melts?
05 [((Cleans glue gun on cardboard))
06 (1.3)
07 That's what my dad says, you have to pull it back when you are finished
08 (2.1)
09 Shelly: There, now put the glue right there
10 Clare: [You have to have things in so they don't like.
11 Shelly: OK, put the glue right up here, right there.
12 Tim: [Right there?
13 [((Tim leans forward, puts glue gun))
14 Clare: Wait, wait, wait!
15 (2.9)
16 [((Tim glues))
17 Clare: Just put a line across there.
18 Tim: [Right there?
19 [((Moves back))
20 (4.0)
21 Clare Let it dry, Shelly.
22 Shelly: We need it on this side too.
23 (1.3)
24 This side.

25 Tim: [This is melted you guys, you have to hold it (.) together (.)
            where it doesn't do anything.

In this episode, Clare and Shelly indicated how they wanted two pieces to
be glued together and where the joint should be located (lines 01, 09–11,
17, 22–24). In the context of these requests, Tim demonstrated proper glue
gun use. He showed how to hold the glue gun and demonstrated how to
remove it after the joint had been made (line 07). As the girls watched, he
put the glue on the skewer pieces (lines 12–13, 18–19). Shelly wanted to
have another joint on the other side of the piece and began to turn it around.
Tim, however, intimated that they could not yet handle the piece because
the glue was still "melted". He thereby implicitly instructed the two girls
how and how long to hold the two pieces so that the hot liquid glue could
set (line 25).

In this episode, Tim gave specific instructions on glue gun use; Clare, who
was literally at his elbows, observed Tim very closely as he engaged in this
material practice as part of an on-going project rather than as an especially
prepared didactic situation. But it was not as if Tim knew everything or
imposed his own understanding. Rather, the two girls instructed him where
to put the glue bead. They knew where and how to run the bead to achieve
their goal. Tim was an intermediary who assisted in achieving this goal. In
this way, Clare and Shelly drew on the distributed aspects of knowing in this
classroom.

The girls continued requesting Tim to glue for them.

26 Clare: Ok Tim, we need glue again.
27 Tim: Ok, tape it. (.) Where do you wan'it? Where do you wan'it?

As in this excerpt, Tim suggested repeatedly that they use adhesive tape,
but every time, complied with the girls' request. When Tim felt overwhelmed
by the constant requests for glue gun joints, which distracted him from his
own work, he gave Clare permission to use the glue gun. She made her first
joints under Tim's supervision. When he was satisfied that she used the glue
gun appropriately, Tim no longer oversaw her work. Once Clare was using
the glue gun, Shelly also asked for it. At first, Clare taught Shelly, "OK,
you can try it . . . Here, you have to press [the trigger] a couple of times".
Under Clare's supervision, Shelly subsequently made her first attempts. The
community of glue gun users had expanded by networking yet another actor.

Although the girls had begun using the glue gun and developed increasing
competence, they were not independent. That is, they had not yet attained
full member status as glue gun users. When they encountered new problems[6]

---

[6] The notion of "constructing new problems" would be more appropriate because a situation
has to be rendered problematic by the person, that is, constructed as a problem, before it can
be recognized as such. Problems are real, from a phenomenological perspective, if they interrupt
an agent's "concernful" (Heidegger, 1977) activity, that is, an activity towards a goal that the

about proper handling of the glue gun, they returned and asked Tim to help them out. For example, the two girls did not yet know exactly when the gun had to be recharged and how to recharge it.

28 Clare: There is hardly any glue left, Tim.
29 Tim: Yeah, you use it until it's all gone.
30 Clare: Like it doesn't come out anymore?
31 Tim: Yes.
32   (3.3) ((Both continue with their activities))
33   There is still lots of it.
34   [((Leans over and looks at glue gun))

Here, Clare thought that she needed to refill the glue gun. Tim, however, suggested she had to work "until it was all gone" (line 29). Clare followed with a question to clarify Tim's description "all gone" (line 30). Tim confirmed the implied description about when the glue gun had to be refilled (line 31) and, as he inspected the glue gun in Clare's hands, ascertained that there was still a lot of glue left. In situations like this, students learned from each other how to recognize that a glue gun did not need recharging, although, to an inexperienced eye, it might look like it. Other similar experiences with glue guns then permitted students to recharge the tool in a more transparent way. That is, it became a tool ready-to-hand, a tool which afforded certain actions without drawing attention to itself.[7] At the moment, however, there were still many occasions in which the glue gun was a tool present-at-hand to the two girls. They had to focus their attention on the tool and learn how to use and care for it. This took their attention momentarily away from the design task itself. On the other hand, learning to use a glue gun was not a "skill" that they learned in some decontextualized repetitive activity.

Following behaviorist analyses, "doing science" was thought to be hierarchies of context-independent "science process skills". Individual skills were therefore taught in decontextualized activities; students were expected to learn these skills and later apply them in different settings. As shown by experience with the curricula of the 1960s and 1970s, students could not learn

---

agent cares about and therefore wants to achieve. For an extended discussion of the ontology of problems see Roth and Bowen (1993).

[7] In phenomenological theories of cognition (e.g., Heidegger, 1959, 1977) human agents do not normally re-present the world of their experience. The world is its own re-presentation. Therefore, people use tools to do things without re-presenting them. I hammer a nail into the wall without representing the hammer, nail, or the wall. I also plan and do my weekly grocery shopping without representing the car which takes me to the super market. Should the tool be broken or absent, however, it begins to obtrude. I start to re-present it, trying to analyze what is wrong. In the process, the tool becomes present-at-hand. Writing in German, Heidegger actually plays on the relationship between the prefix *vor* (before, in front of) and the verbs *vor*handen (present, being, ready to use) and *vor*stellen (thinking, re-presenting).

decontextualized skills and apply them later in specific settings. Similarly, Gitte and her co-workers had designed "Engineering for Children: Structures" based on the notion of teaching some basic skills first and then engaging students in design activities. First, students were to complete decontextualized joining and strengthening activities (see pp. 45–47) which they were to apply during their subsequent design challenges. As the analysis of the "triangles" in Chapter 7 shows, the approach of teaching the use of braces independent from designing complex structures initially met with little success. Here, in the case of the glue gun, where the teaching/learning occurred in the context of children's concernful activities, as-needed and just in time, learning was rapid and lasting.

Students in this study learned to enact practices during concernful activity, in the course of accomplishing personally relevant goals, e.g., to build a bridge entirely from skewers. In the following episode, Clare and Shelly learned how to recharge the glue gun. They were again taught and closely supervised by Tim.

35 Clare:	Tim, there is no more glue in there.	
36	[((Turns around toward Tim))	
37 Shelly:	[The glue is not working	
38 Tim:	[Yeah, you pull really hard	
39 Shelly:	Yeah, this is as hard as you can get it	
40 Clare:	[I'll get it] [((Helps Shelly to press hard.))	
41	(3.5)	
42	Here, let me do it.	
43 Shelly:	I don't think they're that strong.	
44 Clare:	Tim, serious, there is no more glue in here	
45 Tim:	I put a new one in.	
46	(2.2)	
47	Put a new one in.	
48	(5.8) ((Clare searches for glue sticks))	
49 Clare:	Do I take one of these little ones?	
50 Tim:	Yeah.	
51	[((Shelly still pulls hard))	
52 Clare:	There is still a little teeny tiny bit in there	
53 Stan:	This won't come out!	
54 Tim:	Just push it in with another one.	
55	[((Stan attempts to wrestle glue gun from Shelly))	
56 Clare:	Sta::n	
57 Stan:	Push it in with another one.	
58 Clare:	I will, I will.	
59 Tim:	Hold the trigger, the trigger down	
60 Stan:	It's coming out	

This episode began when Clare and Shelly told Tim that the glue gun was not working and suggested that the glue gun might be empty (lines 35–37). Tim suggested that the problem had arisen because they did not pull hard enough on the trigger (line 38). But the two girls insisted that they already pulled as hard as they could (lines 39–43). Clare reiterated that in her assessment, there was no more glue left. Tim proposed that she refill the tool with another stick of glue (lines 45–48). However, it was not obvious to Clare how to do this or whether she should take one of the shorter pieces of glue stick lying around (line 49). Clare then noticed that there was a bit of glue left in the gun which, as Stan stated, did not come out (line 53). Here, Tim provided instructions of how to insert a shorter piece by making use of another piece of glue stick, "Just push it in with another one". Clare did not appear to understand so that Stan wanted to do this operation for her and tried to wrestle the glue gun away from her. Tim, who supervised the recharging procedure provided one more instruction whose implementation was necessary to guarantee a successful recharge. The trigger of the gun needed to be held down so that the glue stick could be placed appropriately behind the nozzle and, consequently, that the gun could be properly operated (line 60). After Stan's comment that the glue was running again, the two boys went about their own design leaving the girls to themselves. From that point on, Shelly and Clare recharged glue guns on their own.

By now, Clare and Shelly had developed enough competence to use the glue gun in a variety of situations and to recharge it when necessary. However, they were still not core members in the glue gun network. Their trajectory of learning was not over, for the practice of properly operating a glue gun goes beyond gluing skewers and recharging a glue gun. Towards the end of the same lesson, Tim followed Clare's request and showed her how to clean the glue gun. By making use of a skewer (which was part of her building materials), Clare was able to push the short, still unused piece of glue through the back and out of the glue gun. One week later, Clare and Shelly experienced yet another breakdown; their glue gun was not working.[8] Clare told Tim about it and illustrated this fact by showing that no more glue oozed from the nozzle. Tim, who just returned to his seat near the power bar, suggested that the circuit blew (there were three other glue guns on the same outlet). Shelly did not understand what was wrong with the circuit, but Tim merely reiterated that the circuit must have blown as one could infer from the fact that no glue gun on the power bar was working. The two girls then decided to take the glue gun to the other outlet.

Clare and Shelly continued to increase their competencies in appropriate glue guns use. One week after learning to recharge the glue gun and during the same bridge construction project, they learned from Chris what to do when burnt by hot glue. Earlier, on the day Clare first learned how to use

---

[8] A more detailed analysis of this episode is provided in Chapter 9 (pp. 266–268) which focuses on the networking processes by means of which individuals and groups form networks that are constitutive of knowledge-building communities.

the glue gun, she had already burnt herself with hot glue. Then, she waited until the hot glue had dried, and then wiped it off on her sweater.

Clare: I burned myself. It doesn't hurt anymore. That's where I touched it. It's OK, it was just the hot glue. I went like that, and there was hot glue right there.

Now again, Clare wanted to wait until after the glue had cooled so that she could pull the hot glue off without burning her other hand. But Chris, who pointed out that she was an experienced glue gun user, urged Clare to immediately wipe the hot glue off with some paper because, as she argued, the burn would be more serious otherwise.

   This extended case shows how Clare and Shelly were networked into a community. They learned the practice of using a glue gun within a growing culture of glue gun users. Here, they interacted repeatedly with more experienced and more competent peers who showed, talked about, and demonstrated the proper use of the tool and how to manage problems associated with the technology. The two girls began to participate in the circulation of a tool and its related material practices. Over time, they required decreasing support from their peers. Thus, their learning can be understood as a trajectory that began with a peripheral status in the small community of those who were competent users of the tool. Then, with time, the two girls became increasingly competent users themselves as they co-participated in appropriate use with old-timers (such as Tim and Chris). Because the old-timers worked in the proximity, they served Clare and Shelly as easily accessible resources whenever they experienced a problematic situation. The two girls progressively moved into the core of the glue gun community. Such learning trajectories, whereby newcomers to a community co-participate to achieve the status of old-timer, has been termed *legitimate peripheral participation* (Lave & Wenger, 1991). Here, the term has to be understood relative to the classroom community rather than some scientific or engineering community. Students such as Tim, because of prior out-of-school experiences, were already part of a wider network of glue gun users. Within the class, they formed a small core community of experienced practitioners. Through their interactions with newcomers, and by allowing the newcomers to co-participate in the practice, the network extended.

   This case study shows that co-participation increased as the newcomers became more competent in working with the tool. The close proximity of the glue guns, forced by the limited number of electrical outlets, may actually have fostered rapid development of this culture of glue gun users. Newcomers worked side by side with and at their elbows of old-timers where they received the requisite support that allowed them to become old-timers themselves; and they could maintain a sense of what *everyone else is doing*. This learning is therefore appropriately described as legitimate peripheral participation.

From an apprenticeship perspective, aspects of Clare's and Shelly's trajectory to becoming competent glue gun users can be divided, heuristically, into the three stages of modeling, scaffolding, and fading (Collins *et al.*, 1989). At the elbows of their more competent peers – including Tim, Stan, and Chris – Clare and Shelly first observed appropriate glue gun operation (modeling); then they tried the practice, accompanied by instructions from the person who oversaw their initial attempts (scaffolding); and finally they glued on their own, unsupervised, and returned for help only when they identified a problem (fading). Here, they first learned to glue two pieces together; later they learned how to recognize that a glue gun still had enough glue and when it needed refilling; and still later they learned how to clean the tool and how to ready it for storage. In each case, I could roughly divide the girls' relationship with Tim into the three stages. That is, the entire trajectory required that these stages be completed recursively as the girls learned more and more aspects of the glue gun practice. But, new learning episodes overlapped with others where they already engaged in competent practice. So it was not as if there were three major phases in their learning. Rather, they learned proper glue gun use in the process of achieving their goal so that their "curriculum" was not broken into little bits of component skills to be acquired in a step-by-step fashion. When they deemed them relevant, students developed competent practices in an as-needed and just-in-time fashion. The best way to describe and theorize such practice is in terms of heterogeneous competencies and asynchronous development.

6.2.5.1. *Unsuccessful Circulation of a Practice*

Students' instructions such as those Tim provided to Clare and Shelly were not always successful even if they took great pains to move from pure instruction (telling), to modeling and scaffolding the practice. In the following episode, Tim attempted to teach Andy how to make sleeve joints for producing longer beams (from straw) needed in the construction of their tower.

61 Tim: Andy, go make some more of these.
62   [((Shows double length straw beam))
63   Andy you are the one who Andy, you just cut a thing in the middle of a
64   [((He tapes two straws))
65   straw, one of the straws, and then you have a whole straw, and then guess
66   what you do, you stick it inside and roll it up and tape it together.
67   Ok you go like this

68   [((Takes straw, makes a lengthwise cut into one end. Andy
            watches))
69   You go like that
70   [((Gets another straw))
71   You go like this go like this
72   [((Sticks another straw into first, using the cut as an easy entry
73   point; separates the two again. Andy watches him.))
74 Andy: ((Picks up the two straws; joins them by inserting one into the
75   other as illustrated previously. Then taps it on the table and
            on Tim's head.))

Tim had already prepared five double-length straws. Tim then asked Andy
to help him (line 61), and, when Andy did not react, began to explain how
to make the long beams. However, Andy did not react to Tim's explanation
(lines 63–66). Tim then picked up two straws and began to demonstrate his
joining technique (lines 67–73). Although watching, Andy did not respond
to Tim's explanation or demonstration. He finally picked up the straws
prepared by Tim (lines 74–75), inserted one into the other, and began to
manipulate them. Andy demonstrated that he could do at least part of the
construction, the joining of two short beams after the cut had been prepared.
But he did not prepare any long beams of his own nor did he demonstrate
the competence to do so.

Here we have an instance where a practice is "taught" to another student,
so that the potential for the technique's circulation was laid. (A similar
episode can be found in Chapter 7, p. 161. There, Tim teaches Renata how
to stabilize two-dimensional structures by forming triangular subsections.)
The excerpt shows that the explanation in and of itself was insufficient either
to interest Andy, or to help him understand how to prepare and make the
joints for long beam construction on his own. In this case, instruction is
simultaneously adumbration. Tim resorted to a step-by-step demonstration,
he showed where to make the cut, and how to insert one straw into the other
at the cut end. Andy's successful attempt to insert one straw into the other,
and thus partially construct a long beam, was the first step in the circulation
of a practice. His use of the prepared pieces to execute part of the whole
joint construction shows similarity with the descriptions of teaching used to
support an apprenticeship metaphor (Rogoff, 1990). Here, a more able
individual scaffolds the first efforts of another by modeling the skill, then
taking on parts of the activity to leave the easier part to the other.

6.3. CULTURAL PRODUCTION AND REPRODUCTION IN A COMMUNITY OF PRACTICE

6.3.1. *Embodiment*

As they were adopted, practices changed. In Chapter 5, we already saw how Andy tried to copy Dennis' "spaghetti stretch" (p. 126–127). As he discovered the plan for making a spaghetti stretch in his own repeated actions, Andy actually enacted it for the first time and transformed the practice. When other students joined Tom in the use of a glue gun, they created new joints in addition to the stock of existing ones. Tom's joints became a shared resource not simply by a process of copying. Newcomers had to go through the experience of making these joints, a process by which the joint-making was transformed – before the joints appeared as part of a new tower. Andy did not merely copy the "spaghetti stretch" but transformed it through his experience. He enacted and produced a sequence of actions the first time. That is, he re-produced and embodied a practice through his own experience. In this course of events, he had to produce the appropriate action sequence a first time on his own. Clare and Shelly learned to operate the glue gun in the same way. They first observed Tim (and Stan) use it, and then had to engage in their own production. This set them up for discovering the practice in their own sequences of actions. This first production and embodiment of a practice by a new member is likely to vary from individual to individual. In the process, the practice itself changes. Cultural reproduction therefore always also includes, to an extent that depends on the contingencies of each situation, cultural production and transformation of a practice. Glue gun practices were at the same time produced and re-produced. This process required time and was heterogeneous and asynchronous. That is, individuals exhibited varying levels of competence in varying aspects of gluing along their varying trajectories of learning.[9]

Thus, knowing a material practice is more than knowing its description in written or graphical or spoken form. As most readers know from experience, the possession of a set of instructions does not guarantee success in assembling furniture, utilizing a software program, or preparing a new dish. The same cookbook recipe leads to an almost infinite number of tastes, odors, and visual presentations. With some cooks, the dish tastes, looks, or smells awful and unappetizing. With others, the same recipe is a culinary and aesthetic marvel. And in few situations, the re-produced dish is better than that on which the recipe was based. The important element missing in the notions of copying, transfer, and diffusion of knowledge is that these do not

---

[9] In a Piagetian language game, this would be both horizontal and *vertical decalage*. However, the evidence provided in this study suggests that such decalage is the norm rather than the exception: Competencies are heterogeneous, distributed, and asynchronous within and across individuals. This is one of the reasons, why a Piagetian framework is inappropriate to describe the kinds of learning that happen in open, knowledge-building communities.

address the embodiment of practices; it is through such embodiment that descriptions came alive in Andy's "spaghetti stretch". It is in the experience of "several-times-through" that instructions become embodied in actions (physical, linguistic-mental) which make the knowing/acting come alive in a new place. It is only through the work of embodiment that a material practice could be said to include a new member. After they become embodied skills, practices can be compared across different people or sites. Embodiment therefore is one element in the earlier described transformation of practices. On the other hand, objects and facts such as a Canadian flag (which Kitty produced for Melinda and Brigitta) or liquid glue-based joints (produced by Tom and Tim) may circulate in a community with little transformation. The practices needed to produce the resource are "frozen" or "blackboxed" in the object.

Such transformations of practices as newcomers enter a community change the very community and contribute to its transformation. While Tom's joints did not change much as long as he held the monopoly on glue guns, the circulation of the practice to newcomers changed the available practices and resources so that new joints became available. The circulation of the resource and its associated practices both enlarged and transformed the community-constitutive network.

The adoption of material practices implies knowledge beyond that which can be communicated in a journal article, technical report, or written proto-col, because plans and descriptions always underdetermine the action they describe. Instructions are inherently adumbrations. Any fit between plan/in-struction and action has to be constructed retroactively. Thus, Andy's plan for the "spaghetti stretch" underdetermined his future actions – it took him about 20 trials to achieve a successful match. Only through his (repeated) experience did the planned actions take form in competent practice and become embodied. While it is possible to take a course or a focused work-shop, the best route is to learn material practices on-site from competent members while they engage in actual practice, for "There is no way to acquire [a practice] other than to make people see it in practical operation or to observe how this *scientific habitus* . . . 'reacts' in the face of practical choices . . . without necessarily explicating them in the form of formal pre-cepts" (Bourdieu, 1992, p. 222, *italics* in the original). The close proximity within which our glue gun old-timers and newcomers worked permitted this form of teaching which Bourdieu calls "pedagogy of silence" (p. 223). It is quite clear, then, that if material practices are to be taught, the relevant tools need to be available in the classroom in sufficient numbers to set up favorable conditions for networking more than a few target students. How-ever, here as in most science classrooms (and in spite of the central impor-tance of innovative tools and apparatus in most scientific research), material practices and occasions for learning to solve related problems were not the most important aspects of the curriculum. Yet it was in the context of material practices that I observed many instances of peer teaching among students.

6.3.2. *Evolving Networks of Practice*

The impact of the glue gun on the classroom was considerable. Glue guns were initially available only to a few students who monopolized the tool. Opportunities for learning in the class were thereby limited. Once glue guns became more widely available, they transformed the classroom in a physical and social sense. Students and projects gravitated towards the two electrical outlets which led to areas of high pupil density. In these areas of high student concentrations, the level of inter-group exchanges expanded learning opportunities so that newcomers to glue gun operation could learn this material practice at the elbows of a more competent peer, an old-timer. Material practices, however, were not simply copied, but underwent change as newcomers transformed their observations (during modeling) into sets of coherent and competent actions. The extended case study of Clare and Shelly documented a trajectory of legitimate peripheral participation in the practice of operating a glue gun within a growing community of competent users. Their increasing participation in using glue guns enlarged the network of practitioners and moved them toward the core of this community. The close proximity of the glue guns, forced by the limited number of electrical outlets, may actually have fostered this rapid development of a culture of glue gun users. Newcomers worked side-by-side with old-timers where they could receive the necessary support to become old-timers themselves.

The socio-technical evolution surrounding the glue guns (as evidenced by the changes in the community and the resistance to adopting one curricular objective) were significant. In part, these transformations were related to the power of the glue gun technology to solve many construction-related problems. However, a technology is only powerful within a corresponding network of social practices which, associated with the instrument itself, leads to the necessary transformation of the objects on which the instrument operates (Law, 1987). In the present situation, the glue gun in and of itself was not a powerful tool, for in the hands of newcomers (like Shelly and Clare in the beginning), it achieved little. But, by relying on the expertise and guidance of others in the initial stages, the two became part of the network (culture) of glue gun users. Thus, Tim's instructions (where to glue particular pieces, how to recharge the glue gun, and how to clean it after use) were important elements in enlarging the network and transforming the classroom culture in which the glue gun technology replaced most other forms of making connections.

The success of the glue gun as technology, and the culture of glue gun users could not be derived solely from the glue gun (technology) or social factors. The appeal of glue gun joints, the glue gun's amenability to the needs of many different student groups, physical and geographic constraints, availability of materials, social interactions over access, and the nature of the task are but some of the influences that shaped the actor network and with it, the emergence of the practice. The glue gun's future, its developmental trajectory, or its future consequences could not be determined or predicted with any degree of certainty from any finite set of elements and conditions. The glue gun technology was both indeterminant and indeterminate. It was indeterminant in that it is not autonomous and self-determining; that is, it cannot be used to explain its past changes retroactively purely by reference to technology itself and the knowledge it constitutes is insufficient to explain its future course as a projection from its present and past.

Technology is also indeterminate; that is, it is next to impossible to reliably predict the form and function of a technology from any set of ex ante social conditions. The question is how to take into account the simultaneous presence of several interacting elements (actors) of comparable weight that are associated with the transformations. This problem is not trivial as any changes in one actor ripples through the network (community) to affect all others which in turn affects the original actor itself.[10] Such actor networks cannot be understood by studying each actor in isolation, because a system's characteristic features depend on the interaction, and are not reducible to individual component actors. The actor network approach used here was developed to model complex systems such as technology constituted by heterogeneous actors.

Invention, innovation, and circulation of existing norms into which school children are to be enculturated require networks through which accomplishments can be sustained and propagated. The open questions in student-centered classroom communities are, "How can 'desirable' ideas be sustained and promoted among students?" and "How do we promote the acceptance of these ideas?" A network approach provides some answers because it allows the combination of heterogeneous elements, describes the establishment of systems without decision-making centers, and encourages investigators to shift perspective by taking the viewpoint of a different, even non-human actor.

In the present case, the network of glue gun users established itself without a central human decision maker. From the glue gun's perspective, in order to be successful, it had to be able to satisfy the needs of all the various student groups (that is, to have different meanings in different settings) in order to become the dominant technique.[11] Because of its flexibility, the glue

---

[10] In systems with non-linearities, minor influences can change the state of the system dramatically. Such systems are chaotic so that their future states can be determined only probabilistically. These systems cannot be modeled on the basis of classical linear structural relations.

[11] It may sound strange to take a glue gun's perspective on a community's change. However, it is but one example of the ethnographic technique to learn about phenomena by making the familiar strange. Latour (1992) used this technique to understand the failure of the ARAMIS project (an individualized urban transport system) despite 18 years of development work, billions

gun satisfied Ron's need for strong joints, Maggy's demand for a neater assembly, Tom's desire to produce miniature joints, and Stan's requirement for a pin that could not slip. The interpretive flexibility of tools was therefore an important aspect (see Chapter 8). The glue gun had to be able to bring together other elements as well. There had to be receptacles, additional power bars, fuses strong enough to support the electrical load, materials that could be glued with this technology (as students readily told me, not all glues work with all materials; some just do not stick, others destroy the material), students willing to share the tool, enough science money in the school to buy glue sticks (Tammy had spent the small amount of money available to the entire school on her fall science project alone), and the teacher's willingness to abandon tight control over the activity and to renegotiate the imposed gluing techniques. Remove any of these actors, and the network begins to shift. As a consequence, I would have observed rather different student-produced artifacts which here, as elsewhere, were used by teachers to document students' achievement for parents, other students and teachers in the school, and even the public at large (through an exhibition in the local science museum).

The actor network approach models not only the successful apprenticeship, but equally importantly the heterogeneous nature of Clare's bridge. First, Clare (and her partner) had a set goal of building a bridge. To achieve this goal, she enrolled various allies, including materials, tools, contingencies of the physical setting, numerous peers, and even teachers. That is, by enrolling Tim, she translated his competence to advance her project, and coincidentally developed a new set of practices. In this way, "her" bridge was not simply the product of her mental and physical activity, or a product of the joint activity with Shelly, but an artifact that emerged from the interplay of numerous actors. In Chapter 8, I will deal in greater detail with the heterogeneous and distributed aspects of knowing, learning, and designing, and the products that result from these processes.

Actor networks also model unsuccessful circulation of resources and practices. Despite Tim's request for participating and despite his explanations and demonstrations intended to demonstrate how to construct double-length beams, Andy did not contribute to further long-beam construction. That is, the network of long beam constructors was not extended and long-beam construction practices were not circulated. In actor networks, opposing forces and recalcitrant actors are used to model phenomena that impede network construction. In the present case, there are several possible explanations. First, Tim's explanations were overlaid with Serge's talk, "It's dying . . . make

of dollars spent, and high potential gains for several parties involved. He used the same technique to investigate the role of a door closer in socio-technical relations (Latour, 1988a). Taking the perspective of a non-human actor is consistent with Latour's research directive to follow people and artifacts, and investigate how they take part in various relationships. Following this directive reveals the social nature of many aspects which in psychological theories showed up as individual characteristics (Latour, 1987).

that, it's alive . . . Andy it's a alive", thus competing for Andy's attention. Here, Andy's interest in learning how to make long beams and to contribute to their construction may not have been strong enough; Tim could not get him interested and enrolled. Second, there were only a few more beams needed. When Andy did not react, Tim took on the task to complete the last construction. A third possibility is (though there is not enough evidence to support it) that the joint was difficult to make. In a similar situation with a similar joint, Maggy did not participate in making Renata's sleeve joint because it was too difficult ("Our biggest problem was cutting the straws . . . - because you have to cut just part of [the straw] without cutting the whole straw, and then cutting the little straw").

The developments surrounding glue-gun use were largely student-driven. Critics may have concerns about the quality and quantity of knowledge students acquire in student-centered classrooms. The challenge remains with teachers to make use of existing individual practices and resources and to foster a classroom climate in which these are adopted by other students. In industry, this is facilitated when companies utilize existing knowledge (represented in papers, computer programs, mechanical parts) and know-how (embodied practices of their engineers, managers, and workers) to create new combinations. That is, by bringing different people into contact with different expertise and by providing access to various information sources, new resources and practices are produced. However, the difference here with some of the uses of actor networks in the science and technology literature lies in the virtual absence of economic and political forces. These often shape scientific and technological networks in decisive ways (Latour, 1992). Thus, if teachers want children to form more autonomous networks, they have to provide for other, equally powerful forces to achieve specific curricular objectives (content knowledge). Providing students with opportunities to set and pursue their own goals and agendas often brings about such forces. Bernard, the teacher in Moussac, certainly experienced students both producing and reproducing French culture as they co-participated with and in the community. My own teaching experience in open-inquiry and open-design learning environments certainly lends support to the tremendous learning observable in student-centered classrooms.

Given the present data and analyses, the notion of knowledge "diffusion" appears inappropriate – unless diffusion also implies transformation. That is, the "production" needs to be emphasized as much or more than "re-production". The re-production of practices always entails transformations as the practices are appropriated and embodied by newcomers. A growing community of practice therefore implies a transformation both of the individual (who embodies a new practice) and the community (which is larger and has changed in the available practices). Learning therefore is a process of individuals *and* the collectivity in which they are members. The actor network view proposes the notion of "translation" (Fujimura, 1992; Latour, 1987) to capture the forms of learning I just described. The translation model identifies how a technological innovation is subjected to innumerable transformations as it is promoted, adopted, adapted, discarded, and hybridized by different collectivities. Such change processes also embody the possibility for a break-up of a parent community into several daughter communities (such as the progressive splintering of the natural sciences over the past 300 years). Yet, as the often-used notion of scientific community suggests, these different daughter communities, despite their widely differing material and discursive practices, are often considered as one. As a consequence, the referents of "community" or "community of practice" has to be specified in the context of using these terms.

7. EMERGENCE AND CIRCULATION OF DISCOURSE PRACTICES

Discursive practices – which those in a Wittgensteinian tradition commonly refer to as language games – are constitutive of the ways we see and interact with the world. When a civil engineer talks about stabilization in terms of braces and plans to use a triangular brace she is engaging in a practice characteristic of her field. Appropriating the language games of canonical science and technology and populating them with one's own intentions has long been a goal of science education. However, as this chapter illustrates, teachers sometimes – traditional teachers often – focus more on the acquisition of a discursive resource (being able to *recall* the name and definition of a practice) rather than the practice itself. Such a situation is explicit in the following account of students' trajectories of competence in one aspect of an engineering language game ("triangles"). The second section of this chapter is devoted to teacher questioning that scaffolded students' story telling so that they told increasingly complex engineering design stories. Finally, in the third section, evidence is presented that illustrate how whole-class conversations served to build and extend engineering design discourse.

7.1. TRAJECTORIES OF COMPETENCE

The objectives for the "Engineering for Children: Structures" unit are not formulated explicitly as has been the norm in science education practice ("The student will be able to . . ."); nor did the stated objectives outline hierarchies of skills that students should learn. However, Gitte and Tammy had some tacit objectives. They wanted students not only to engage in science and engineering activities to have some fun,[1] but also to develop material and discursive practices important in engineering design. Thus, students should

- know about material properties and how to change them through strengthening materials;
- know about and using joints that interface the same and dissimilar materials involving a variety of joining techniques (glue, adhesive tape, rope, pins); and

---

[1] This word is used far too often by teachers and often means that students do not have to engage in learning. While I believe that learning should be associated with positive affect, I do not believe that activities should be just for fun. Furthermore, I have not seen much evidence about how putting your hands on some material and equipment changes cognition, but some of my research has attempted to show just that (Roth *et al.*, 1997a). This book, and especially Chapter 8 was written to address this issue.

- achieve structural stability by means of triangular configurations (braces, tripods, staves).

Throughout the "Engineering for Children: Structures" unit it was quite clear that no other objective was more important to Gitte and Tammy than that of using triangular configurations (braces) to achieve structural stability. To introduce this practice to the community, they designed and taught a special introductory lesson in which children were to learn that the addition of braces to form triangular arrangements stabilized material arrangements in the form of pentagons and hexagons or three-dimensional structures such as cubes (e.g., Figure 2a, p. 162). In subsequent lessons, they assisted students in their design activities to allow the triangular braces to emerge. In many ways, this approach has overtones of traditional curriculum design. First, one teaches the skills – here stabilization by means of braces – in some decontextualized "general" activities. Second, children are to apply what they learned. The problem with such a conception lies in the assumption that *knowing that* acquired in the first step will be converted into *knowing how* in the course of the second step of the sequence. What such assumptions neglect is the inherently unspecifiable nature between *knowing that* and *knowing how*, that is, the precarious relationship between plans and situated action (Suchman, 1987).

The following two subsections account for the results of the teachers' effort to bring a discursive practice into circulation: The first subsection provides a cross-sectional view of an emerging community of triangle practice six weeks after this practice was first introduced; four levels of competence are identified. The second subsection shows the trajectory of one student through these levels of competence over the course of the "Engineering for Children: Structures" unit.

7.1.2. *Snapshot of an Evolving Community of Practice*

Six weeks after the initial lessons on bracing and at the end of the bridge project, 11 of 14 projects in the classroom featured triangular configurations (e.g., Figure 1b–d), and 7 of the 13 available written project descriptions mentioned triangles[2] as important features of the described structure. During a more detailed investigation, I realized that in two of the 11 projects, the triangular configuration had no structural significance; students had added them for aesthetic reasons because *everyone else used* triangles and because the teachers emphasized the presence of triangles. A third group employed a number of triangular configurations of structural and aesthetic nature, but highlighted the aesthetic triangles in their written descriptions and during whole-class presentations (e.g., Figure 1d). Some students employed no triangles at all. Among these students, there were two groups (one with

---

[2] "Triangles" is used as a short-hand for "triangular braces". Triangle is the term developed by and used in the community and therefore constitutes an emic term.

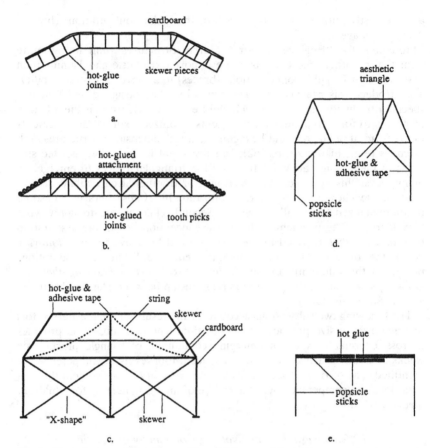

Fig. 1. Five solutions to the bridge problem. (a) Peter and Paul did not use triangles because they did not need them (b) Tom and Andy planned structural triangles as a basic component of their design. (c) Tim and Stan developed the "X-shape" in response to the initial instability. (d) Brigitta and Sylvia initially included only aesthetic triangles, but after interactions with Gitte, included structural braces. (e) Dennis and Patricia and Anna laminated but did not use triangles.

aesthetic triangles) who argued that they did not use triangles because their materials, joints, and building techniques did not necessitate the (teacher-desired) configuration (Figure 1a). Two other groups provided no explanation for not using triangles (e.g., Figure 1e).

The summary in Table I shows that four groups either used triangles as structural features of their bridges by design or as responses to structural weakness and two further groups had developed alternate solutions which did not require the use of structural triangles for improving strength or stability. Six groups began to include structural triangles after significant

TABLE I

Use of structural triangles six weeks after the initial instruction

Level of Competence	Description	Triangles		Frequency structural (aesthetic)
		Planned use	Unassisted use	
1	Students planned triangles and used them as structural features. (e.g., Figure 1b)	Yes	Yes	2
1	Students initially planned to use triangles but, in the course of their projects, found solutions that were strong and therefore did not require bracing. (e.g., Figure 1a)	Yes	No	2 (1)
2	The use of triangles (or "X-shape") emerged as a contingent response to structural problems but without teacher help. (e.g., Figure 1c)	No	Yes	2
3	The use of triangles was planned. However, triangles to improve structural strength and stability were included only after teacher scaffolding. (e.g., Figure 1d)	Yes	No	6
4	Despite teacher scaffolding, students did not include structural triangles to improve strength or stability. (e.g., Figure 1e)	Yes	No	(2) 0
4	Student had left class to attend another school so that he could not benefit from teacher scaffolding. (similar to Figure 1e)	No	No	(1) 0 (0)

teacher scaffolding and two projects did not feature any triangles. Thus, although students employed the word "triangle" in their notebooks or during talk about their projects and although they could say the "magic word" when prompted by a teacher, many students did not build them into their structures by design six weeks after the first instruction. Some groups began to strengthen and stabilize structures by means of braces when provided with sufficient scaffolding. Other groups used them without scaffolding, but they emerged out of the specific situation in which the children developed their constructions in response to structural weakness. Some individuals or groups did not employ triangular configurations at all.

Figure 1 and Table I re-present a community of practice in transition. Its members, who appropriate the discursive and material practice surrounding triangular configurations at different rates, are found at various points along their individual trajectories to competence. (In the following section, we follow Tim in his trajectory.) As a cursory heuristic to mark different regions of these trajectories, I developed a four-tiered categorization.[3] Students who engaged in discursive and material practices related to triangles, or replaced them with other elements, had achieved the highest level of competence (Level 1). Students were classified in Level 2 when they did not plan triangles for their bridge designs but developed them later as contingent responses to structural weakness or instability as part of the design process. Level 3 categorized those who used triangles with teacher scaffolding which included probing questions, suggestive questions, and suggestions to use triangles. The beginning of the trajectories, Level 4, was used to categorize those students who, even with teacher help, did not engage in the material practice of stabilizing and strengthening structures by means of triangular configurations.

Cross-sections of communities as the one depicted above – while they illustrate different levels and degrees of competencies and characterize the extent to which they are distributed in a community – are not suited to theorize the development of competence. The following case study shows how an individual student's competence in using triangles for design changed as he participated in the "Engineering for Children: Structures" activities.[4]

---

[3] These four levels are for coarse descriptive purposes only. Human competencies in a domain are not homogeneous so that one can establish clear classification. As Piaget noticed already and conceptualized in the notion of *horizontal decalage*, competencies are heterogeneous: an individual is better on some aspect of a practice than on another, and may be near newcomer performance on a third. A sports analogy may help. Some soccer players are outstanding in the defense, but only mediocre forwards, and entirely useless as goal keepers. In sports, the heterogeneous competencies of a soccer player appears to be common sense. Schools still insist, however, that competencies are homogeneous (written tests are used almost exclusively for assessment purposes) and can be determined through singular measures (a single grade).

[4] Readers should note that here, too, practice has at least two aspects: material and discursive. Using triangles by design to make a stable structure is an aspect of an engineering discourse; manipulating tools and materials to stabilize structures are aspects of a material practice. But in this unit, where children's designing integrated the two (as will be clear from Chapter 8), the two often co-emerged.

7.1.3. A Trajectory of Competence in Triangular Bracing

7.1.3.1. A Traditional Lesson about Triangles

This lesson was intended to teach children the design practice of triangular bracing as an ideal geometric configuration to achieve structural stability. (Unlike other lessons in the "Engineering for Children: Structures" unit, these tasks were much more specific than the design challenges.) That is, triangular braces were to be introduced as resources that the children designers could use whenever they wanted to design stable and strong structures. The teachers' intent was therefore, to begin a network within which a specific discourse could recur and be circulated. Tammy and Gitte had planned to first demonstrate the concept and then allow the children to practice and extend what they had learned. At the beginning of this lesson, children had a few minutes to write a paragraph about stability in their engineering logbooks. Tammy then asked students to read their paragraph to the class. The following episode illustrates how the teacher flagged the "magic" word as soon as Ron read it from his notebook (line 03). (As Tammy later admitted, she was surprised and pleased that the word had come up so quickly.)

01 Ron: I think stability is something strong, keep things up like bones,
02   beams and triangles.
03 Tammy: I like that last word. You guys think that's very funny but that's
04   very, it's excellent.
05 Jeff: I think stability is something that helps something else to be
06   stronger like gravity or forestay on the mast.
07 Tammy: Good for you. And what kind of shape is that forestay?
08 Jeff: Triangles.

In the subsequent exchange, Jeff made reference to forestays (lines 05–06) which provided Tammy with another opportunity to "draw out" the term triangle (lines 07–08). After asking two girls to read out there paragraphs, Tammy came back to forestays and triangles.

Jeff mentioned that a forestay on a sailboat gives stability to the rest of it. So, if you're looking at the sailboat here ((draws a sail boat on the blackboard)), this is called the forestay and this is the wire that supports, that's attached very very firmly to the boat and attached to the mast. And the mast is attached to the bottom of the boat. So, you got yourself what, Ron was talking about it too, another triangle, the triangle.

Tammy then showed a square from straws and asked how to stabilize it. Jeff immediately suggested to add another straw diagonally to form two triangles. Because he had uttered his response under his breath, Tammy asked him to repeat his response loud enough for all students to hear. Jeff repeated the word "triangle", and Tammy followed with the introduction of a new task for the students:

As Ron and Jeff mentioned, we stabilized this shape ((holds up straw and plasticene square)) by making two-dimensional triangles. How many triangles would it take, straws would take it to make this shape ((Holds up pentagon)) stable?
((Children move to their construction sites))
Let's just use this one ((holds up a pentagon from straws and plasticene)). Could you try and show me your best stabilization of this shape ((holds up and shakes pentagon)) show me, use your toothpicks and your marshmallows to show me how you stabilize the shape. Make the shape first. I don't think I need to tell you this, but, put a straw through. How many triangles would it take, straws would it take to make this ((holds up pentagon)) stable?

Here, Tammy, who had discussed the lesson previously with Gitte, used a technique that is widely known as "application". She had shown and talked with students about the square from four straws pinned at the ends which was stabilized by adding a fifth straw across one of the diagonals. She said that she had shown the concept and therefore transmitted it. Now, she wanted students to apply the concept to a new situation, pentagons from toothpicks and marshmallows. As if she was afraid that the children would not produce what she expected, she added "How many triangles would it take?" From my perspective, there was no reason to assume that anyone had seen a concept, understood what was meant by "triangles", or was in a position to engage in the engineering practices of talking about and building braces. It was not surprising that Tom, a very eager and willing student, provided a negative response to Tammy's last question ("How many triangles would it take, straws would it take to make this stable?").

09 Tom:	Beats me ((looks at his materials))
10	(2.0)
11	I don't know how to make this.
12 Tim:	I know, you put one across there and then one down
13	[((gestures link vertically and horizontally
14	between pairs of corners))
15	And then it stables it that way, and then that way.
16	[((gestures same link vertically and hori-
17	zontally between pairs of corners))
18 MR:	Why don't you show it?
19 Tim:	I'm gonna try ((Begins to add braces to his pentagon))
20	(2.1)
21 Tim:	A little bit too big.
22 Tim:	That's stable
23	(1.2)
24	It can't hardly move
25	((He squeezes the pentagon, and it begins to fold))
26 Tammy:	((In the background)) °The least number of straws.°
27 Tim:	I only have used two so far. Mine is stable now. I got it

Tom's first reaction was that he did not know, and maybe one that signaled disinterest in following a prescribed activity (lines 09–11). Tim, on the other hand, quickly indicated that he had found a solution, and gestured the connections to be made (lines 12–17). He explained that the connections he indicated with his gestures would stabilize the pentagon. When I encouraged him to show how he wanted to stabilize the pentagon, he began to add two braces to his pentagon (Figure 2a). Tim emphasized that the pentagon "hardly" moved (line 24) and that he had used only two braces (line 27). His subsequent effort was directed to change his original solution from a "hardly moving" to a stable one without increasing the number of braces. In the background, Tammy could be heard talking to another student about triangles, but Tim gave no indication that he had heard this comment. He remarked, however, that he had now found the solution. His pentagon was stable. However, about one minute later, while talking to himself, Tim suggested a different solution, "It took three". Renata responded, "I think it takes about four, because there is one sticking out here". But the following episode shows that he was confident about his two-straw solution and showed Renata how to do it.

Renata had asked herself, under her breath, how to stabilize and make rigid a pentagonal shape, "I wonder if I can make it more stable?" Tim instantly began his explanations:

28 Tim: You just take, you only have to put one on the top. You
29   remember she said make a triangle. If you make a triangle, if
30   you make a giant triangle, it keeps it stabeler.
31   If you make one giant triangle
32   [((points to her pentagon))
33   It only takes two sticks to make a giant triangle.
34   Like you see.
35   [((points to the pentagon which he had glued into his
           notebook))
36   I got the shape and then I got two pieces of spaghetti.
37   ((Renata watches him))
38   So it's a giant triangle.
39   [((Traces a triangle in the air over his pentagon))

Because Renata did not arrive at a two-straw solution, Tim explained what to do (lines 28–33). He suggested that she should make triangles just as the teacher had suggested. However, Renata made no sign that she understood Tim's explanations. Tim tried again. He used his own pentagon from spaghetti and mini-marshmallows to outline where to put the two pieces (lines 34–36), and how this would yield the triangle he talked about (lines 38–39). This explanation and the accompanying gestures were sufficient, for Renata subsequently stabilized her own pentagon in the way the teachers wanted.

In the context of this lesson, Tim had shown competence. Not only did he stabilize his own pentagon but he also taught another student how to do it. When his verbal descriptions did not help Renata to understand, he used

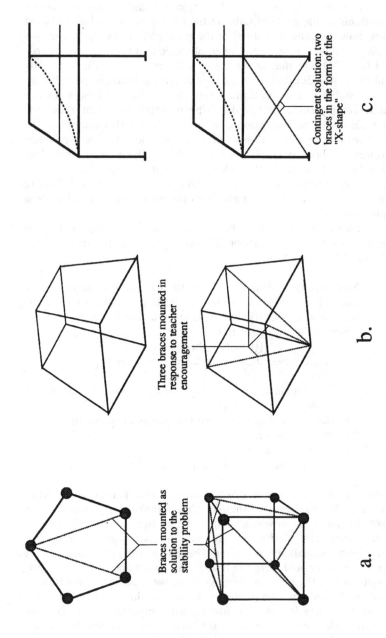

Fig. 2. Tim's activities related to "triangles". (a) During an initial lesson, he solved the stability problem for a pentagon and a cube. (b) His initial "cube" for the tower did not contain any triangles. He added them after repeated teaching attempts by Tammy. (c) He developed and included the "X-shape" after his first test of the bridge's strength.

the pentagon in front of them as referent for gestures that supported his explanation.[5] Tim, as several other students, went on to build a cube which he stabilized by forming diagonal braces (Figure 2a).

In this way, Tim and his peers achieved the curricular objective of stabilizing pentagons, hexagons, and cubes. Some of the students arrived at the teacher-desired solution on their own (Jeff), some with a little encouragement (Tim), some with the help of their peers (Renata), and a few with more direct instructions by the teachers. Teachers take such data often as indicators for the success of a lesson. They assume that children saw the concept (Tammy demonstrated how to stabilize a square), had practical experience (children stabilized pentagons and hexagons), had an extension experience (children stabilized cubes), and were reasonably successful on the task. However, as the following episodes show, the students were far from forming a community in which the practice of stabilizing with triangles was common place.

7.1.3.2. *Significant Teacher Scaffolding*

Over the next several weeks, Tammy and Gitte scaffolded students' designing in a way that would allow students to appropriate the triangle practice.[6] They read to the whole class examples of student reflections that mentioned triangles and used the technique of drawing out to make students talk about triangles during their project work and class discussions. For example, on the day following the first lesson, Tammy praised Kathy for her descriptions of the stabilization activity.

I just think it was just the amount and quality that Kathy wrote. OK when we were talking about stability, if you remember, and we first gave a definition of stability. And then we talked about what we did and what made us think of what we did, and sort of gave the process of what we went through. ((Begins to read from Kathy's notebook)) "A thing has to be stable to stay up that's what stability is, the roof is stable. I think that stability is mostly triangles that you put on a shape. I did a dome and something like a cube and two things like a pyramid. I did six things, I worked on Carla's ideas, and she worked on my ideas. When I think of stability, I think of triangles, because when I put a triangle on all my shapes, they were stable . . .".

Through commenting and reading from students' notebooks, Tammy repeatedly reiterated the connection between stability and "triangles".

A week later, students began to design their towers. As children worked on their own projects, Tammy and Gitte talked to all groups about stabilizing their design artifacts. The following episode, recorded during the tower

---

[5] In Chapter 8 (pp. 231–232), I provide a detailed analysis of how and why the artifact assisted Tim in his effort to teach Renata the stabilization of a pentagon by creating triangles.

[6] Competent triangle discourse does not simply mean being able to say the word, but using it as part of a discourse of designing that specifies bracing as a structural feature. Triangle-sayers are not necessarily triangle-users.

construction, provides an indication of what Tammy understood as "drawing [the concept] out" of students.[7]

40 Tammy: OK, do you want that to be like this, or do you want that this
41   to be more stable?
42   [((squeezes the tower which bends in all
                    joints))
43 Simon: [We have, we are]
44 Tim: Stable.
45 Tammy: You want to stabilize?
46 Simon: [We are, with a beam
47 Tammy: Which shape does one have to make if one wants to stabilize
                something?
48 Tim: [Triangle
49 Tammy: Interesting.
50 Simon: Triangle.
51 Tammy: [How are you going to make the triangles, there?
52 Tim: Like this.
53   [((Gestures across the diagonal of one of the straw cube faces
54   [((Simon holds straw across face
55 Tammy: Does this make a triangle, Simon?
56 Simon: No, this is a.
57 Tammy: But you could elong, how could you elongate it?

Andy, Simon, and Tim had begun their tower. Despite the initial lesson about stability, repeated reiteration that triangles and stability are linked, and other previous instruction (4.5 hours) and one hour of constructing towers, the three boys had not used triangular configurations to stabilize their tower. In her conversation with the three boys, Tammy bent and twisted the structure repeatedly (Figure 2b, top), and then asked them whether they wanted their tower to be like this (lines 40–42). Simon suggested that they could stabilize with a beam (line 46), but Tammy was not satisfied. The students had not yet uttered the magic word. (In some conversations with other students, Tammy actually asked for the "magic word" when the desired term was not forthcoming.) Here, Tim uttered the desired word clearly to Tammy's satisfaction (line 48–49) and Simon repeated the word once more (line 50). She then followed up by asking how (and possibly where) the boys wanted to make the shape. Both Simon and Tim pointed across the diagonal

---

[7] Given that this class was part of a French immersion program, Tammy tried to speak in French with the students as frequently as possible. This conversation was originally in French. But because neither Tammy nor the students were native speakers, their French sounded like directly translated English which made my reverse translation a lot easier. However, because of the translations, the overlaps are only approximate.

on one of the square faces in their structure (line 53–54). Tammy left the group after she had ascertained that Simon understood that he needed to take a beam long enough to reach across the diagonal. However, she briefly returned several times. One minute later, she commended the boys, "You started with the triangles, nice". Four minutes later she returned to ask again about triangles. At that point, the three students had added one brace to form a square rather than a triangle. But at the end of the period, only one of the ten rectangular faces of their tower construction was stabilized with a triangular brace (see also Figure 1c in Chapter 8, p. 202). This lesson ended with a whole-class conversation. Several students talked about their projects and pointed out triangles in their structures. During the following double period (on the next day), Tammy returned to the group three more times to talk about and "draw out triangles" before Andy, Simon, and Tim had finally used a sufficient number of triangles to stabilize their tower (see Figures 2b and 3).

This episode was typical for the competencies which I classified as Level 3 (Table I). At this stage, students did not design structures on the basis of triangular braces as elements that provide structural strength and stability. A lot of scaffolding and "drawing out" was necessary before students engaged in designing with braces. At this level of competence, students sometimes indicated that their structures were complete although, from the teachers' perspectives, they were structurally weak and did not include the prized configuration. Both teachers encouraged such students to do further stabilization work and eventually incorporate triangles. Both teachers recognized the amount of teaching, drawing out, and questioning they had to employ so that students at this level could achieve sufficient structural stability for their towers.

7.1.3.3. *Contingent Emergence of Triangles*

In some cases, students appeared to know about triangles but did not use them *a priori* as design features. When they used them without teacher help, they did so in response to locally-emerging problems. In the case of the triangles, these problems were of structural nature. Students attained this Level 2 of competence when they used triangles not as a constituent design feature but in response to structural weaknesses and without teacher help.

Prior to beginning their bridge, Tim and Stan showed Tammy the triangles in their individual plans for a bridge ("Look at all the triangles in here!"). But, these triangles were of an aesthetic rather than structural nature; both used them because they knew that triangles were expected. However, Tim seemed to know that structural stability could be achieved by means of triangles. When Shelly asked Tim a little later how to stabilize a structure from skewers and glue (line 58), he told her to use triangles:

58 Shelly: How are we supposed to get that stand up?

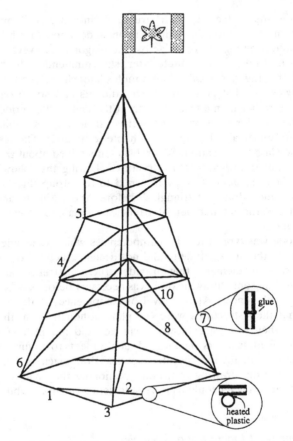

Fig. 3. The tower designed by Andy, Simon, and Tim as it was during their presentation about two weeks after the lessons. To facilitate the reading of this drawing, the main beams are drawn in heavier lines than the cross beams, although they were all from the same material. The numbers index those points that the three referenced during their conversation.

59 Clare: We *will*. We *will*.
60 Tim: It's easy, you just make a triangle out of that stuff.

Clare did not want any help (line 59), but Tim responded, "It's easy, you just make triangles out of that stuff" (line 60). However, during the first two lessons of their own bridge design, Tim and Stan did not talk about triangles or include them as a structural feature in their design (Figure 2c, top). Not until three lessons after telling Shelly what to do that Tim began to use triangular shapes. When he used a triangular configuration of building materials, it was in response to repeatedly experienced structural weaknesses of their bridge (lines 61–63).

61 Tim: Stan, it's still not good enough, see it's still wobbling.
62   [((He pushes the side of the bridge
63   which moves in all the joints))
64 Stan: I know.

A few seconds later, Tim asked, "Stan, why don't we connect some straws like this?" and gestured one diagonal beam (Figure 2c, bottom). First, they used straws for the diagonal beam. But their first test showed that the straws did not hold up ("We've got a bridge and it is all stable. But Stan broke it". They then decided to go with skewers and negotiated with a teacher for permission to use them. Stan got two skewers, but Tim suggested they needed more.

65 Tim: We still need two more skewers.
66   (4.8)
67 Stan: Why do you need two more?
68 Tim: Because we put them the opposite direction (on each side?).
            See look.
69   (15.4) ((Tim tapes two skewers in V-shape)) ((Stan gets more
            skewers))
70 Tim: The wrong way Stan, put them.
71   [((Stan holds new skewers parallel to the others))
72 Stan: I don't (these doesn't?).
73 Tim: And now we're gonna make it criss-cross.

Tim had started to work two skewers into the structure to stabilize the piers. He announced that they needed two more skewers (line 65) but Stan did not understand why. Tim suggested that they could put them in opposite directions. When Stan returned, he held the new skewers parallel to those Tim had just attached. Tim's response shows that this was not how he had intended his "put them in the opposite" (line 70). He had meant, as his later elaboration shows (line 73), to put the skewers between each pair of piers below the deck so they "criss-crossed" the existing skewers and thereby formed two Xs (Figures 1c and 2c, bottom). Later, he called the resulting shape the *X-shape*.

At the end of this project, I took the account presented earlier in this chapter. Tim and Stan had employed the braces without teacher scaffolding. But they had not planned the use of braces *a priori*. Rather, when their bridge was not stable, and because it could not carry as much weight as they thought it should, Tim began to gesture triangles. At first, he proposed and used drinking straws for one of the diagonals between bridge piers (Figures 1c and 2c, bottom). But when these broke, he switched to wooden skewers and, in the process of building, added a skewer to the second diagonal to form an X-shape. As I showed in Chapter 4, this invention was to become an important aspect of the classroom discourse about creating stable structures.

Eleven children included the term in their glossaries of important engineering design terms. As other aspects of the design discourse, the X-shape was widely circulated. With this wide availability, other students added rationales for the stability that one could achieve with this configuration. In his glossary, Andy included the X-shape. Here, he suggested that the X-shape was so stable because it made four triangles. This provides evidence that Andy was a competent member in the triangle network and therefore should be classified at Level 1 of my heuristic.

Another feature of Tim's bridge also indicated his status as a newcomer to the practices related to structural strength. He (and Stan) had included string so that his design looked like a suspension bridge. However, the string had an aesthetic function and did not carry any load. As with the aesthetic triangles used by other students, recognizing the suspension as an aspect of bridges and using it to strengthen a design are different things.[8]

Another student who discovered the triangle practice in his exploration was Jeff. Having set himself the goal to establish the class record for the load a bridge could support, his bracing technique emerged after he assessed early designs as exhibiting structural weaknesses (see Figure 7, Chapter 4 and Figure 4, Chapter 7). Here, he did not begin to design his bridge based on the bracing practice. Curiously enough, his pre-unit drawings had already included triangular configurations in the pillars of a bridge and as features of a dome (see Figure 3 in Chapter 5). This may be an indication that in the early drawings, he included the braces as decoration and because of concerns to be accurate in his re-presentation. My observations of his design process on the bridge suggests that he had not been a competent triangle-user. Tim also appeared to have shown competence in the triangle practice by suggesting to Shelly that she should stabilize her structure by designing triangles with her materials. But his own triangle practice only emerged from the considerations of weakness later in his project. This further suggests that one needs to be very careful in assessing a practice. Early assessments appeared to indicate that Jeff was a competent designer of bridges, but his later design project revealed that braces were not resources he drew on as a matter of course in designing for stability.

7.1.3.4. Competent Practice

The highest competence level was achieved by those students who used triangles as design features or who explicitly rejected the use of triangles because they had developed alternate means of achieving structural stability. During the bridge project, Tim did not reach this competence level. How-

---

[8] Bridge design is not an easy thing, even for much older and more experienced university engineering students. An interesting account of the practices of such a students was provided by Gal (1996).

ever, during the following project in which students designed "mega struc-tures" (structures large enough to contain all four members of a design group) from newspaper rolls and masking tape, Tim included triangular braces in his design drawing and built them, together with his three partners, into his structure.

During the bridge project, students such as Tom and Andy or Chris and Arlene already competently used discursive and material practices related to triangular braces. Tom and Andy, in their independent design drawings, had included structural triangles. They highlighted these as positive features in the evaluations of their designs and, one lesson later, repeatedly emphasized – during the class discussion about the rules for building bridges – that stable structures needed to include triangles or braces as constituent parts. They began their material activities by gluing structural elements, including braces, from toothpicks and glue (Figure 1b). They explained their process, "We sort of made the bottom one, and the bottom crosses, and the triangles in here". That is, they used structural triangles as the basic building unit. In the description of the bridge for an exhibition at the local science museum, the two emphasized again that their bridge's structural strength was achieved by using triangles ("It could hold 40 blocks. We put triangles to make [the bridge] stronger"). In this way, their bridge ultimately supported 154 wooden blocks.

Other students explicitly argued against the use of triangular configurations on their second bridge design. Chris and Arlene did not use these configur-ations, despite both teachers' repeated encouragement to achieve structural stability by means of triangles ("What can you do to stabilize that?", "How are you going to get a little more stability?", "Are you going to have any sort of structure on the bridge?", and "You have to stabilize it somehow before testing it".) and despite the teacher's admonition not to use the glue gun ("Can you guys try and stay away from the glue gun?"). The two girls argued that they did not need triangles. They could achieve structural stability by other means. The joints they had constructed using glue guns and a laminating technique made their bridge strong and stable. I could further-more ascertain during their first bridge project that Chris and Arlene had reached Level 1 competence in their triangle-related discursive and material practices; they had included several bracing techniques during their previous tower project. In both projects, they planned the use of braces, and built them into their emerging design. Because of unrelated problems, they decided after several days to abandon their first design and to build a new bridge from newspaper.

I classified both explicit use as well as explicit non-use as triangle practice at the highest competence level. Andy and Tom had used triangles as *a priori* design features, critiqued their designs in terms of the presence of these as building blocks, and implemented them from the very beginning of their actual constructions. Arlene and Chris had planned their first bridge on the basis of triangular configurations. They explained the lack of triangles in their final project by the presence of other features in the design and materials that made triangular configurations superfluous.

7.1.4. *Actor Network Approach to Changing Discourse Practices*

The process of learning to design stable and strong structures by means of triangular configurations of building materials shared similarities with that of becoming competent in the material practice of using the glue gun. In both processes, members appropriated new practices and developed competence in them by actually participating in activities. Both trajectories were characterized by the decreasing amount of support newcomers needed for participating in the practice. The two trajectories differed, however, in the driving force of the transformation. The collective transformation of the material practice involving glue guns was by and large student-driven. The collective transformation of the discursive practice around triangles was related to the effort both teachers spent to scaffold students' actions.

The two transformations and, therefore, the individual trajectories differed. The transformation of the collectivity in terms of its discursive practice around triangles was relatively slow. After six weeks, there were four groups who had achieved the highest level of competence. Six other groups also engaged in the practice, but only with considerable and continuing teacher assistance. The glue gun practice, as other material practices, was appropriated much more quickly by the newcomers, so that the community as a whole transformed much more quickly. That is, the network of glue gun users was much more rapidly enlarged and the practice more rapidly circulated than the network of the discursive practice.

But one can distinguish even for the notion of triangle a heterogeneous development. The notion of triangle as a mere fact (resource), was recognized by students rather quickly. When cued with questions such as "What does it take to make things stable?" or "What is the magic word?", most students responded appropriately within two lessons. The triangle word was circulated rather quickly as a fact. But the associated design practice was not circulated at the same rate. Students became rather quickly triangle-sayers while they took much longer trajectories to become triangle-users.

This difference between triangle-sayers and triangle-users expresses that between a resource (triangle as a fact) and a discursive practice (triangle as a practice). Repeating facts about stability and strength – students as triangle-sayers – and actually engaging in construction with the purpose of making stable and strong structures – students as triangle-users – are entirely different things. I believe that these differences reflect much of the current disjunction of school knowledge from its everyday use and the practice of "teaching concepts first and then having the learners apply them". Teachers are all too often satisfied to hear the triangle words, or see the suspension strings; and are little concerned whether students enact the practices indexed by these words. The fault does not rest entirely with teachers. Schools, grading systems, standardized examinations and the associated phenomenon of accountability do much to diminish learning to the memorization of facts (resources) and do too little to encourage the development of authentic material and discursive practices. My case study of the triangle showed that direct teaching had only limited success if one considers the amount of time actually spent.

In many ways, these observations remind me of those in another Grade 4 class (Newman *et al.*, 1989). There, children were taught the intersection scheme (finding all pairwise combinations of a collection of objects, substances) in a lesson similar to the first triangle lesson. However, when asked to complete a chemistry task that required testing all pairwise combinations, the children did not draw on the previously "taught" intersection scheme to organize their activity. The intersection scheme was not, as the teachers and researchers had hoped, a resource to children's future actions. Rather, if the intersection scheme was used at all, it emerged from the physical and social contingencies of each investigation and student group. Similarly, the children in my study appeared to develop competence rather quickly when they had a reason for designing. New material and discursive practices frequently emerged, without teacher help, from specific situations in which the children responded to problems they framed such as structural weakness, instability, and weak joints.

In my study, students traversed long trajectories before they used triangles as design features, and in some cases, students did not implement triangles without having better alternatives. When I took account at the end of the bridge construction, the distribution of competence can be interpreted as representing a community in transition, with members at various levels of competence. In contrast to the earlier described almost effortless circulation of resources and material practices, the circulation of the triangle practice cannot be understood independently from the teachers' efforts. Conversely, when the effort was less, the network necessary for the circulation of the practice did not come about. I observed such a situation with the notion of "catastrophic failure", which describes the engineering technique of bringing structures to their breaking point to study weaknesses in their structural designs. Despite teachers' repeated use of the words in whole-class and small-group situations and despite teachers' encouragement to test structures to their breaking point, the notion did not become part of the students' discourse and practice until the last week of the unit (see Chapter 4 for the extent to which this word entered the children's glossaries). "Triangle" and "catastrophic failure" were therefore resources, but resources that the children did not associate with the canonical material and discursive design practices these words refer to in adult communities of designers.

Thinking in terms of the actor network approach about the difficult transformation of, and resistance to change in, the classroom community affords both teacher and student perspectives. While for the teachers, some fixed curricular objectives were the goal of the unit, students' primary interests were the projects. To construct an alien, tower, bridge, or mega structure, the students enrolled various allies (materials, tools, peers, teachers, etc.) and in the process, transformed their practices. These transformations, in turn, affected future practices and changes thereof. From the students' perspective, the practices which were the teachers' main interest (making triangular braces to achieve structural strength, checking for weak spots in buildings by causing "catastrophic failures") were seldom constructed as useful allies in their projects.[9] Sometimes, students' emerging practices did the same or better work than the braces suggested by the teachers such as when Arlene and Chris constructed strong and stable structures based on laminated and hot-glued joints. For most children, however, the triangle practice was not an easily accessible and readily available resource and ally.

Here, the needs of practice, as defined in relation to the community's changing activity systems, were more important and appropriate regulators of learning than teachers' beliefs about the value and importance of bits of concept-related resources and practices. This is an important issue, for design activities are powerful contexts for learning exactly because they allow students to frame problems and solutions, develop authentic practices (meaningful in *this* activity), and define community-specific activity systems that regulate learning. Tammy and Gitte's enormous efforts to "teach" the triangles appeared, at times, out of place and heavy-handed.

From an actor network perspective, teachers can be viewed as powerful Machiavellian actors engaged in building networks; here, these are networks in which the discursive practice surrounding triangles are circulated and used by a large fraction of the community. The teachers tried to get everyone (that is, students in the classroom) to participate in some practice. The role of the network-building actor can be expressed in an advertising metaphor that includes the notions of *recruitment, enrollment, attachment*, and *intéressement*.[10] In this metaphor, teaching is like advertising intended to expand the network of loyal customers. To expand the network, teachers attempt to recruit and enroll students into the network and encourage them to "become attached" to their discursive practice. The teachers had to capture students' interest to use triangles which they attempted to do by pushing or squeezing the children's structures to make structural weakness and instability observable. Once children were dissatisfied, teachers probed for the word "triangle" as an aspect of offering a better solution. The notion "intéressement" captures this phenomenon. Literally translated, intéressement means

---

[9] How the children's artifacts emerge as products of the interaction of multiple, heterogeneous, human and non-human actors is the topic of Chapter 8.

[10] These are my noun versions of the French verbs "recruter", "enrôler", and "fidéliser" used by Latour (1992). I developed this advertising metaphor in the context of another study about the emergence of discursive practices in high school physics classrooms (Roth, 1995a). Intéressement, one of the notions Latour (1992) introduced to the sociological literature remains untranslated in the Anglosaxon literature in science and technology studies (e.g., Fujimura, 1992; Star, 1991).

to get company personnel interested in the well-being of their organization through the allocation of a bonus. Getting students to "buy into" a point of view may be a good way of translating this term. In our present case, teachers did everything to sell the children on triangles so they would continue using them and thereby demonstrate mastery of the practice. In this way, teaching bears resemblance with advertising.[11] Gitte and Tammy's insistence on the "magic" word bordered on the kind of brainwashing one can experience while watching big sports events on TV (like Olympic games) where the same advertisements are repeated over and over again, ad nauseam.

7.2. LEARNING TO TELL ENGINEERING DESIGN STORIES

In knowledge-building communities, newcomers learn to tell their biographies and stories of their work in community-specific ways (Lave & Wenger, 1991; Orr, 1990). For example, the trajectories of newcomers to Alcoholics Anonymous are constituted by an increasing likeness between the newcomers' histories of their diseases and the histories of the disease told by old-timers. These trajectories are brought about and facilitated by participating in general meetings where old-timers tell hour-long stories of their previous lives as alcoholics. These stories are not ad hoc inventions but often develop over months and years. Newcomers also participate in smaller, discussion meetings which focus on aspects of what will be a reconstructed life story.

In the present study, the process of telling an engineering design story was also a trajectory described by increasing competence in accounting for a design artifact and the history of its making. However, in contrast to the Alcoholics Anonymous example, a central feature of learning to tell an engineering story included question-answer sequences with the teachers, particularly Gitte.

Questioning is an important aspect of teachers' daily practices. However, despite the prevalence and importance of questioning practices, fine-grained analyses that construct details of these practices are rare and many open questions remain. I conducted the analyses presented in this section to understand teacher questioning in the context of an innovative child-centered engineering curriculum. There were a number of indicators that the learning described in this book was, in part, precipitated by Gitte's skillful questioning techniques. These were increasingly adopted by Tammy as the course progressed (Roth, 1996d, 1998). These questioning techniques constituted an important practice that scaffolded students' accounts of their work, and therefore, the resulting stories about their design processes. This and the following section of Chapter 7 present evidence for the increasing com-

---

[11] This is particularly true for conceptual change teaching. Conceptual change teaching works on the assumption that students will buy into a new discourse when they are dissatisfied with their old one, and when this new discourse fills their needs for understanding.

petence of students to engage in telling engineering stories and talking engineering design.

Gitte's questioning reveals an interaction of questioning sequence and
question content. Her questions were nested such that questions related to
knowledge about the final design product, resulted in students' talk about
materials, engineering techniques, and issues related to the development
and testing of the artifact. Thus, Gitte may have asked a question about
performance that led students to talk about design concepts: "What happens
if you have an earthquake?" or "Have you taken into account the idea
that there might be an earthquake?" These performance-related questions
embedded other questions about design concepts, driven by Gitte's concern
that students report their work by means of an engineering discourse about
structure-stabilizing configurations. Thus, the question implicitly asked for
design principles, although it was couched within a context pertaining
to more global issues of design. Gitte also used consumer-producer relationships to couch questions and thereby draw out children's knowledge related
to any topic including new ideas: "Why should I be interested to buy your
structure?", "What might be a benefit to me if I bought this kind of tower?"
or "What is the selling feature here?" These questions embedded more
specific questions related to operating performance and ultimately, design
practice and material properties. In the following episode, Gitte encouraged
Andy, Simon, and Tim to construct an account of the outstanding features
of their tower. The question was phrased to elicit feature that stand out from
the students' point of view rather than the ones Gitte might have addressed.[12]

01 Gitte: OK, I'm a fellow engineer who might want to buy this tower.
02   How can you sell it to me? (.) What are the features that I
03   would want to buy this one (.) rather than someone else's?
04 Simon: Earthquake-proof
05 Andy: [See how you can work it.
06 Gitte: How do I know it is earthquake-proof?
07 Tim: Because if you go like this (.) it won't fall down.
08   [((shakes desk))
09 Gitte: Is that right?
10 Simon: Yeah
11 Tim: [We don't know yet because we haven't tested it yet
12 Simon: [we haven't tested it yet
13 Gitte: OK
14 Tim: [Because it wasn't strong=
15 Gitte: =What else might you sell me on this?

Gitte opened the interaction with questions that allowed a contingent query.

---

[12] In this way, students' talk was "concernful" (Heidegger, 1977) in that they talked about
issues relevant to, and understood by, them.

Her question as to the selling feature of the tower allowed students to specify the current topic of talk. Gitte used Simon's response as a spring board for further questions about the engineering design aspects (line 06). Simon had claimed that the tower was earthquake-proof. Gitte's utterance can be heard as questioning the answer and challenging the group to justify this claim. Such justifications are potentially in engineering terms, so that Gitte could achieve her purpose. However, Tim's initial response to this second question (line 07) was not about the tower's structural features but about the possible outcome of an earthquake test which he hinted at by shaking the table (line 08). However, he addressed the existing gap between the possible and actual tests in his subsequent two utterances where he pointed out that the test was not conducted (line 11) because of some structural weakness (line 14). Rather than following up on the issue of the structural weakness, Gitte elicited from the three boys further outstanding features of the tower. Because Gitte and the boys had already talked about the structural problems earlier in the conversation, such a topical shift constituted a reasonable move by the teacher.

Gitte's opening questions initiated a conversational topic and requested information about important features of the design as they were identified by the children. As such, Gitte could not know in advance what the answer would be. Rather, she used the students' first responses as a platform for further inquiry. Gitte followed up by requesting an elaboration of the initial answers. For this reason, I termed this form of questioning technique "contingent query" (Roth, 1996e). Gitte could not know students' responses but framed later questions in terms of previous answers. Her questions implicitly encouraged students to structure their answers in terms of knowledge relating to the properties of materials or design concepts. Consistent with her interest in developing children's discursive practices, Gitte frequently asked the students to name the items. In this way, technical items which students made available in their discourse in an indexical manner – by pointing, uttering "this" and "that", or gesturing – found an equivalent in children's discourse. As with the initial request, Gitte could not know the answers to her questions, but appeared genuinely interested in finding out about children's work.

When the students had exhausted their topics of interest, Gitte began to point out specific features in the structures and relate her requests to them (much in line with her intent to have children learn about specific engineering techniques). In this way, she allowed students to talk about and elaborate their structures. Her questions scaffolded students' efforts so that a complete account (or at least a sufficiently elaborate one) could emerge. For example, she asked Andy, Simon, and Tim questions such as, "I like the lightning rod on the top. What is its purpose?", "Your tower has a leaning effect. What is the purpose of that?", or simply, "What's that?" At other times, she directly addressed the feature of interest, "How do you think about these little feet (stabilizers)?" and "I am interested to know why you put [braces] going that way like that instead of just this?" In this way, Gitte made sure that the children's accounts covered a wide range of topics and therefore, developed their engineering design discourse.

One question sequence in particular was so successful that first Tammy, and then the children themselves began to use it. At first, it prompted students to talk about one of the problems they had encountered. As a follow up, students were asked about how they resolved the problem for which they had just provided an account.

16	Gitte:	Any problems so far?
17	Tim:	One. We had, we tried to make it just from pins. And it didn't
18		work and we ended up it was flopping around. And so we had
19		to, had to go back to square one.
20	Gitte:	And what did you do to solve it?
21	Tim:	We put pins for the columns and tape for the beams, so it
22		would be stronger.

After Gitte had encouraged the three boys to specify one of the problems encountered during tower design, Tim talked about the joints constructed from pins which were not very stable (lines 17–19). Gitte's follow up question then afforded Tim to provide an account of the solution (lines 21–22). Later, students began to ask questions in the same way: "What was your problem and how did you solve it?" These developments ultimately led the presenting students to address problems and their solutions without being prompted. Thus, the story that accompanied the bridge Jeff and John had designed included the following account of a problem:

We only used glue, thumbtacks, and tape for the joints. Our difficulties were to strengthen it. We solved that problem by using braces. Luckily we didn't get any catastrophic failure.

Gitte suggested that distributed knowledge would be integrated by children when they could share their experiences in whole-class sessions. Thus, she specifically planned show-and-tell sessions so that material and discursive practices of individual students could be made public and thereby appropriated by others. She planned these sessions in such a way that by calling on specific students, the presentation and surrounding discussion focused on those specific engineering techniques that she had previously identified. Her questions in whole-class sessions served to "pull out" significant experiences and point out similarities in experiences across groups ("I have, at the end of these activities, drawn out those techniques over and over again").

Students' contributions to conversations became increasingly independent of the teachers' scaffolding-questions which led to sustained student-centered conversations. This was especially apparent during whole-class discussions. Here, an additional element was introduced to the conversations: student questions. These lengthened the exchanges which were not simply responses to teacher questions, but self-sustaining conversations to which Gitte and

Tammy contributed on an equal footing with students.[13] An example of sustained student-centered discussion is featured in the following episode. This episode was part of the whole-class session in which Andy, Simon, and Tim presented their tower (Figure 3, p. 166). (This is the account the three students provided. My account of the tower's development and its intermediate stages appears in Chapter 8.)

Andy began this sequence with his account of the solution to a stability problem they had encountered. At this point, Week 6 of the unit, there is still considerable scaffolding observable. (The numbers in double brackets, e.g., ([9]) serve to index those points of the tower on page 166 that students touched or pointed to in the course of the conversation.)

01	Andy:	We had a new idea of how to make it more stable, because we
02		have this little thing here. And we put this thing across, because
03		then it would bend sort of, and these ([1]) two things are pinned
04		in, so it makes it so it can't bend and stuff like that.
05	Tammy:	It's way more stable.
06	Tim:	Like the shape, we were trying to make it stable and trying to
07		get the least amount of straws to make it. And so we used that
		idea for this ([2]) part.
08		'Cause it was really, it moved a lot, it ([3]) would bend.
09	Tammy:	The top would be more unstable than the bottom was, I remem-
10		ber that it was really unstable.
11	Tim:	And now, it won't sway as much.
12	Andy:	And we also have lots of cubes right here ([4]) and some
13		triangles that help it to be stable.
14		(2.6)
15	Tammy:	What did you find difficult?

Before Tammy asked her question (line 15), Andy and Tim talked about the process of tower construction interrupted only by two commentaries (lines 05 and 09–10). The core issue in this first part of their presentation was stability. Tim's comment about using the least amount of straws (line 06–07) resonates the instructions that the teacher had provided for the earlier activity of stabilizing two- and three-dimensional shapes. After Andy used the magic word triangle in the context of stability (line 12), there was a long pause (line 14). As Tammy could interpret this pause to mean that Andy and Tim could not continue their tower story, she provided another question to which the students could respond. This question therefore, from Tammy's perspective, allowed students to continue their account. It as a kind of scaffold to assist students to construct a more elaborate account.

---

[13] Gitte's interactions with students cannot be studied independently because of Tammy's presence. However, Tammy had shifted to similar patterns of interactions due to their collaborative efforts to improve questioning.

15 Tammy: What did you find difficult?
16 Andy: This part ([5]).
17 Tim: [This part ([5]).
18 Gitte: Why was that?
19 Tim: Because it would break easily, trying to put some pieces on
20   'cause it was really fragile when we started to make it. Pieces
21   would fall apart. Like they would fall off.
22 Andy: And at the start, we used pins and it would keep on going, it
23   would keep on moving.
24 Tim: [sideways]
25 Gitte: And how did you change that? What did you do?
26 Tim: We used tape, a tiny bit of tape for a few corners and
27 Andy: [Tape
28 Tim: we put pins to attach them together (.) when we put them on.
29 Andy: 'cause when we attached them with tape, it would always just
             come apart.
30 Gitte: Interesting.
31 Tammy: So you made your joints from a mobile joint to a stable joint.
32 Tim: [stable [Yeah
33 Gitte: That's interesting, your's seems to be one of the few structures
34   that managed to survive over a period of two weeks.
35 Tim: Well, it broke last week and we had to fix it.
36 Andy: But it wasn't too broke.
37 Simon: It wasn't that broke.
38 Andy: 'Cause all this ([5]) wasn't there but it was the bottom ([6])
39   part that collapsed, that's when we made that ([2]) part.

In the context of this conversation with the tower directly available to be
pointed at, Tammy's question (line 15) could be answered with a laconic
"this part" (lines 16–17). It took Gitte's follow up question to get a longer
explanation from Tim (lines 19–21). Whereas the earlier part of the conversation was about stability, Tim and Andy elaborated here the problems they
had framed about joints. Gitte's question (line 25) constituted another scaffold for extending the explanation Andy and Tim had begun (lines 26–29).
Gitte's comment about the extraordinary stability of their tower encouraged
the three boys to elaborate even further on the emergence of their tower,
especially what they did after the initial design had collapsed. On the other
hand, Tammy reiterated the link between stability and the joints by insisting
that the mobile joint was changed into a stable joint (line 31).

It is important to note that the presence of the tower artifact allowed
many indexicals into the discourse without creating misunderstandings. For
example, Andy pointed to three distinct parts in their tower (lines 38–39)
while indicating which part did not exist, collapsed, or was made after their
tower had come apart. Also, in response to Tammy's question about their

difficulties, both Andy and Tim simply pointed to a place in the tower uttering "this part".

Gitte all of a sudden shifted the topic (line 40). She identified in the issue of using the least number of straws (lines 06–07), one of those moments where she could link children's ideas, discourse, or design practice to canonical engineering. As she usually described it, she was "drawing out the engineering part". However, with the hindsight of the analyst, there was not enough evidence that the purpose of Tim's utterance was that of canonical engineering. Rather, as earlier indicated, his comment may have referred to the lesson about stabilizing about two weeks earlier. Nevertheless, Tim's utterance provided Gitte with a reason to talk about constraints engineers face when they design (lines 40–44).

40	Gitte:	I really like that idea in the way you that you're talking about
41		using the minimum number of straws to make it stable, that's
42		really important, because that's a really critical engineering
43		technique, not using too much material to make it too heavy.
44		So, that's something you should be thinking about when you
		are doing the bridges.

It is not clear whether and to what extent these comments were constructed as relevant by the students. In the present case, Simon continued the description of what they had done when the tower had come apart (line 45–46). From a pragmatic perspective of talk as situated action, this means that Gitte's comment had little effect on the conversation.

45	Simon:	I was trying to put these ([7]) parts back together. There are
46		pins in here so I had to take them out and Tim had to put
		some glue in here.
47	Tim:	And it wrecked a bit so we had to start all over again.
48	Gitte:	How did it wreck it?
49	Andy:	And because
50	Tim:	[Because when you would take a pin out, it would move
51		and pull another apart and we had to start again to make it.
52	Andy:	We cut, to make these ([9]) things. We cut a little bit of straw
53	Tim:	[The tape is too short
		([10])
54	Andy:	and then we opened it more so we could put that last straw in
55		and then we glued it to make it more.
56	Tim:	So it wouldn't come apart again.
57	Andy:	Come off.
58		(7.5)
59		I think that's all.

The entire sequence was about what the students had done during their reconstruction of the tower. One problem they had framed connected the long beams that were not as solid as Tim had previously thought. Readers will recall that Tim had developed a special joint for his long beams after Tom refused to glue them (see Figure 1a, p. 134). Simon suggested here that they removed the pins and, when they put the long beams back together, glued them. Here, they used the technique Tim had already proposed to Tom (Figure 3, insert, p. 166) – which provides us with yet another account that joint constructions changed when glue guns were available.

It is not certain whether the boys needed Gitte's question (line 48) to continue their account of the reconstruction work on their tower. She had posed it without waiting for the boys to continue their account on their own. A different situation occurred, when there was a very long pause in which nobody spoke (line 58). A pause of 7.5 seconds, in a conversational situation, is a moment of silence considered very long.[14] Andy's comment "I think that's all" provided a definite marker that they had ended. This marker was necessary because the pause between two utterances normal for a Western culture was overextended. At this point, Gitte asked another question. This question then, as others before, provided another scaffold so that the boys could extend their tower story.

60 Gitte:	I just want to ask one more question if you don't mind. I'm
61	interested in how you came up with the idea to stabilize the base in the way you did?
62 Tim:	We just wanted to add something on (.) so we just did something
63 Gitte:	[Was that the first thing you did or did you have to try a couple
64	of things?
65 Tim:	This ([2]) was the last thing we did just before it was time to clean up.
66 Simon:	[Something broke]
67	So we had to put it back together.
68 Tim:	We put this ([8]) on first and then we put this ([2]) on for fun.
69 Gitte:	What an incredible job to make it stable.
70 Andy:	It took us a long time 'cause Tim's glue gun got stuck in, and
71	we couldn't get the glue out.
72 Simon:	It was like a rusty bullet.
73 Andy:	A rusty bullet and no glue was coming out.

Gitte's questions (lines 60–61 and 63–64) now changed from being about the material basis of designing to being about the process of designing, that is,

---

[14] Readers will recall the research on wait time done in the late 1970s and early 1980s (e.g., Tobin, 1984). This research showed that teachers left much less than one second for students to respond to a question and uttered the next question without leaving time to elaborate the content of the first one.

her questions were from the category knowledge about knowledge in the typology of engineering design knowledge (see Table I in Chapter 4). Gitte wanted to know how students came up with an idea (lines 60–61) and whether they arrived at a solution to their problem immediately or after several attempts (line 63–64). The presentation came to its end. Andy and Simon briefly described a problem they had with the glue gun when no glue came from its nozzle.

The official presentation was now concluded. In the early parts of the unit, both teachers invited the comments of other students only after such official accounts were completed. Gitte therefore invited comments which, in the present situation, were quite brief.

74 Gitte: Do you have any comments or questions for this group?
75   (5.4)
76 Tim: Maggy?
77 Maggy: I like the way you put the flag and the triangles on the top.
78 Tim: Thanks.
79 Gitte: How did you come up with the idea of the flag?
80 Tim: Decoration.
81 Andy: All the, and all the other people have it like, the flags.
82   [((looks around and to other towers))
83 Gitte: Seemed like the thing to do.

Maggy's comment is representative of many interactions between students during whole-class sessions, especially in the beginning of the unit (line 77). Rather than engaging in a constructive critique of aspects that could be improved, students most frequently indicated what they liked about the presented artifact. Of course, this form of interaction cannot be understood independently of Tammy's concern that children sometimes put down each other. So rather than providing for opportunities which could have encouraged analysis and therefore engaged students in learning about engineering principles, the "positive comments" often related to issues unrelated to structural engineering. Here, Maggy mentioned the flag and the triangles.[15] However, over the course of the "Engineering for Children: Structures" unit, more design-related questions were asked, and students asked more questions of their own. The whole-class conversation in the next section contrasts what happened in the present conversation. The responses to Gitte's question about the origin of the flag are interesting from the perspective of the issues raised in Chapter 5. Here, Tim and Andy provided two different explanations for their flag. Andy's response, while not directly attributing to everyone else, strongly suggests that this was, as Gitte described, "the thing

---

[15] Given that she did not include triangles by design in the tower or the subsequent bridge project, it is likely that she was not yet competent in the triangle discursive and material practices as they relate to structural issues.

to do". Tim, on the other hand, eluded the question regarding the origin and provided a more basic rationale for including a flag; they had produced the flag for decorative purposes (line 80).

It is interesting that more than half of the time spent talking over the project of Andy, Simon, and Tim was concerned with what they did with the partially collapsed tower after the lesson proper was already completed. The story that emerged here was that of the tower's reconstruction and the practices the boys engaged to stabilize and strengthen it. This is an important design lesson which the three boys learned only after the officially scheduled time for the tower project.

Throughout the conversation, the artifact served as a referent that afforded pointing and gesturing. In this way, even in the absence of a complete engineering discourse that would have allowed them to label each part, students could talk about their design using indexical terms and still be understood. In itself, the absence of words is not a negative aspect of these conversations, for the discourse of engineers and scientists is equally laden with indexicals. In fact, there is evidence that in many situations, scientists and engineers are dependent on the presence of drawings and artifacts to allow constructive and worthwhile conversations to happen (Henderson, 1991; Knorr-Cetina & Amann, 1990). Here, the presence of the artifact allowed tentative accounts to emerge in which indexicals featured a lot; as shown in the following sections, teachers or other students contributed new items for their common engineering discourse that could take the place of an indexical.

Ultimately, this transcription of a whole-class conversation showed that two aspects of the setting were important to the emergence of the students' stories about the artifacts they created. First, the artifacts where immediately present to be talked about and referred to. Any ambiguities that arose could be dealt with by using the artifact as a referent for pointing and gesturing actions. Second, the teachers asked questions in such a way that students could tell their stories. Because these stories were initially rather short, the teachers' questions provided opportunities for students to extend their stories. These questions therefore served as scaffolds that allowed engineering design stories to emerge.

7.3. ENGINEERING DESIGN CONVERSATIONS

Whole class discussions such as that featured in the previous section provided many opportunities for developing an engineering-related language game. In particular, the verbal and visual presentations of student-produced artifacts served as occasions to stimulate such interactions. First, student presenters talked about their design and specific problems and solutions they had framed. Then, questions, comments, and suggestions by peers and teachers created a self-sustaining conversation about engineering topics, and a context for framing and solving structural engineering problems. During these occasions, students, teachers, and visitors introduced new aspects to the existing discourse or they pointed out new phenomena or engineering techniques, extending the current discourse. In this way, the classroom community sustained and modified a public engineering discourse. Student-produced design artifacts served in important ways to anchor these conversations and to provide topical cohesion. Finally, students had opportunities to integrate personal experiences into ongoing conversations. Using excerpts from whole-class conversations that occurred during Week 11, I illustrate here four dimensions that were typical of classroom conversations throughout the unit (presenting the artifact, extending language games, using artifacts as conversational anchors, integrating personal experience, and sustaining student-centered discussions). At this time, students had already developed remarkable competencies in accounting for what they had done.

7.3.1. *Presenting the Artifact*

Whole class conversations often began after students presented their design artifact to the class. For example, Tim and Stan presented their bridge during Week 8 (the design is featured in Figure 1c, p. 156).

01 Stan:	OK, we started out with just cardboard, tape, string and skew-
02	ers. I said, "what can we do with this?" Tim said, "we can
03	start with the base". So we took 3 piers and joined them to a
04	piece of cardboard. Then I said, "we should start the platform".
05	When the platform was finished, we glued it onto the piers.
06 Tim:	I said, "it's too weak". So, we took the skewers and made an
07	X-shape in between the piers. That made it very strong. Then,
08	we made a top part and then we made railings. That is how
	we made this awesome bridge.

Here, Tim and Stan described their project to the others in the classroom. By that time, descriptions like *pier*, *platform*, and *railings* had already been in common use for some time. They also introduced *X-shape*, a technique used to stabilize a rectangular shape with two diagonal supports forming an "X". This was a significant event because it was here that Tim and Stan introduced a new element to the classroom community's discursive and material practices. Readers will recall from Chapter 4 the considerable number of entries about the X-shape in students' glossaries (pp. 85–87). These are evidence that Tim and Stan's invention had brought about a change in the common design discourse. This presentation also provided opportunities for the teacher to introduce a new label for a technique used by the students but which Stan and Tim had simply called *top part*. That is, teachers could assist students in extending their language game. Their notion of X-shape was circulated in the community. It has to be further underscored, however, that there were few opportunities for students to engage in the

material practice related to the X-shape because there was only one project
that followed Tim and Stan's presentation. Although their glossary entries
were evidence that they used the words appropriately in writing, this is not
enough evidence to conclude that students are competent X-shape designers.

7.3.2. *Extending Language Games*

After a presentation, the artifacts and verbal descriptions formed a ground
for subsequent interactions. There were comments, suggestions for improvement, and questions, each leading to further elaborations of the details in
design, engineering techniques, and the problems students framed. As the
conversation evolved, students and teachers constructed the need for naming
objects, events, or techniques and proposed either their own labels (such as
X-shape or *triangle*) or imported them from other discourse communities
(which students experienced in the science museum or by listening to presentations of, and talking in the context of their own work to, the visiting
engineers). In this way, then, the existing engineering design discourse was
extended to include new descriptions of artifacts, construction practices, and
experiences in this classroom. New aspects of the ever-increasing engineering
design discourse were circulated in the network.

About three minutes after Tim's and Stan's formal presentation and some
first comments, Tammy asked all students whether the presenters had used
acknowledged engineering techniques. Tim was the first to respond.

09 Tim: Some of it is held up by the top, like it is sort of braced from
10   the X on the bottom and it is.
11 Tammy: It's called suspended, held up by the top is called suspended.
12   And that's why you have suspension bridges.
13 Tim: We were going to make one because we thought it would be
14   stronger if it put like this instead of rope.
15 Tammy: And what would you attach the rope to?
16 Tim: Here and up to here and attach it up like that.
17   [((Points to bottom of the bridge)
18   [((Points to top of middle pier))
19 Tammy: Kathy, isn't that similar to what Jane and you made; in the
20   idea not in the actual material used? Doesn't that remind you
21   of yours the way the triangles go on top to keep the bridge
            from on top?

Following Tammy's call for other engineering techniques, Tim talked about
his top which, like the *X-shape*, provided support to the bridge (line 09).
Tammy interpreted Tim's utterance as the search for a name and she supplied
the terms *suspended* and *suspension* (lines 11–12). Tim subsequently described how to add string (linking the supporting cable with vertical cables

to the platform) to make the bridge stronger so that it looked even more like a suspension bridge (lines 16–18). Tammy took Tim's description as an opportunity to provide a link to Kathy and Jane's bridge which showed a similar feature (suspension of platform). In this way, Tammy highlighted the potential need for including *suspension* in the shared language. Evidence presented in Chapter 4 illustrates that other students appropriated these new aspects of engineering discourse. For example, Kitty was one of the 11 students who described in the glossary her understanding of *suspension*. Andy also produced a sketch accompanying *pier* which looked very much like Tim and Stan's bridge, including the higher middle pier and the cables for suspending the bridge platform.

During Stan and Tim's presentation, Jeff volunteered two other words relevant to the construction. There was *tension* in the strings (cables) and, when Stan and Tim tested their bridge, the piers were *compressed*. As we saw in Chapter 4, *tension* and *compression* became integral parts of the children's language games. Compression and tension were important items in students' glossaries appearing 22 and 21 times, respectively.

Characteristically, Tammy and Gitte generally did not "force" students to include new concepts in their language. The earlier described cases of triangles and catastrophic failure were notable exceptions. Rather, students could include them when they felt a need for them. Both teachers wanted to develop student-based expansions of the engineering-related language game rather than to inculcate standard definitions. Gitte commented:

Gitte: We didn't give them that vocabulary. The vocabulary came from the kids. But the way it came was through questioning and finding their own voice to describe what they have done. By sharing with each other they came up with the words stay, beam, column, reinforce.

As we have seen in the first part of this chapter, both teachers found out that in the few cases where they "pushed" a particular concept (such as "triangle" or "catastrophic failure"), it could take a considerable amount of time and effort to bring about the desired change: children simply had not identified the need for changing their language game in the direction favored by the teacher. It is important to note that this form of teaching draws heavily on students' informal learning outside schools. The contribution of the teachers in the present study is then to be seen as highlighting and reflecting on those aspects of children's everyday language games that are suitable and viable in an engineering context. "Drawing out" therefore is a way of relying on the cultural resources and practices which the children appropriated elsewhere. The "discovery" rests in marking something students are already doing as significant rather than finding something new or on their own. It is a discovery in students' own enacting cultural knowledge. This may lie at the origin of the phenomenon that students took little ownership for words and discourse (which are re-productions), but more for the artifacts that they designed (which are productions).

7.3.3. *Using Artifacts as Conversational Anchors*

The design language games did not change instantaneously. Thus, although teachers may have suggested a way of talking about an engineering technique, parts of a structure, or a tool, this form of discourse did not become immediately part of the students' common language. I already presented evidence for the teachers' arduous work to bring about change in the students' discursive and material practices related to the use of braces; although the teachers used various teaching techniques to get children to include braces in their designs, only a small number of them actually did so spontaneously and without prompting. Nevertheless, the presence of artifacts allowed students to communicate sufficiently by pointing and gesturing in various ways.

In the above situation, Tammy simply introduced "suspension". About three minutes later, Tim had another opportunity to talk about this aspect of the presented bridge. Responding to Sandy's question, "What difficulties did you have and how did you solve them?" Tim responded,

22 Tim:	We had difficulties with the X right here because we used the
23	glue gun. But it was too skinny to hold so we had to use the
24	tape. And we had troubles here making it even and we had
25	some problems putting in the top part because Stan had to hold
	this and that and I glued here and there.

In answer to Sandy's question, Tim described the problems Stan and he had framed in the course of construction. However, rather than talking about how they had to *suspend* the platform from the string, he referred to the *top part*, *this*, *that*, and *it*. Here, one can see the crucial importance of the design artifact in the construction of a suitable language game about structures, designing, and constructing. Because they still had available their own descriptions and the indexicals which they used in conjunction with pointing and gesturing to explain and elaborate their ideas, the extension of their language game by including new elements was allowed to come about on a need-to-know and just-in-time fashion.[16] This should not be used as evidence against the claim that students developed and appropriated a complex engineering design language. Even scientists' laboratory and engineers' shop talk depend in fundamental ways on the support of artifacts; there is extensive evidence that laboratory and shop talk makes heavy use of indexicals. The presence of the artifacts simply allowed students to use indexicals, even during presentation, until they had appropriated a formal, and in this class-

---

[16] I have described and theorized elsewhere the important function of various artifacts to developing a common language (Roth, 1995a, 1996f; Roth & Roychoudhury, 1992).

room accepted, language game about specific issues. Hence my use of "conversations over artifacts" that necessitate the presence of the referent so that "top part", "this", "that", and "it" become meaningful. I will return to this issue in Chapter 8 with detailed observational and theoretical descriptions of the role of artifacts in design, cognition, and discourse.

7.3.4. *Integrating Personal Experiences, Classroom Discourse, and Formal Engineering*

Students and teachers linked conversations to personal experience outside school (bridges and towers they knew from personal experience), and to common experiences including the presentations by visiting engineers, the visit to the science museum, or the films. In this way, the different aspects of the curriculum were not separate from each other, but were linked through their inclusion in the classroom engineering design language.

In the following excerpt, Gitte attempted to expand the context of Stan and Tim's bridge by linking the classroom discourse about foundations to the talk Karen, a visiting geotechnical engineer, had given about the same topic:

26 Gitte:	That part of the bottom, that cardboard base part, Karen talked
27	about that. She had made something similar when she was
28	working as an engineer. Does anyone know what she called
	that? It was a type of foundation.
29 Stan:	A base?
30 Gitte:	Think about it, it looks very similar to the object whose name
31	she gave it, it's a very difficult question.
32 Tim:	Is it a dam she was talking about?
33 Gitte:	The cofferdam? She made this thing in the cofferdam (2.7),
	Jeff?
34 Jeff:	That sheet that hooks together? The sheets?
35 Gitte:	That was one thing in the cofferdam for sure. What she was
36	referring to was a raft foundation. Remember that? It looks
37	like a raft that you float on.
38 Tim:	Like if you go to Kelowna, there is a bridge that floats on the
	water.
39 Tammy:	That's right.

While Gitte's questioning technique was here very traditional in its quest for specific words,[17] this conversation had the important function of tying the engineer's visit to aspects of children's designing. Here, Stan and Tim's bridge provided an opportunity to link children's activities and inventions to

---

[17] This was one of those situations in which Gitte wanted to elicit one specific word as she had done with "triangle" and "catastrophic failure".

those of engineers, thus supporting Gitte's effort in authenticating students' work. Once *raft foundation* (line 36) was uttered in the presence of an audience, Tim immediately contributed an example from his personal experience with the potential to change his language game. In the process, the children talked about *cofferdams* which was another important item in children's repertoire – as evidenced by its prominence in the glossaries. These *cofferdams* are part of an engineering technique which allows the construction of piers under water.

In this, the conversation also illustrates the dangers of seeking to elicit specific words from children. Here, Gitte did not realize the legitimacy of the children's talk about *foundations*, but persisted in the elicitation of *raft foundation*. This episode further illustrates that "formal" engineering talk was not always readily appropriated by the students. Here, this was not surprising because there were no opportunities for making the notion of raft foundation relevant in children's projects and therefore in their engineering-related talk. Underwater construction of piers was not one of their concerns so that there was little incentive to appropriate this discursive (and material) practice. Students did not have their own intentions which could have populated this form of engineering talk; they saw no need for this engineering talk and possibly had little inclinations to appropriate it.

Gitte's questioning for specific canonical descriptions was consistent with her belief that children spontaneously invented many engineering techniques, or knew about them but that it was her duty to "draw [this knowledge] out". By *drawing* this knowledge out over and over again, that is, by getting children to talk engineering, Gitte believed the children would eventually develop canonical engineering discourse. This *drawing* out is evident in the previous episode. Gitte tried to elicit specific answers – consistent with her pedagogical discourse that once this knowledge was made public in the classroom forum, it could be appropriated by all students – and evaluated children's responses in those terms. The initial remark left no doubt about the fact that the answers to her question could be found in a previous conversation ("That part . . . Karen talked about that" "Does anyone know what she called that?"). She also indicated that she knew the hidden answer to the question, which itself was "difficult" (line 31). Subsequent student responses were evaluated in terms of these still hidden solutions. Thus, it was clear that Tim's response (line 32) was not the one she wanted to hear, for it was not the cofferdam but "this thing" (line 33) she expected. Here Gitte elicited another response as a way of indicating that Tim had not provided the prefigured correct answer. In these cases, Gitte made her evaluations implicit in the indication of incompleteness in such comments as "Yes that's one, what's another one . . .?" or "Good, there is another technique that you are not getting yet?" On the other hand, Gitte felt justified because, as she argued, it was through such recitation that the students actually learned so much engineering design-related content knowledge in this class. Gitte was supported in this by Tammy who observed that the students in this study had learned so much more than those to whom she had previously taught parts of the unit.

7.3.5. *Sustaining Student-Centered Discussions*

Teacher scaffolding during small-group interactions and whole-class conversations led to increasingly independent student-student exchanges and increasingly subject-focused engineering design discussions. These student-centered discussions provided a great deal of opportunities for students to participate in talking and arguing engineering design, and thereby increase their discourse competence.[18] For an example, we return to a conversation from which excerpts were presented in Chapter 4. This conversation arose from the presentation of a bridge design by Jeff and John that sustained, by far, the greatest load. Through repeated testing, catastrophic failure, redesign, and the contingent emergence of strengthening techniques, Jeff and John had arrived at a bridge from straws that withstood a load of 347 wooden blocks suspended from the center (Figure 4). The following excerpt comes from the 13-minute presentation and discussion of the project, wherein

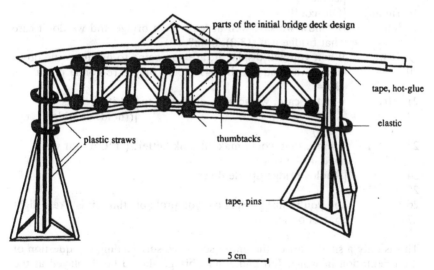

parts of the initial bridge deck design

tape, hot-glue

elastic

plastic straws

thumbtacks

tape, pins

5 cm

Fig. 4. The design Jeff and John proposed for a bridge that can withstand heavy loads. In a load test, this design supported 347 wooden blocks suspended from the center of the artifact.

---

[18] Although the opportunities should have been there for everybody, it was unfortunate that they were mostly taken by the boys. Despite their efforts, Gitte and Tammy could not significantly increase the participation rate of girls in these whole-class conversations. In this sense, the unit had failed, though both made every effort to spend more time with small groups of girls than with groups of boys.

several students expressed concerns with the procedure Jeff and Jon had
chosen to test their bridge (they placed it upside down).

01 Ron: Did you write anything in the description thing about (0.8)
02   about that it held that many, but upside down?
03 Jeff: No, we didn't write down 'upside down.'
04 Tammy: It would be an interesting thing to add. (1.4) Doug?
05 Doug: Is this a car or a people bridge?
06 Jeff: We don't know. We were just concerned about the bridge.
07   (2.5)
08   To us, it doesn't really matter if it is for people or cars.
09 John: Yeah, [before
10 Tammy: [It's the idea of a bridge, right?
11 Jeff: [Yeah
12 John: [Before we put the braces on,
13   we were gonna make people through there and cars go up on
14   top, but it wouldn't be strong enough, so we put the braces=
15 Jeff: ='Cause, 'cause like that it would be like an elevator to bring
              people up
16 Tammy: [Oh, cool]
17 Jeff: But now we say, well at least it's a bridge and we don't care
18   what bridge it is. (3.2). Stan?
19 Stan: When you will display it, are you going to put it upside down
20   or the way you got the 347 blocks or you're gonna put it? (.)
              The=
21 Jeff: =In the case, we actually are going to put it like this
22   [((holds bridge up-
              right))
23   I think, that would make it look better. (.) 'Cause if we
              have it like this
24   [((holds bridge upside down))
25
26   and you're just passing by, you probably think it is like a flight
27   control or something.

This is only a small part of the entire sequence surrounding the question of
the orientation in which Jeff and John's bridge should be displayed at the
local science museum. Throughout the presentation and discussion of the
project, students brought this topic up repeatedly. Here, both Ron (lines
01–02) and Stan (lines 19–20) were concerned about the discrepancy between
(a) claiming the bridge's strength when it was tested upside down and (b)
displaying it right side up during an exhibit in the local science museum.[19]

---

[19] Jeff's test in which the bridge was placed upside down had emerged as a contingent achievement during the series of tests he and John had conducted. As they increased the load, the tests

Tammy's utterances (lines 04, 10, and 16) always came during, what may have been interpreted by some participants as embarrassing or tense situations; their effect was to diffuse any problematic interpersonal situations. The important questions, however, were posed by students themselves (lines 01, 05, and 19). In terms of content, questions and answers addressed issues related to testing procedures, the design's purposes and user specifications.

In response to the question about his bridge's purpose, Jeff suggested that he was concerned about the bridge design (line 06) rather than its ultimate purpose for people or cars. He immediately reiterated this concern by accounting for the addition of braces in response to the earlier experienced weakness (lines 12–14). Doug's question (line 05) itself appeared to be a sideline in the questioning sequence by several students concerning the legitimacy of the testing procedure. Stan returned to this issue. He wanted to know whether Jeff would exhibit the bridge upside down or right side up when it was put on public display in the science museum. Whereas Jeff acknowledged that the bridge could be displayed either way, he suggested that in the upside down position, it left too much freedom for the interpretation of any onlooker – here, the shape of the bridge (see Figure 7b, p. 89) could be interpreted as a flight control unit.

Later, during the presentation, Ron returned to the question of the bridge piers ("legs") and their relation to the strength of the bridge. Again, the effect of his interrogation was that it questioned Jeff's claims about the strength and, in addition, details about the structure that would result in different strengths if the bridge design was tested right side up or upside down. Issues central to the design of a bridge to produce strength are the topic in this extended student-centered exchange.

28	Ron:	I got a question. If you took the legs off, if you cut them off
29		around the bottom of the bridge, do you think it would hold
30		the same as upside down?
31	Jeff:	Yeah.
32		(1.6)
33	Ron:	Is it build the same?=
34	Jeff:	=Well actually, if you cut the legs off, you might, you might
35		wanna try it putting it down like this (.) without the legs. See
		how strong it would be.
36		[((holds bridge upright so
37		that the deck rests on his hands))
38		(1.4)
39		You might try that

showed that the piers were the weakest part. The two, less concerned about the piers than about the bridge deck decided to turn the bridge upside down. As the deck was initially designed symmetrically (see Chapter 8), this procedure appeared feasible. The symmetry was broken in the later stages of the development when Jeff and John added the bent beams to the structure.

40 John: [It would still be strong, sorry
41 Jeff: But we don't think it would hold any more, 'cause
42   (1.4)
43 Ron: Is the bottom and the top built the same?
44 Jeff: The bottom and the top, yes. 'Cause originally this was it,
45   [((center of bridge deck))
46   and then we built this under the bottom, to make it stronger.
47   [((points to additional beams below deck))
48 John: Initially we put in this.
49   [((points to braces))

Jeff's response to the initial question (line 31) was read by Ron as another claim so that the latter followed up with another question. Here, Ron made a connection between "holding the same" (line 29) and being built the same (line 33). Jeff then provided the description of a possible test if the bridge held the same in both positions. That is, one can hear his statements, "You might wanna try it" and "See how strong it would be" as acknowledgments that he did not actually conduct the test that would support his earlier claim. John chimed in also asserting that the bridge would be as strong as in the orientation they tested it; Jeff added that it would not hold more. Ron was not satisfied with these responses, and queried whether the top and the bottom parts of the bridge deck were actually the same (line 43). Although the bridge did not appear to be symmetrical, which may have been at the source of Ron's questioning, Jeff maintained his claim that the bridge deck was designed symmetrically (line 44). He continued to explain that they added some straws below the deck, and John pointed out again the braces which they had included after the bridge collapsed during first test.

This conversation during Week 11 of the unit attests to the increasing complexity of the presentations. At issue here were basic principles of engineering design, the strength of a structure, and symmetry. Stan and Ron questioned the legitimacy of testing the bridge in an upside down orientation but presenting it right side up, which can be heard as a pretense that the record was brought about in this orientation. In the process, Ron questioned the symmetry of the design which provided Jeff and John with opportunities to point out how they achieved structural strength by means of reinforcements and braces. The conversation was authentic because driven by the goals of students. Here, they addressed issues that really interested and concerned them. The data provided here and in Chapter 4 showed that these conversations toward the end of the unit allowed canonical engineering and science to emerge. In addition to the engineering issues that were the focus of the above excerpts, the episodes presented in Chapter 4 showed how Jeff explained the strength in terms of the direction that stresses and strains acted in his bridge. Finally, and most importantly, the conversation no longer needed teachers' questions to be sustained over considerable time.

Here, we are able to see the emergence of a community whose members engaged in recurrent conversations and how they produced, re-produced, and circulated an ever evolving engineering discourse. Although there existed attempts on the part of the teachers to push certain issues despite students' reluctance, the learning conditions they fostered, by and large, facilitated the emergence of this community of engineering practice.

7.4. TEACHERS AS NETWORK BUILDERS

In the present setting, the most visible aspect of children's learning – besides their development of material practices – was their competence to talk about topics related to structural engineering design. This development of children's discourse was made possible primarily by the competent questioning techniques Gitte contributed to the community which Tammy adopted as the "Engineering for Children: Structures" unit evolved. The contingent queries which were frequent during initial stages of the projects began to fade as children's (hi)stories of their designs addressed an increasing number of engineering-related issues and thereby became more elaborate. These contingent queries functioned like those used in effective tutoring. In tutoring, questions are employed to coach students through a solution path.[20] The tutor queries students to assess their knowledge, confidence, understanding, and current status of the problem, or to focus students' attention on specific aspects of the problem text, their own writing, and so on.[21] These questions assist students' efforts in selecting and taking certain paths leading them to solutions, and thereby constitute scaffolds that permit (aid) task completion. In the above conversations with Andy/Simon/Tim and Stan/Tim, Gitte's questioning can be understood as scaffolding. Its purpose was to allow children to construct stories of and about engineering. During the initial stages of students' designs, questioning was more frequent, purposefully addressing many issues related to material and discursive practices of engineering (pertaining to both canonical practices and those idiosyncratic to this classroom community). This allowed students to develop those discursive competencies that overwhelmed the visiting elementary teachers (for their comments, see Chapter 4, p. 75).

Gitte decreased her support as students' accounts of their work and plans became longer and more complete. This fading was apparent from all transcripts and is also observable in the episodes used here (e.g., compare the early presentations of Andy/Simon/Tim and Simon/Tim with those featuring Jeff/John). Interacting with this fading support was associated with a change in the questions' content. Because students already talked about many issues

---

[20] A detailed analysis of question and answer sequences during tutoring was provided by McArthur et al. (1990).

[21] It is not self-evident that students should see phenomena and see them in canonical ways that would allow them to develop appropriate observational and theoretical descriptions (Roth, 1996f; Roth, et al., 1997a, b). In fact, considerable effort has to be expended so that students see the world and parse it into foreground and background similar to scientists (Roth, 1995a).

without being prompted by Gitte's questions, these questions were also from fewer domains of the typology of engineering design knowledge (see Table I in Chapter 4). This was especially evident from the whole-class discussions to which Gitte often contributed but a question or two; the remainder of critical issues were covered either by the presenters or addressed in questions by other students (such as the issues surrounding Jeff and John's controversial testing procedure).

The view of teacher questions that scaffold the networking process underscores the social nature of individual accounts, and, because of co-participation, the common engineering-related discourse. That is, questions constituted teachers' contributions to the social construction of the discourse and networks, and the circulation of stories. While the projects and the conversations over and about the artifacts arose from children's interests, the competence to provide longer and more detailed engineering-based accounts cannot be understood independently of teacher questioning. This does not mean that the teachers were the only or key contributors of new words for the common language game. Rather, many of the terms originated with the students and were therefore from the domain of everyday discourse. The teachers' work consisted in providing opportunities so that items originating with individual students could become resources for others and therefore could be circulated in the network. During whole-class conversations and with the teachers' insistent questioning, many of these student-generated terms were appropriated by an increasing number of members. That is, the teachers became network builders who did their best in enrolling members.

A content analysis of Gitte's questions underscores the extended nature of her engineering design-related discursive practice.[22] This discourse was not simply fact based, in which case she would have asked questions about the properties of materials and names of engineering techniques. Rather, it combined topics from five different domains of knowledge in engineering design related to materials, design practice, development and testing practice, performance of final products, and knowing and learning through design. In this, the subject matter covered included dimensions nearly identical to those that had been identified for innovative engineering designers, though the extent of the engineering discourse was admittedly much more limited (see Chapter 4, especially Table I). Gitte's experience as curriculum designer, teacher, and observer of many construction sites and interactions with engineers had enabled her to scaffold children's engineering design-related conversations so that they achieved great competence. On the basis of my data, the claim that good question techniques require a great deal of competence in the discursive practices of the subject matter domain appears justified.

At this time, it is worthwhile to return to the notions of doing it first, copying it, and everyone else is doing it. These were useful descriptors in

---

[22] I provided a detailed analysis of Gitte's questioning techniques, and particularly the content of her questions elsewhere (Roth, 1996e).

the context of student-produced knowledge, facts and artifacts, material practices, or discursive practices. I did not observe situations where these descriptors would fit when they concerned the practices and resources introduced and legitimized by the teachers (mostly discursive practices). Although the teachers believed that students originated the triangle practice (the case study illustrated that Ron and Jeff uttered "triangle" and Tammy flagged it as an important term), neither student ever claimed ownership in the same way they did for other resources and practices. This difference may reflect the varying degrees of ownership students felt over their learning, and resemble the differences between learning in-school and out-of-school settings. There are a number of ethnomathematical studies that indicate continuous and effortless learning as people participate in meaningful activity, structure their interactions with the world, and change their interactions according to local contingencies (e.g., Lave, 1988; Saxe, 1991). Similarly, learning appeared to be effortless in a Grade 8 classroom where students controlled the content of their activities (Roth & Bowen, 1995). In all of these settings, the agents were in control of their learning, which was always related to practice and goals with which they could identify. My observations in this study indicated such effortless learning when students participated in meaningful collective activity and in pursuits with which they were truly concerned. Because of the distributed nature of knowledge, different students could be inventors while others copied or followed trends. The triangle example showed that learning teacher-determined engineering practices was much more akin to traditional school learning. Here, the teachers wanted all students to learn what they had legitimized: a certain form of engineering practice. This in itself is not of negative value because the cultural reproduction of resources and practices has the potential to provide students with even more resources for dealing with design problems. However, in their concern for transmitting legitimized knowledge they become "the means by which representatives of academic disciplines constitute legitimate knowledge and what lies outside its boundaries" (Lave, 1993, p. 11). Ownership over learning has to do with the learner's status with respect to "legitimate knowledge". Here, students are no longer in control of their learning which lies in the teachers' hands. This lack of control may be associated with learning and with it teaching that is no longer effortless participation in changing practices.

A follow-up study which investigated cognition in a unit where students learned about simple machines by designing their own machines provided evidence that the issue of ownership is not related to who introduces and circulates some discourse (McGinn et al. 1995). My research team showed, for example, that although the notion of mechanical advantage was introduced by the teacher, words like advantage and disadvantage are current in everyday discourse. Advantages and disadvantages already were common aspects of the students' language games before the teacher made mechanical advantage the focus of the science class. Possibly because the shift in language from "advantage" to "mechanical advantage" is small – the associated mathematical discourse requires the comparison between effort and a reference load – students made the transition in discourse rather quickly. Even if students did not calculate mechanical advantage at first, they could easily assess whether their machine provided an "advantage" or a "disadvantage". With repeated use during whole-class conversations over and about students' models and the teacher-designed pulley configurations and transparencies, students quickly appropriated the new language game. The teacher had provided the classroom community with an opportunity to develop a "special network of recurrent conversation" (Winograd & Flores, 1987, p. 158). Across situations – small-group and whole-class conversations, design activities, and structured investigations – mechanical advantage became a central aspect of the language game about simple machines.

PART III

NETWORKING ACROSS INTERSTICES

In Part II, the transformation of a community was presented as network building and circulation of resources and practices. This process of network building is reflexive because the circulation and building go hand in hand. As new members appropriate a new discourse, they form a new node in the network. At the same time, they also begin to participate in the circulation of this discourse. Part III focuses on integrating and networking across two gaps often created by researchers disciplinary concentrations: the gulf between humans and non-human and the gulf between Self and Other. In Chapter 8, I show how knowing, designing, and acting have a material basis. In other words, I show how individuals acting in material contexts give rise to what we recognize as cognition. That is, I am shifting the unit of cognitive analysis to a holistic individual-acting-in-setting. In Chapter 9, I take a similar tack, but take a closer look at how individuals network into groups and into culture more generally.

8. NETWORKING HUMANS AND NON-HUMANS

In this chapter, I show how children's design processes were characterized by their *heterogeneous* nature wherein multiple elements including tools, materials, community standards, teacher-set constraints, current states of the design artifacts, individual preferences, and past discursive achievements were networked to allow the emergence of the design artifacts.[1] At any one moment, the design artifacts enfolded past achievements with affordances and constraints for future developments. Although individuals and groups identified themselves (and were recognized) as the creators of the resulting design artifacts, these artifacts embodied heterogeneous origins and were more than the sum total of individual resources, practices, and competence.

Tools, materials, community standards, teacher-set constraints, current states of the design artifacts, individual preferences, and past discursive achievements did not have any absolute ontology – that is, singular meanings and applications – but were *interpretively flexible*. In this environment, children learned to recognize and exploit in productive ways the interpretive flexibility of materials, discourse, tools, design artifacts, and teacher-set constraints. Rather than being stifled by the open-ended and ill-structured situation of designing, students used the flexibility of their situation in creative ways. Problems did not exist in any absolute sense but emerged as interactive achievements. As a consequence of this interpretive flexibility, the situational choices made by the principal actors (members of a design group) appeared to be inherently unpredictable.

The students' discursive and material activities were over, about, on and with the emerging artifacts. The students' discourse was about, and their practical activities were directed toward, the artifact; the activities were over and with the artifacts in that artifacts became tools and media for designing. These artifacts supported students' cognition, that is, their sense making, planning, thinking, and testing of ideas. In this, design artifacts became *self-reflexive structuring resources* which, in their development, provided opportunities and constraints for their own future development.

---

[1] In this chapter, I draw examples from the tower project by Andy, Simon, and Tim. However, the article in which I first developed my ideas about situated cognition during designing (Roth, 1996a) provides a range of complementary examples from other students and groups.

8.1. HETEROGENEOUS DESIGN PROCESSES AND DESIGN PRODUCTS

In this section of Chapter 8, I attempt to provide an answer to the question, "What elements of the learning environment contribute to the design processes and products?" My answer will elaborate the following argument. Designing, and therefore the cognition involved, is a situated activity in which human actors (designers) network resources and practices. Designing is a heterogeneous process. The artifacts resulting from design constitute the residues of these networking activities. The cognitive activity of designing cannot be reduced to individual mental processes. Rather, the cognitive activity of designing has to be understood in terms of holistic relations including various human and non-human, individual and collective actors that together make for the experienced setting.[2] Decisions taken by principal actors, or owners of projects, need to be modeled in terms of the networking of tools, materials, artifacts, teacher-set rules, interdictions, and constraints, and the emergent properties of collective, discursive and material activity and design artifacts.

When the children's creations were "celebrated" during whole-class conversations and show-and-tell periods, individual children or groups of children claimed and were accorded sole authorship for artifacts. However, there were other aspects of the setting which contributed in fundamental and non-negligible ways to the art and artifacts of designing. Designing in this classroom did not occur in a vacuum but in a studio-like atmosphere that provided a learning culture and the associated network of relationships among psychological, sociological, and material actors in specific settings. I use the genesis of one artifact, the tower constructed by Andy, Simon, and Tim to show the contribution of various actors to the project. Because some of these actors are non-human, this case study documents the material basis of knowing and learning. That is, cognition is not the sole product of individual brains (in the way some have interpreted and used radical constructivism) or of societies (in the way some have interpreted strong theses of socio-cultural theories). Rather, there is an important material basis to cognition, and consequently to designing and problem solving.

---

[2] I emphasized the experienced setting because much of my research has shown that different students do not experience a setting, such as a science laboratory, in the same way. They see and describe different things as relevant, and hear their peers' and teachers' utterances in different ways. Consistently with phenomenological perspective (e.g., Heidegger, 1977), I therefore try to understand not just what people do, but the worlds in which they do what they do. This is a radical departure from traditional psychology and cognitive science where researchers use their own worldviews as norms against which "subjects'" behavior is judged. The approach where researchers use their own worldview as a norm lacks an appreciation for the consistency of action within the worldview of the respondent.

8.1.1. Design History of an Earthquake-Proof Tower

Readers are already familiar with one account of the tower constructed by Andy, Simon, and Tim. This account of the tower emerged, as shown in Chapter 7, from the interactions between the three boys and their peers and teachers. That account of the towers history from the boys' perspective, was constructed in a teleological way in the sense that it described aspects of the tower at hand. The tower as it presented itself at the end of the second day of designing (Figure 1m) also constituted an account of what had gone on during the design process. Both accounts, however, deleted much of what shaped the towers design and the design process from which it emerged. To construct an understanding of the design process, we need to know designing as it emerges from ongoing activity. I therefore re-construct here key aspects of the towers historical trajectory. Andy, Simon, and Tim worked on their design for approximately two hours spread over two successive days (Day 1: 9:42–10:23; Day 2: 9:07–10:18). Several individual stages of the artifact are re-presented in Figure 1. (During the following week, the three had to reconstruct the tower which led to several reconfigurations and changes of joints leading to the tower as re-presented in Figure 3 in Chapter 7. Here, however, I account only for the initial two days of designing the tower.)

Day 1, 9:42. Andy, Simon, and Tim chose as their topic to design an earthquake-proof tower. Andy received the materials they were to use: straws and pins. Simon and Tim scrounged for additional "tons of string" and cardboard. However, Gitte suggested "Tammy and I are saying 'no' to any extra cardboard". Simon and Tim returned to their work site.

01 Tim: We could make cubes and then build them up.
02 Simon: Ok, cubes, make cubes.
03 Tim: Make cubes, and then we can attach them all together.

Andy joined his team members and the three boys began to construct cubes from straws that they cut to different lengths. Simon and Tim each talked about making the base or making the top. Tim suggested the possibility that they could design their tower like an earthquake-proof building in the city, where a larger top was sitting (actually hanging) on a narrower base.

Day 1, 9:49. The boys began to argue who was making the top and bottom parts. Tim's cubic artifact (Figure 1a) was smaller than the piece Simon was working on (Figure 1b). Tim suggested, "We are making the bottom though, like the bottom part, because we can make like a smaller part and a bigger one, like the earthquake building downtown". Simon retorted, "But then you have to turn it over, so it can stay on the ground. So I am actually building the bottom". Tim then proposed to try both options, "Yeah, we

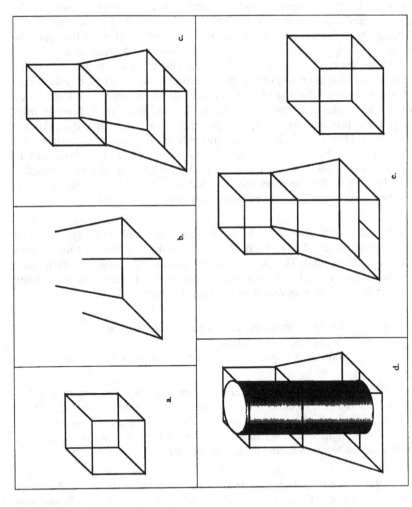

Fig. 1(a–e). Stages in the development of the tower for which Andy, Simon, and Tim received credit. There were many other human and non-human, individual and collective, material and discursive actors who contributed to the artifact as it finally presented itself. Some of these actors are indicated in the final plate (Figure 1m).

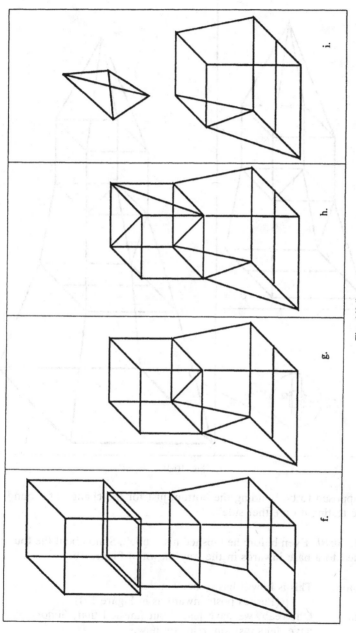

Fig. 1(f–i)

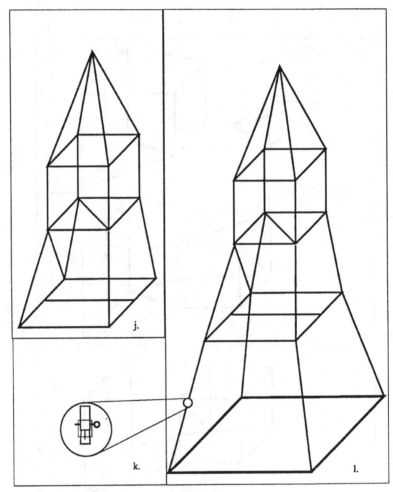

Fig. 1(j–l).

are supposed to be building the bottom just for the change of ground, and we are testing it on either side".

Day 1, 10:10. Even before he finished his "cube", Simon bent the four posts attached to a base inwards in the way shown in Figure 1b.

04 Simon: This is how it has to be, Tim.
05   [((Pushes posts inward as in Figure 1b))
06 Tim: Oh, yeah, we have just about finished that, Simon. You just
07   have, let's just join mine on now.

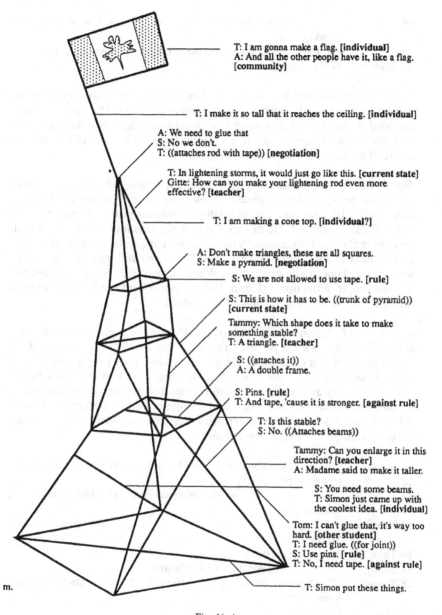

T: I am gonna make a flag. [individual]
A: And all the other people have it, like a flag. [community]

T: I make it so tall that it reaches the ceiling. [individual]

A: We need to glue that
S: No we don't.
T: ((attaches rod with tape)) [negotiation]

T: In lightening storms, it would just go like this. [current state]
Gitte: How can you make your lightening rod even more effective? [teacher]

T: I am making a cone top. [individual?]

A: Don't make triangles, these are all squares.
S: Make a pyramid. [negotiation]

S: We are not allowed to use tape. [rule]

S: This is how it has to be. ((trunk of pyramid)) [current state]

Tammy: Which shape does it take to make something stable?
T: A triangle. [teacher]

S: ((attaches it))
A: A double frame.

S: Pins. [rule]
T: And tape, 'cause it is stronger. [against rule]

T: Is this stable?
S: No. ((Attaches beams))

Tammy: Can you enlarge it in this direction? [teacher]
A: Madame said to make it taller.

S: You need some beams.
T: Simon just came up with the coolest idea. [individual]

Tom: I can't glue that, it's way too hard. [other student]
T: I need glue. ((for joint))
S: Use pins. [rule]
T: No, I need tape. [against rule]

T: Simon put these things.

m.

Fig. 1(m).

08   [((Tim puts his cube on top of Simon's pyramid
             trunk))
09   We have to use tape, reinforced tape. ((Begins to wrap))
10 Andy: Perfect! Stable!

Here, Simon attached his module to Tim's cube even before finishing his
own and without producing the agreed-upon cubic shape. The present artifact
afforded inclined posts that would be directly attached, resulting in a unit
with the shape of a pyramidal trunk. Here, the present state of the artifacts
afforded a particular, unplanned design move.

Day 1, 10:15. Simon discovered another group across the room had utilized
one of the heavy cardboard pipes, "Look, they are making these rolls, from
cardboard". In response, Tim immediately walked to the supply area, got
two rolls, and returned. He dropped one of the rolls into the center of their
existing artifact and suggested, "Here, Simon, put it in the middle. And then
attach them to the top to make it work". As one roll was too short, Tim
went to ask Tom to join two pieces. When he returned, he dropped the roll
into the center of the tower (Figure 1d); his subsequent test instantly showed
that the previously "wiggly" tower was stabilized.

Day 1, 10:18. Tammy joined the group. During the conversation she said
about the rolls, "It's not for use. You can't use it". She also suggested, "I
would like that you do not use any more masking tape. I want you to try it
without masking tape". At this time, the three boys had one cross beam,
but it did not form a triangular brace (Figure 1c). As we saw in Chapter 7,
Tammy insisted on teaching the use of triangular braces. Referring to Tim's
solution that stabilized the tower with the two cardboard rolls, Tammy
reiterated the rule, "This is not for use".

11 Tammy: Ok, do you want that to be like this, or do you want that this
12   to be more stable?
13   [((squeezes the tower which bends in all
             joints))
14 Simon: [We have, we are]
15 Tim: Stable.
16 Tammy: You want to stabilize?
17 Simon: [We are, with a beam
18 Tammy: Which shape does one have to make if one wants to stabilize
             something?
19 Tim: [Triangle

Day 1, 10:22. All three boys had completed their "cubes". Two cubes were
joined and included two braces, the third one, a little bigger than the current
top cube lay next to it (Figure 1e). Andy suggested a new design possibility:

20 Andy:	Take this one off
21	[((Touches current top cube, Figure 1e))
22	because this one is bigger than that one.
23	[((Points to the single cube, Figure 1e
24	[((Points to current top one, Figure 1e))

He then tested the alternative to his proposal by putting the larger cube on top of the presently existing tower without actually attaching it (Figure 1f). In a way, this produced the previously discussed option in which a larger module was to be placed on top of a smaller one.

Day 1, 10:23. Tammy returned to the table and questioned the boys about the tower's flexibility. They affirmed her first question about an earthquake-proof tower's flexibility.[3] Tammy was not satisfied:

25 Tammy:	You want to say that the building should move a bit?
26 Tim:	A little bit.
27 Simon:	Yes.
28 Tammy:	That much?
29	[((She bends the structure in several joints))
30 Tim:	No.
31 Tammy:	What else do you need?
32 Tim:	Just like this.
33 Tammy:	To have in a building like this?
34 Tim:	You can use a lot of triangles.

It is evident that this excerpt is one of those examples previously discussed where Tammy wanted the "triangles" to emerge from the children's activities. However, at this point, it took considerable prompting (lines 31 and 33) until a student, here Tim, uttered the magic word. Nevertheless, subsequent to this conversation, the boys added only two triangular braces. At the end of the day their tower design had emerged to the state shown in Figure 1g.

Day 2, 9:07. The three boys began their second day of designing by deciding to "put more triangles on". The addition of two more triangular braces made the artifact more sturdy and less flexible as Andy's test revealed. But at the same time, the tower also became "unshapely" because the additional braces

---

[3] During earthquakes, dwellings that show some degree of flexibility actually hold up better than rigidly designed concrete buildings. Thus, traditional post and beam constructions that can give will stand up to considerable tremors. The building which Andy, Simon, and Tim considered as design precedent was one of the well-known architectural sites in the city: it suspended the main building from a central post so that the building could give during an earthquake. Most of the children in this class knew of and had seen the building.

put new strains on the structure (Figure 1h). Tim decided to make a cone top and began its construction.

Day 2, 9:14. Simon commented that the tower looked "demented" which Tim suggested was "because [they] added too many things on". Andy first proposed to add even more braces, but then joined forces with Simon in taking the top off the current structure. They ended up with a complete pyramidal base that included Simon's original unit (Figure 1b) plus the bottom of Tim's cube (Figure 1i). Through this operation, they produced the module Simon had originally agreed to build but it was in the shape of a pyramidal trunk.

Day 2, 9:18. Tim proposed his "cone top" (Figure 1i). Both Andy and Simon rejected it because of its triangular base. Although Tim suggested that it could be attached to the structure by means of supporting beams across the top face, the other two suggested that "it wouldn't look good then" and "it is too difficult". They then negotiated a solution.

35 Andy: Just cut down the bottom one
36 Simon: [Just make a pyramid.
37 Andy: Just cut down the bottom and don't worry about (?)
38 Simon: [All you have to do, just make
             a pyramid.
39 Andy: The top, the top, the bottom needs to be square.
40   [((Points to the bottom of Tim's cone))
41 Tim: Square?
42 Andy: Look, all these are squares.
43   [((Touches tops of two "cubes"))
44   (1.8)
45 Tim: You just need one thing there. ((Cuts open the base of the
             cone.))
46   (8.4)
47   It's gonna be a small square.
48   (2.8)
49 Andy: So we make a lot of cubes and make them all smaller.
50   (1.5)
51 Tim: OK, you guys get started and I'm gonna make this.
52 Andy: And I'm making cubes. ((Picks up construction materials))
53   (3.7)
54 Simon: Make it 3 layers high, not
55 Tim: Let's make it sort of like the Empire State Building.

Andy and Simon provided descriptions of complementary courses of action. Andy suggested to "cut down" the triangular base (lines 35 and 37), whereas Simon proposed a new design for the cone top, a pyramid (lines 36 and 38).

Andy reiterated that the bottom of the cone top needed to be square, pointing to the piece in Tim's hands (line 39–40). Tim's question (line 41) suggested that he had not understood the point Andy and Tim had tried to make. As soon as Andy pointed out the square faces on the other two modules, Tim began to rebuild his top.[4] Following his suggestion that the square would be small (if he used the same size pieces as he already had), the three began to describe a new design of the tower: several cubic modules that would get successively smaller (line 49) so that the structure would resemble the Empire State Building (line 55). Tim reiterated that the last cube needed to be really small, but Simon gestured with his hands the same pyramidal shape that the present artifact had.

Day 2, 9:24. Tim had taken off the base of his top, and added another straw so that his artifact looked like four tent poles standing up. He proposed, "Why don't we just attach it on like this, it would be a lot faster". Approximately two minutes later, Andy asked Tim to attach a square base.

56 Andy: Now, you make a square.
57 Tim: No, I'm gonna attach it on someone's cube, like this
58   [(((Holds   "tent"
            onto a cube))
59   (3.0)
60 Andy: Perfect size, because you can switch it around different sizes.

Here, Tim proposed to attach the half-finished module to speed up the design process and used the existing pieces to model the result. Andy, on the other hand, recognized in the same course of action a solution to their earlier problem of fitting differently-sized faces of joining modules (line 60). Because the module was not completed, he could attach it to any cube without a prerequisite for size of the part that would become the interface to the next module.

Day 2, 9:32. The three boys attached the three modules (Figure 1j) and presented it around the classroom as their finished product. They conducted an earthquake test by shaking vigorously the desk on which the tower was placed.

Day 2, 9:35. At this point, the tower design had evolved to include three levels (Figure 1j). Tammy joined the group and, after requesting and observ-

---

[4] I return below to the sequence between lines 39 and 42 to show how the existing artifacts assisted the boys in making sense of the situation and thereby opened the possibility for a mutually-agreed upon course of action.

ing a test whether the tower met the specification for being earthquake-proof, engaged the students in the following exchange.[5]

61 Tammy: Can you enlarge it in this direction?
62   [((Gestures up))
63 Simon: No.
64 Tim: [No.
65 Tammy: Why not?
66 Simon: Because it is very unshaky and out of the ways.
67 Tammy: Could you stabilize that with triangles?
68 Simon: Yes.
69 Tammy: Or could you add something there on the bottom?
70   [((Points to the bottom part of
            structure))

Tammy wanted the children to make the tower taller and underscored her utterance with a gesture above the tower that could be read as stretching. But the students were opposed because they found that the tower was not stable enough to add another level (lines 63–64, 66). Tammy explicitly suggested taking care of the stability problem by adding triangles and then continued with the topic of tower size. She did not initially specify (lines 61–62) where she wanted students to add another level, but designing a taller tower could mean adding more on top or bottom. In her next utterance that took the form of a question (line 69), Tammy implied adding a new level to the bottom of the existing artifact. Although this utterance was in the form of a question, the students' subsequent actions showed that they took it as an instruction to make the tower taller. Andy would later explain that they had added a new level because "Madame said to make it taller because it was only that size ((gestures first three levels of tower))" and Simon had heard the utterance as "it would be a good idea to make the tower bigger".

Day 2, 9:52. Andy, Simon, and Tim finished their newest addition to the tower. In the course of adding the new bottom level, they had encountered a problem. Tim had found a way to join two straws by making a cut with the scissors in "the middle" of one, and then inserting it into another one (see Figure 1k and Figure 1a in Chapter 6). This yielded straw beams almost twice the length and fit the bottom unit. However, Tim wanted to fix the joint by means of hot-glue. Tom – who presently owned the only glue gun in the class – said that he could not do the job Tim requested, because it was too difficult. In the course of the morning, Tim would add some tape to prevent the beams from slipping in and out, and, in some instances, even added a pin (see Figure 1k and Figure 1a in Chapter 6). This yielded a tower

---

[5] This conversation was in French except for the word "unshaky" in line 66. The episode could be translated literally because its structure was that of a word-by-word translation from English.

four stories high (Figure 1l). As we saw from their account, when the tower partially collapsed during the following week, Tim, now in the possession of his own glue gun, produced the joints in the way he had earlier requested (Figure 3, insert 7 in Chapter 7, p. 166). At that point, he also evolved new joints which required small spots on two straws to be melted which then fused as he held them together.

Day 2, 9:54–10:18. The three boys added a few more braces, encouraged by both Gitte and Tammy. In addition, the following brief exchange resulted in a new design direction for the top, which also provided a possible solution to Tammy's previous request for making the tower taller. Earlier on, Gitte and the boys had already talked about the pyramidal top as a lightning rod. Here then, Gitte raised another question about the top.

71 Gitte: How could you make your ah: lightning rad (.) rod even more
             effective?
72   (0.8)
73 Simon: Up.
74 Tim: [Just like this.
75   [((Holds a straw on top of cone))
76   (1.1)
77 Gitte: Good Simon. ((Walks away))

Following this exchange, Tim attached first one straw, and then added more to make this part increasingly longer. He wanted to design the tower "to make something different. I make it so tall that it reaches the ceiling". However, the long rod (about 1.5 m) bent, collapsed, and then came apart. Tim rebuilt a rod from just two straws. A few minutes later, he announced that he would make a flag. Once designed, he mounted the flag on the existing rod. In the meantime, Andy and Simon had added another few braces so that, at 10:18, they announced the completion of their project (Figure 1m).

This history of the tower, while far from complete, includes many human and non-human actors that contributed to its final shape. This history not only accounts for the final design, but also includes some of the design options not taken or abandoned for a variety of reasons. Figure 1a–m shows the emergence of the tower's design. Figure 1m also indicates the origins of some of its aspects as they related to tools, materials, community standards, teacher-set constraints (rules), current state of the design artifact, individual preferences, and past discursive achievements. This allows us to recognize the tower as a heterogeneous assemblage that emerged from the coming together of various resources and practices. The tower emerged as a product of the interaction between these elements, with the three students, the designers, as principle actors (Figure 1). Despite the observed heterogeneous origins of the artifact, Andy, Simon, and Tim were celebrated in the end as its designers. It was *their* tower. This attribution was asserted by (a) the children through accompanying descriptions assigning ideas and work to themselves and (b) the teachers when they praised the designers.

Although a variety of tools, artifacts, materials, individuals, and rules existed in the classroom at any one point, not all of them were of equal importance in accounting for the emergence of an artifact. This is, because for each person and group there exists an interpretive horizon of events which delimits the openness that surrounds artifacts.[6] Elements which lie outside this horizon, although they might exist for other people in the same setting, are not relevant aspects in the design history. For example, although a glue gun was present in the classroom, it did not influence the design of the tower in its early stages. Only later, with a growing sense in this classroom community that glue guns make strong joints, did Tim consider the use of this tool in his own activities. That is, none of the actors identified – tools, materials, rules, negotiations, current state of design artifact, etc. – has an absolute ontology so that they are therefore not available to all designers in the same way. In the following subsections, I show how various aspects of the setting, as identified by the designers and therefore within their horizons, influenced the emergence of the children's designs. In addition to the tower designed by Andy, Simon, and Tim, I include examples from other projects. In the subsequent major section, I address the varying ontology of the non-human actors.

8.1.2. *Material Basis of Designing and Design*

Designing, and the artifacts emerging from this activity, have an important material base. That is, designs are shaped by available resources (materials, tools, rules) as they exist within the interpretive horizons of designers. For example, if a glue gun is interpreted as lending itself only to making a hot-glued joint, it shapes the design in a particular way. If, on the other hand, a designer uses the hot tip of the glue gun to make holes into plastic straws in which other straws can be lodged – as Tom had done to produce the joint in Figure 1b, Chapter 6 – new design affordances emerge that may lead any design into unforeseen directions. Here, I elaborate some of the ways in which designing had a material base. In other words, I illustrate how student designers networked various non-human actors into their emerging artifacts.

8.1.2.1. *Networking Tools*

As we saw in Chapter 6, the availability of certain tools changed the design process and affected the artifacts produced. With an increasing number of

---

[6] The openness surrounding us is filled with views of appearances of things which in our representing are objects (Heidegger, 1959). Our horizons, which are a product of our experiences, common sense, and the situation at hand, delimit what and how we can see, and act, in any situation.

glue guns available for circulation, with the expansion of the network of glue gun practices, and with the circulation and development of new glue gun practices, the design culture changed. New joining techniques based on this tool emerged and were circulated. The strength which the materials and joints received from this technique was so great that some students found specific engineering techniques unnecessary, in particular some of those that the teachers wanted students to "discover" in their own activities, such as the triangles (see Chapter 7). In the following examples, the glue gun changed the task of designing. Most frequently, the tool afforded the construction of artifacts impossible to achieve with other techniques. In other situations, it prompted students to pursue courses of actions that led into blind alleys.

The tower Andy, Simon, and Tim constructed collapsed during the week between the termination of the project and their presentation of the artifact to the class. During this week, Tim had brought a glue gun to class. While trying to repair the tower, Tim serendipitously discovered a "welding" technique in which two plastic straws could be joined by softening each with the glue gun tip (see Figure 1c in Chapter 6, and Figure 3 in Chapter 7). When the softened pieces were brought quickly together, a new joint formed. Drawing on this technique, his team repaired several braces and thereby achieved a very strong and stable structure. As we saw in Chapter 7, this tower received Gitte's approval. In particular, during the boys' presentation she pointed out the strength and stability of the tower which had made it last much longer than those of any other group.

The absence of a tool also obtruded and influenced the design of Andy, Simon, and Tim's tower. That is, the tools became present-at-hand in their absence so that students had to attend to alternative designs for instances where they felt the tool was missing.[7] After Tammy had asked the three boys whether they wanted to enlarge their towers, Tim decided to construct long beams to design a large base unit. He immediately proposed cutting one straw "down the middle" and inserting it into another straw. To make the seam solid, he wanted to hot-glue it. Tim asked Tom to glue the joints of his long beams. Tom, however, refused and suggested that he use tape. Initially, Tim did not follow this advice and designed the tower as one that could be easily assembled and disassembled. Later, when these joints came apart, he first added masking tape and later pushed pins through both straws (see Figure 1m and Figure 1a, p. 134).

The use of a tool allowed students to assemble stable and strong structures without drawing on the triangular braces as resources in the design process. Arlene and Chris built their second bridge from paper. At first, they rolled newspaper tightly around wooden dowels which were later removed. Using

---

[7] This makes reference to the work of Heidegger (1977), who suggested that tools are normally ready-to-hand ("zuhanden") in which case they are used in a transparent way. Tools become the focus of attention, and therefore are re-presented when they are broken or absent: they are present-at-hand ("vorhanden").

these rolls as basic design elements, they constructed the bridge deck and piers by laminating and bundling the materials. Finally, they joined piers and deck using the glue gun. The hot, fast-setting glue, and the additional materials used to overlap deck and pier produced firm joints. As a consequence, they did not need triangular braces.

In some situations, the increasing use of glue guns led students into blind alleys. Sandy and Dennis were among those students who abandoned their old practices too quickly and tried to apply hot-glued joints to any materials. This created many problems while they worked on their bridge which was to be made from spaghetti. Sandy and Dennis attempted to glue individual pieces of spaghetti to form bridge deck, piers, and braces. However, due to the brittleness and small size of the material, all their initial attempts to design the bridge were thwarted. For example, in attempting to make a bridge deck, they laid about eight spaghetti flat on a table touching each other and tried to join them by means of a bead of hot glue at right angles, across the spaghetti pieces. They found out quickly that their plan did not work because the bead was not strong enough. They made progress with their design only after they bundled the spaghetti, firmly securing the bundles with masking tape. These bundles then constituted the new basic design elements which could be hot-glued. In this situation, there existed forces and reticent actors that intervened with Sandy and Dennis' efforts to network the individual spaghetti into a spaghetti bridge.

8.1.2.2. *Networking Materials*

Materials shape design products and designing processes. But it is far from evident what kind of designs evolve given a particular set of materials. Although material properties appear to be given *a priori*, they can be changed through various engineering practices (laminating with different materials, bundling, layering). Furthermore, different structural arrangements of materials afford different load capacities, stresses and strains in lateral and longitudinal direction, and torques. Often, it is not even possible to evaluate whether or not a particular geometrical arrangement would have evolved given different material. For example, Jeff and John initially received spaghetti as building materials. Even before they began to design, they negotiated with Gitte to exchange the spaghetti for straws. Given the openness of the design process, we cannot even hazard a guess about how their bridge design would have looked had they used spaghetti.

The materials used for joining the structural elements also played an important role in design. Tim asked Simon to get the masking tape because it would afford stronger connections between beams than did the pins they had used up to that point. When the community discovered the strength of hot-glued joints, the designs took new and unforeseen directions. Hot glue afforded to Peter and Ron the construction of a bridge from pieces of wooden skewers without drawing on the triangle practice. On the other hand, indiscriminate use of a joining material also created problems. Dennis and Sandy tried for nearly three periods to construct posts by making bundles from hot-glued spaghetti. Here, their fixation on a particular joining material led to a dead end in the designing process. It took Gitte several conversations with the two boys before they abandoned the process and shifted to creating bundles from spaghetti and masking tape. They could then return to their favorite joining technique to fasten the various posts and beams created by means of the bundling technique.

8.1.2.3. *Networking the Current Artifact*

Children's ideas literally took form in the emerging design artifacts. As the ideas took form, the artifacts constrained and limited other design options and afforded new and different ideas. In this sense, children's designs became self-reflexive artifacts that embodied past activities and, simultaneously, foreshadowed affordances and future constraints. The future constituted an open horizon so that the number of possible trajectories grew exponentially with each design move. However, in the evolutionary process of design, only one of these trajectories was realized. The emerging artifact embodied and enfolded the actual design trajectory, decisions, conversations, and so on. In this way, design projects summed over their own history: any new idea took the present state of the artifact as starting point (in more than 50 projects there was only one counter example where a design was abandoned in its current form). Thus, existing artifacts had normative character such that designers rarely returned to an earlier form unless, as in the case of Andy, Simon, and Tim's tower, previous design decisions led to an "unshapely" design.

Andy, Simon, and Tim began their tower (Figure 1) from cubic modules. Nobody asked how the cubes were to be joined or whether the smaller or larger cube should form the base (see their negotiation in this situation below). They framed their alternative designs as "the earthquake building downtown" (also "West Coast Energy") or as the "London Tower" which are narrower on the bottom or top, respectively. After the decision was made to go with the "London Tower" plan, Andy/Simon/Tim continued with this shape, although they encountered stability and strength problems so that they reframed their project as "Tower of Pisa". Later, while they discussed how to attach Tim's cone shape to the existing artifact, Tim suggested several intermediate cubic modules to produce the shape of the "Empire State Building". Finally, the group agreed that Tim could attach his "cone top" if he made a square base for it so that they could make an "Eiffel Tower" shape. At each of the decision points, future design states were discussed in terms of the present artifact and individual or common prior experiences. Had the students decided to use the "West Coast Energy" design with a base narrower than the top, subsequent framing of their design problems as

"Tower of Pisa", "Empire State Building", and "Eiffel Tower" would have
been less likely.

In a similar way, Jeff and John designed their bridge through several cycles
of testing and adding new features. Although their initial design included a
bridge deck constructed from one level, they continued to evolve the deck
into a double beamed-structure held together by straight beams. They then
added braces between the uprights and finally included arched beams. Rather
than redesigning their bridge after the first test of its limiting load, Jeff and
John designed further additions to the existing artifact to increase its structural strength under heavy loads.

8.1.3. *Social and Psychological Basis of Designing and Design*

The networking of materials is thoroughly tied to the networking of human
agents and collectivities. Individual actors proposed ideas, but these emerged
from engagement with materials, other individuals, culture, teachers, and so
forth.

8.1.3.1. *Networking Individuals*

Design artifacts emerged as material witnesses of the collective planning,
constructing, and negotiating of team members. During these negotiations,
designs changed and "master" plans emerged, but they were not the sum
total or least common denominator of the key actors' initial inputs. Rather,
negotiations constituted a changing context such that individual proposals
were transformed. New plans emerged with or without bearing any likeness
to the originating individual ideas. For example, Tim first suggested and then
constructed a "cone top" for his group's tower. His design had a tetrahedral
shape, that is, a triangular base. Simon and Andy objected, claiming that
the triangular base could not be fitted to the square of the cubic shapes that
they had produced. After Tim's solution to the lack of fit was rejected on
both technical and aesthetic grounds, for it consisted of joists across the top
ceiling (Figure 1i), he changed the triangular base into a square base. The
resulting top of their tower emerged from their interaction rather than being
the idea of any single individual. Here, neither Tim's pointy and triangular
"cone top", nor Andy and Simon's "cube top" became the common plan.
They designed a much wider pyramidal top that fit the top level of the
existing tower artifact.

8.1.3.2. *Networking the Embedding Culture*

Throughout Chapters 5 to 7, we saw that the classroom showed similarities
with design studios in which knowledge and information circulate freely
among members. Especially the increased density of students in the vicinity
of electrical outlets – once a larger number of glue guns were in operation

– facilitated interactions between different groups. The nature of the classroom as design studio facilitated networking with other design groups and the emergence of community standards and trends. Such standards and trends contributed to individual designs and could be recognized in resemblances among artifacts belonging to different groups and the associated explanations such as "everyone else is doing a Canadian flag". Furthermore, students directly networked with students from other groups. Through these interactions, outsiders to a design project became themselves actors who shaped the emerging artifact of another group. We already saw how the Canadian flag emerged as a common resource so that some students added flags (or had the intention to do so) to their structure because everyone else was doing it; we also saw how the design conditions changed inside the classroom where everyone else quickly knew what others did, whereas outsiders (Andy and Dennis) were not influenced by these changes in the culture.

Other design artifacts similarly embodied contributions from passing groups and students. In Andy, Simon, and Tim's tower, a pinned-sleeve joint for producing double length straws was used for the base module (Figure 1k, and Figure 1a in Chapter 6). Tim developed this joint after Tom refused to hot-glue the straws and suggested using cello-tape. When the tape failed to produce a solid joint, Tim added a pin resulting in a pinned sleeve-joint.

8.1.3.3. *Networking Teachers*

The teachers constituted important actors who contributed to the artifacts as they were later presented. My historical account of the tower constructed by Andy, Simon, and Tim included several interactions between the students and their two teachers. As we already saw in Chapter 7, the teachers' key goal was to encourage the emergence of a triangle culture. Consequently, they geared their questions and directions to students in a way that they hoped would bring about triangle-related discursive and material practices. Especially Tammy, much less familiar with the curriculum and engineering practices, tended to ask questions related to structural weakness much more bluntly and, to speed up interactions, frequently directly instructed students to add braces, and gestured where and how to add them. But even when she asked questions such as "Can you enlarge the tower in this direction?", Andy, Simon, and Tim often acted as if she had given the instruction to enlarge the tower in a specified way. After Tammy asked this question, the boys added another level to their tower. Similarly, after Gitte asked how they could make their lightning rod more effective, Tim designed a rod to go atop his current tetrahedral "lightning rod".

Even when teachers were not interacting with a group, the norms they established for a task influenced what and how students brought about their designs. Throughout the unit, we recorded student utterances related to interdictions such as "I don't think we are allowed to do a bottom thing [from cardboard]", "You're not supposed to do that", "we're not allowed

taping [the structure] to the ground", or "we are not allowed to use [tape]".
However, what these teacher rules meant for a particular design was not
something that could be determined *a priori*. The rule "No additional cardboard pipes" meant that Andy, Simon, and Tim did not include this element
as a center in their tower. At the same time, a cardboard pipe was part of
the designs submitted by Jeff and John (a crane) and Damian, Dennis, and
Doug (a tower with a crane).

In this situation, we saw that the meaning of a rule was context-dependent.
Whether or not a particular rule shaped a design in a particular way could
only be established after the completion of a project. That is, rules did not
have an absolute ontology but were interpretively flexible so that their meanings were brought forth in each moment of enacting design. But such ontological uncertainty was not limited to rules. Materials, tools, artifacts, plans,
and utterances were equally flexible, and their meanings were determined in
actual use.

8.2. ONTOLOGY OF RESOURCES

In the foregoing section, I provided an answer to the question, "What
elements of the learning environment contribute to the design processes and
products?" I identified a number of human and non-human, material and
discursive, and individual and collective actors that were networked into
children's design processes and design artifacts. The focus question of this
section is, "What is the ontological status of the elements?" My answer
elaborates the following point: The relative importance of tools, materials,
community standards, teacher-set rules and constraints, current state of the
design artifacts, individual preferences, and past discursive achievements is
indeterminate because these elements do not exist in any absolute sense.
Meanings are determined through the flexible use of these elements. The
degree of the elements' salience is contingent on specific local developments.
Because of the differences in background and experiences between any
pair of human actors, tools, materials, artifacts, teacher interdictions and
constraints, history of activity and design artifacts, or plans do not have
stable ontologies. Different meanings that elements can have for different
actors lead to different design actions and ultimately, learning.

In the engineering design lessons, children learned to recognize and exploit
in productive ways the interpretive flexibility of materials, discourse with
group members, tools, design artifacts, and teacher-set rules and interdictions. Rather than being stifled by the open-ended and ill-structured situation, students used the flexibility of their situation in creative ways. Problems did not exist in any absolute sense but emerged as interactive
achievements. As a consequence of this interpretive flexibility, situational
choices made by the principal actors (members of a design group) appeared
to be inherently unpredictable. There is ample evidence in the data that all

actors discussed in the previous section and highlighted in Figure 1m were subject to variable ontology.

8.2.1. *Interpretive Flexibility of Plans and Artifacts*

In this classroom, it made sense to understand the artifacts as outcomes of first drafts rather than as the fabrication and implementation of complete and standard architectural plans. From their perspective, the children treaded novel design territory and engaged in design innovation. Children's final artifacts were products of a designing process, and their "experiments" were more like engineers' sketches rather than their final plans, models, or actual structures. Such activities are of that kind which, according to Anderson (1990), have not yet yielded to modeling efforts in terms of "rational" processes. The failure of such modeling rises in part because "problem" is frequently used as an ontological category such as in "word problem". In the present context, the interpretive flexibility of materials, artifacts, tools, or rules/constraints led to the simultaneous characterization of situations as problematic and unproblematic. That is, one student may have seen a problem to be resolved before the overall design work could resume, while a partner may not have seen a problem at all. Being able to interpret situations in a flexible manner, to *reframe* a situation such that it was unproblematic – that is, afforded an action that would remove a problematic situation – was one of the major aspects of learning. I illustrate this flexibility in interpreting the current artifact in two examples.

Initially, children's ideas and plans were unspecified so that the meaning of an expression such as "earthquake-proof tower", "two towers with sliding object between", or "baseball stadium with removable top" was quite open even among members of a group. As the artifacts emerged as residues of children's designing activities, meanings converged and stabilized. In the beginning of each project, existing artifacts did not constrain the interpretive flexibility enough for group members to have similar conceptions of how the final artifact would present itself. The following episode illustrates that students could begin their design process (including construction) without pre-specifying exact outcomes so that their plans remained open to interpretation to be settled far into the project.

Andy, Simon, and Tim had decided to construct an earthquake-proof tower. They had planned to "make cubes and then attach them all together". At the point of the following conversation, Tim and Simon had already constructed two "cubes" of different sizes (Figures 1a, b). The excerpt was part of a discussion about which shape the tower should take.

78	Tim:	We are making the bottom though
79		(1.3)
80	Simon:	I don't know
81	Tim:	[Like the bottom for it

82		[((Moves both hands up and down counter each other))
83	Simon:	Bottom part?
84	Tim:	[This is for the top (.) the top part
85		[((Points to Simon's straws))
86		(1.1)
87		because we are making the base 'cause we can make like a
88		smaller part and a bigger one, like the one building, like the earthquake building downtown.
89	Simon:	Yeah, but then you have to turn it over.
90		(1.4)
91		So it can stay on the ground.
92		(1.1)
93	Tim:	Yeah
94		(1.8)
95	Simon:	So I am actually building the bottom.
96	Tim:	Yeah, but we're
97		(1.0)
98		supposed to be building the bottom just for the change of the ground.
99		(1.4)
100		So we are going to change it. Or test it on either side.
101		(4.0)
102	Simon:	That's better

Tim thought that he had built the (smaller) base, which would make his tower look like the earthquake-proof West Coast Energy Building downtown (lines 78, 81–82, and 84–88), a landmark which was repeatedly discussed in class. Simon, on the other hand, questioned the contents of Tim's utterances (lines 80, 83, and 89). He felt that the larger cube should make the foundation to assure the tower's stability (line 91). Because Tim appeared to agree (line 93), Simon concluded that he was actually building the bottom part (line 95) rather than Tim. He later admitted to one of the teachers that their differences in interpretation had confused him ("I thought I was going to do the floor, and Tyler said, 'do the top;' and now I am confused"). Tim resolved the issue by suggesting that they could change their earlier plans or test both designs (line 100). He therefore left open what the tower's final shape should be and who had built the "bottom part". Simon agreed with Tim's proposition (line 102).

Each of the three boys had started the "earthquake-proof tower" with his own idea. Immediately before the conversation, all of these ideas were compatible with their activities of constructing "cubes" of different sizes. During the episode it became clear that Tim and Simon interpreted their current artifact in different ways; thus, whether their designs were the "same" depended on the point of view. Children's discussion over and about the artifact allowed differences to come out and to be discussed. The group accepted Tim's suggestion (line 100) and deferred to a future moment the decision that would remove the interpretive flexibility of the artifact. In this way, the design of their tower was repeatedly reinterpreted as expressed by their labels including "West Coast Energy Building" (Vancouver), "Eiffel Tower", "London Tower", "Empire State Building", "CN Tower" (Toronto), and the "Leaning Tower of Pisa".

Similarly, the bridge Jeff and John designed emerged from several reinterpretations of "strong bridge deck". They began with a simple layer of straws "stabilized" by a rhombical center piece (see Figure 4, p. 189). However, this deck collapsed when only a few wooden blocks were attached. In a second phase of the project, Jeff and John added a second layer to the bridge that was connected to the first by means of vertical connectors. Although they increased their load capacity to 56 blocks, Jeff was not satisfied and proposed the addition of triangular braces to the design.[8] In the subsequent stages, the two boys added the bent straws above and below the double-layered deck (see Figure 4 in Chapter 7). This last addition, although it increased their load capacity to 347 blocks, also broke the symmetry which initially characterized their design. This symmetry became, as we saw in Chapter 7, a point of contention between Jeff and several students who questioned the design. The way Jeff and John tested their bridge underwent reinterpretation. Initially, they tested their bridge right side up. However, the piers proved to be the weakest part of the structure. Because the deck was symmetrical, Jeff proposed a test in which the bridge was positioned upside down. In this way, they only tested the bridge deck rather than the piers. However, when they added, in the final stage, the bent beams which broke the symmetry of their design, Jeff and John did not return to the original form of testing.

John and Jeff's way of improving the bridge was a typical example of a conversation with materials and artifacts. As in conversations with people, artifacts talk back. This backtalk allows human actors to change their interpretations and, in response, design new actions. Such backtalk was central to Gal's (1996) description of an MIT student's design of a bridge during a design competition. There are striking similarities between the processes by which the MIT student and John/Jeff developed their bridges. As I outlined in Chapter 2 (p. 31), watching the Grade 4–5 students' interactions with the mathematical modeling program that assisted them in constructing equivalent fractions alerted me to the crucial importance of systems that can talk back to students in ways they understand as regulators of learning experiences.

---

[8] It was at this stage that the triangle practice emerged from Jeff's activity. For this reason, I had classified him in Chapter 7 as being at Level 2 in his trajectory of participation. In the same way, the triangular base of his piers and the triangular staves emerged as he pursued his goal to make the bridge more stable.

8.2.2. *Ontology of Rules*

Both teachers contributed to the students' designs to non-negligible degrees.
These contributions were primarily of three types:
- teachers suggested possible shapes and forms for the students' artifacts;
- teachers provided hints as to how existing artifacts could be strengthened and stabilized; and
- teachers set certain constraints which allowed only certain materials.

Some of the suggestions and constraints were constant across groups such
as when teachers wanted children to be familiar with specific engineering
techniques (triangular braces, testing structures by bringing them to "catastrophic failure"), while others were more design-specific and changed from
group to group. The children usually recapitulated these constraints by tag-
ging each others' ideas with "We are not allowed to . . .". One of these
constraints related to cardboard during the tower construction. These were
the initial constraints under which Chris and Arlene constructed their gondola. Whether and how such "constraints" limited the students' designs was
a matter of interpretation for the situation at hand. In the following episode,
Chris reminded Ron that what had been his original design goal stood in
contrast with his current project.

103 Chris:	Weren't you doing the one that could stand a fan blowing to me this way?
104 Ron:	We were s'pposed to.
105	(1.0)
106	But (.) we decided to change it (.) and we made up the sky dome.

Here, Ron admitted to having changed the original task. He had done so
without asking one of the teachers. However, such situations often resulted
in established facts that had normative order, that is, which assisted students
in turning later negotiations with teachers in students' favor. Thus, although
students were to make the joints of their tower constructions with pins, Tim
had begun to use tape. Later, neither Tammy nor Gitte asked him to take
the tape off his structure. The accomplishments had a nature of *faits ac-
complis* which teachers, in most instances, did not want to ask students to
revise.

   Thus, teacher interdictions and constraints could not be taken in any
unequivocal way because their meaning was established in each specific
situation. Constraints such as "design a bridge by using only a set of materials
provided" meant different things in different contexts. There were situations
in which students acted in a way that might be interpreted as "according to
the rule". In other cases, the same group of students submitted to an interdiction in one situation, but "violated" the same interdiction a little while
later. Finally, students renegotiated the conditions of a project, essentially

changing it in the process. These observations underscore that rule-following is a creative achievement rather than a self-evident act.[9] It is also possible to argue that rules as such do not exist. What a text (pronounced by a teacher, written on a piece of paper) means has to be constructed in each situation so that it makes little sense to speak of a constraint, interdiction, or rule as existing *a priori*. Rather, "rules" are a posteriori descriptions of patterned actions; a "rule was broken" simply means that actions could not be described by the rule. This precarious relationship between rules and actions cannot be emphasized enough. Whereas most people accept that one can establish whether a recipe has been followed during cooking only a posteriori, many science teachers may not accept a similar situations for rules in their classrooms. That is, while it is accepted that recipes are ambiguous, rules intended to regulate school life often are not accepted as such.

In the beginning of the tower project, Tammy and Gitte admonished students to use only straws and pins.[10] Andy, Simon, and Tim did not ask a teacher for permission to get additional building and joining materials, but immediately began searching for cardboard and tape; they were stopped short by Gitte who reiterated her admonition. Later however, although they hesitated at first to add another joining material (tape), Tim's argument that tape would strengthen their structure legitimized their "forbidden" action. Here, despite the teachers' constraints to use only one type of fastening material, the interpretation of the teachers' other constraint, "make it strong", was more important; it was also the grounds for convincing the teachers that changing the rules of the assignment was necessary and thus legitimate. Tim also tried to add a heavy cardboard pipe to stabilize and strengthen his tower (Figure 1d), but Tammy reminded him of the rule. In this situation, he could not convince her that the additional materials constituted a legitimate solution to his problem of stability and strength of the tower. Although two other groups used cardboard pipes, Tim was not allowed to use them.[11] In the same way, although Gitte had reminded Andy, Simon, and Tim of the rule that no additional cardboard was allowed, Chris negotiated cardboard to be used in the design of her gondola. In both situations, the rule that "no additional materials are allowed" did not have a stable ontology. In the case of Tim's group, it meant that they could not

---

[9] The most important work on rule following was established in the domain of ethnomethodology. The important studies include: how people classify hospital records on the basis of instructions (Garfinkel, 1967), how people follow operation instructions for photocopiers (Suchman, 1987), and how students follow instructions in science laboratories (Amerine & Bilmes, 1990; Lynch *et al.*, 1983).

[10] Gitte, the curriculum designer, wanted students to re-discover through this activity that triangular bracing could produce stable structures even when only pins were used for joining plastic straws.

[11] The photographs of the designs show that in both cases, the cardboard did not primarily serve to provide structural strength. Jeff and John had added the pipe to their tower as a way to model the cabin of a tall crane. Damian, Dennis, and Doug used the pipe as the central part of their hinge that permitted their crane to pivot.

add cardboard or the cardboard pipe; for Chris, Jeff, and Doug (and their groups), the rule allowed the addition of materials other than plastic straws.

Because of the flexible ontology of rules, it was quite difficult to predict how students would interpret teacher-set rules and constraints because this differed not only across groups but even within groups. For example, each group was allowed only limited materials for their bridge designs. When Tim and Stan, who worked together on a bridge from cardboard and glue, wanted more material, they negotiated with one of the adults. They convincingly argued that the skewers provided additional strength to their project. Only moments later, they requested more cardboard, but Gitte refused. Immediately afterwards, Tim and Stan cut up the cardboard pad which they had previously used to protect their table top; the pad became a source for additional materials despite the interdiction. Here, they reinterpreted "cardboard pad" to mean "cardboard as a source of needed material".

Learning can be thought of as a process of entering and establishing a relationship with objects, be they physical or mental (Turkle & Papert, 1992; Wilenski, 1991). A detailed description of how such relationships are established was provided in a study of Faraday's notebooks, especially the records pertaining to the work that yielded the first electromagnetic motor (Gooding, 1990, 1992). This account of Faraday's design process illustrates the pervasiveness of ontological ambiguity, and the agency through which this ambiguity is made to disappear. Ontological ambiguity is not a nuisance but

allows free movement between actual and possible worlds, enabling new phenomenal possibilities to be constructed. . . . The ambiguous status of manipulated objects is essential to the creative development of thought experiments as well as real ones. (Gooding, 1992, p. 72).

The earthquake-proof tower constructed by Andy, Simon, and Tim underwent a series of conceptualizations such as "Tower of London", "Eiffel Tower", "Empire State Building", "West Coast Energy Building", "CN Tower", and "Leaning Tower of Pisa". Here too, the boys constructed their tower's phenomenal possibilities before an open horizon of possibilities. In the end and through the boys' agency and skill, one of these possibilities was realized as a material tower.

Ontological ambiguity is not a nuisance but has important creative potential. "Bricoleurs" (Turkle & Papert, 1992) create artifacts and theories in this movement between the actual and possible by arranging and rearranging materials, associating ideas, and interacting with the social and physical environment. From the work of bricoleurs emerge artifacts whose properties are not pre-planned. Designing as understood here is an evolutionary process which contrasts with professional normal design that is constituted by routine implementation of pre-specified plans. Through their evolutionary emergence, children's design artifacts integrate over their own history. Previously tentative design moves become "facts", and this factual nature shapes future design moves. Thus, the emergence of artifacts is associated with the reduction of interpretive flexibility characterizing tentative plans such as Andy/Simon/Tim's earthquake-proof tower. The addition of new elements to the artifact is constrained by the present state of the artifact. This "concretion" of ideas into artifacts also reaches into the future by constraining the horizon of possibilities. In this way, decisions emerge rather than have to be taken. In the end, among the virtually unlimited number of possible "earthquake-proof towers", only a few are realized in the artifacts presented by the children.

8.3. ARTIFACTS AS STRUCTURING RESOURCES IN INTERACTION

In the previous sections, I provided descriptions of the nature and ontology of the elements that were networked into students' designing and designs. In the course of students' activities, artifacts emerged. Although there are claims by many educators that designing constitutes an ideal learning environment (Harel, 1991; Kafai, 1994; Schank, 1994), there is little if any evidence to show how the artifacts that emerge from children's activities contribute to cognition. In this section, I therefore elaborate an answer to the question, "What is the role of design artifacts in collective designing?" My argument runs along the following lines: Artifacts resulting from the children's designing not only bear the marks of their situated, or heterogeneous origins, but also have important functions in the design process itself. They reflexively constitute their own context. There are two important functions of artifacts in the process of designing:

• design artifacts afford the integration of "thinking" (abstract) and "acting" (concrete; and
• emergent design artifacts support communicative processes by reducing interpretive flexibility.

That is, the design of artifacts is not merely a context in which children learn to design and learn about the materials they use, but these artifacts reflexively influence the way children learn and interact. Children do not *just* design, but the design artifacts structure designing; and children do not *just* talk about their designs, but the design artifacts shape the communicative interactions from which they emerge.

Earlier studies revealed the importance of physical objects for high school physics students' planning of experiments (Roth, 1994a). By means of material objects, high school physics students structured phenomena which allowed them to frame meaningful and important research questions. Furthermore, the objects they used constrained the interpretive flexibility of talk in such a way that they could make sense of each other's ideas. Design artifacts in the present study had the same functions. They were not only products of children's design activities, but more importantly, allowed children to structure initially ill-defined design problems; and they allowed children to negotiate the interpretive flexibility of tools, materials, teacher-set constraints, and the present state of the artifacts. In this way, design artifacts are reflexive because they are both goals of children's activities and they structure these activities.

8.3.1. *Inextricability of Thinking and Acting*

This and other formal studies of children's designing (e.g., Harel, 1991; Kafai, 1994) attest to the power of design activities to engender various forms of learning. One may ask, what is it about designing artifacts that leads to demonstrable learning outcomes? Is it that design artifacts themselves contribute to children's cognition and interaction with others during the design process? I contend that design artifacts allow the re-integration of "thinking" and "acting" that traditional schooling (based on traditional psychological theory) separated into abstract and concrete modes, and after denigrating the concrete, supported thinking and acting to different degrees (Roth, 1995b; Wilenski, 1991). The present study shows that design artifacts are important aspects of learning because they allow thinking through manipulating; they are objects student use to integrate "acting" and "thinking". Children's manipulations of materials, tools, and artifacts were part of an integrated designing activity. Thinking and manipulating tools, materials, and the artifact were intricately intertwined. Thus, designing could not be understood as fixation of ideas that came from the head and literally gelled in and as the emerging artifacts. Rather, designing was an integral activity which had mental, material, practical, and social aspects that make "thinkering" a suggestive metaphor. Thinkering decenters the traditional notion of cognition to include, as the following examples show, the manipulation of non-mental entities.[12]

There were many situations that showed how the artifacts themselves were used as part of designing. For example, Tim had produced a cubical module (Figure 1a). Simon, also wanted to make a cubical shape but stopped when his piece still lacked four straws that would have completed it (Figure 1b). He moved the four open straws around, then folded them inwards and suggested that they should fix his piece to Tim's cube which was a little smaller. In this way, they avoided a potential problem that could have arisen at the interface between square faces of different sizes (in the way they framed the interface between the cone top and top cube as problematic). In the present situation, the state of the artifact Simon had in his hands allowed him to move the uprights. When they were bent inward, he recognized that in this way, they could construct the link between his and Tim's pieces. A

---

[12] In *Double Helix*, Watson (1968) provides an interesting account of how important insights about DNA structure arose from manipulating physical re-presentations of the four bases involved. As he moved the cardboard pieces around, he "realized" that pairs of these bases fit together such that the two resulting pieces were identical. He had "found" the rungs of the DNA ladder. In this case, material objects were things to think with that integrated the "concrete" and "abstract".

similar situation occurred later when Tim connected his cone top – which consisted of four straws held together by pins at one end – to the existing topmost straw cube rather than finishing his pyramid. Andy described the situation as "Perfect fit". Again, Tim's putting the open ends of his artifact to form the corners of differently-sized squares encouraged them to make the perfect fit rather than trying to construct an interface for two pieces of unequal size. In another situation, Andy placed his cube on top of the smaller, existing top cube (Figure 1f). His considerations of alternatives included the design artifacts. These afforded tests of the implications of his design move.

My second set of examples comes from two Grade 5 girls. Although Chris and Arlene constituted one of the two groups that used drawings as part of their planning (which were not required), their design work was characterized by the union of mental and material actions. While Chris manipulated some straws in designing a tower, she repeatedly uttered, "I am just experimenting". The design artifact grew out of these experimentations. She later explained, "I just started out to make an experiment of the frames and then Arlene taped it; and she said to make it more stable". This experimental nature of the designing process, the union of manipulating artifacts and ideas was also evident in the following transcription from the first day of Chris and Arlene's first bridge design.

107 Chris: I got an idea too, we can just, this could be the bottom one,
108   we tie it around like this, like I might use a knot.
109   [((begins to tie))
110   If it's still too wobbly at the end,
111   [((she moves string and structure which "wobbles"))
112   we can put a knot.
113 Arlene: [And then we go around like that and tie it
114   [((loops string around post))
115   [((ties loop around second post))
116   Is this a little far off? We should put it down a bit
117   [((lowers the loop))
118 Chris: Yeah, a bit down.
119 Arlene: Put it down a bit, right here
120   [((lowers string))
121   before we tie it.

After deciding to make a bridge from string, they began tying a string around two posts "which could be the bottom one" (line 107). Chris uttered, "if it's still wobbly . . ." while simultaneously testing the structure (lines 110–111); and Arlene's planning for the next design move coincided with her tying the string into a loop (lines 113–115). Later, Arlene again reacted to the current state of the artifact, especially to the result of an earlier move; she tested her question "Is this a bit far off?" by lowering the loop of string (lines 116–117). The answer to her question was provided by the artifact. It is in this sense that Gal (1996) and Schön (1987) talk about conversations in which designers engage with their drawings and models. Designers make a move and assess the response from drawings and models.

These transcriptions shows how "thinking" and "acting" co-occurred as children manipulated the materials and artifacts in front of them. My examples show that in children's designing, cognition arises from the interplay of material with social and individual worlds. The design and construction of objects in the world allows individuals to engage in a relationship with their ideas and the knowledge required in their construction. The construction of objects requires children to make explicit decisions about various aspects of their knowing. Designing towers and bridges thereby affords children opportunities to get close to their own ideas. Rather than simply talking about braces, children explore ways of using them, explore their strengths and weaknesses, and even develop new ways of building that make braces dispensable. Through these explorations, children construct understandings of design elements such as braces, stays, deadmen, or forces that act within structures or on structures (stability). They "discover" these in their own actions and interactions. It is in the relationship that children establish with materials, tools, and artifacts that designing artifacts became "concrete", for concreteness is not a property of an object, but of the relation a person establishes with objects. Through this relationship, an action-talk back situation, students had continuous opportunities for interacting with their own ideas as manifested in the artifacts: physical artifacts talk back to children in ways that many other systems do not. (In Chapter 2, p. 31, I wrote about the mathematical modeling program that talked back to the students in ways their notebooks had not done.)

The collective design of artifacts therefore is a powerful context of learning because it emphasizes children's cognition as arising from their actions in material and social settings: cognition, rather than being abstracted from individuals' settings is co-extensive with their acting-in settings. It is in building and playing with physical devices that the incompleteness of re-presentations is exposed. Capitalizing on the holistic relations of agents-in-settings thus allows completeness and universality of cognition to arise from the uniform performance of corresponding physical systems.

8.3.2. *How Artifacts Constrain Interpretive Flexibility*

Based on the analysis of the data, I argue that the emerging design artifacts have important social functions: They make designing collective activity. Design artifacts act as conscription devices in the sense that they
- permit students to engage one another by making direct reference to the artifact (pointing, gesturing),
- focus participants' attention and communication, and

- serve to represent the knowledge negotiated and constructed in the pursuit of some goal.

In this way, design artifacts provide a setting and a backdrop to students' design talk. The presence of the artifacts reduces the interpretive flexibility of students' utterances about their design ideas: Artifact and utterances convergently reduce their respective ontological flexibilities. This conscriptive quality of the artifacts encourages and facilitates extended student conversations and is an important aspect of the activity structure that supports the emergence of meaningful engineering design language.

Design artifacts also provided contexts that allowed students to incorporate new elements into their language without loss of intersubjectivity when they used them in inappropriate and ambiguous ways. Sometimes, they appropriated a word such as "span" from a teacher's discourse on the span of bridges. At first, they often used these words incorrectly. Nevertheless, despite incorrect and ambiguous use (e.g., using span to refer to the piers of a bridge), others understood what they said. Their manipulations of and gestures to the artifact helped listeners to make sense of their utterances. It was only over time that students developed a consistent language (e.g., to refer to the piers of their bridge as "piers" or "posts"). The following two examples illustrate how design artifacts facilitated communication by constraining the interpretive flexibility of design talk.

Our first example takes us back to the situation in which Andy tried to communicate to Tim that the base of his cone top was triangular which made it too difficult to attach it to the square face of the current topmost module (p. 208). (The two modules are shown in Figure 1i.) Andy appeared to interpret Tim's behavior as if the latter did not understand his explanations.

39 Andy: The top, the top, the bottom needs to be square.
40   [((Points to the bottom of Tim's cone))
41 Tim: Square?
42 Andy: Look, all these are squares.
43   [((Touches tops of two "cubes"))

To further reduce the unquestionable interpretive flexibility of this transcript, Figure 2 includes key positions of the actors as shown by the videotape.[13] Figure 2a shows Andy (on the right) as he bent forward to point to the bottom of Tim's cone top (line 40) while uttering what can be heard as "the bottom of the cone top needs to be square". Tim (on the left) looked at Andy and questioned, "Square?" (line 41, Figure 2b). Andy, who had faced

---

[13] Here, interpretive flexibility is used in a reflexive way. It describes not only the relationship of Andy's utterances to the state of affairs he described (square and triangular faces), but also the readers relation to the transcript I reproduced. Interpretive flexibility therefore acts at two levels: At the level of described phenomena, students need to reduce it to achieve common ground and an agreed-upon course of action. At the level of this book, where I attempt to help readers see what I had seen in the situation and on tape.

a. Andy (right): The top, the top, the bottom needs to be square.

c. Andy: Look, all these. . . .

b. Tim (left): Square?

d. . . . are squares.

Fig. 2. Shifting gazes, hand movements, and utterances over and about the artifacts allowed the emergence of common ground between Andy and Tim. The artifacts, networked by hand movements and eye gazes, constituted the material basis of communication.

Tim shifted his regard to the cubes in front of him, and while uttering "Look, all these are square" touched their top faces (lines 42–43, Figures 2c, d). That is, Andy acted as if in response to the question, "Why does the cone top have to have a square bottom?"

In this episode, eyes, hands, and utterances were interconnected to result in communication during which the artifacts acted as background to and topic of the conversation. Andy's hand movements and pointing toward the bottom of the cone top assisted Tim in foregrounding it as the topic of Andy's utterance. Andy then shifted his gaze to see (literally) if Tim had understood. But Tim's rising tone of voice in uttering "Square?" gave reason to believe that he had not understood. Andy shifted his gaze again to the square tops, as if he was asking Tim to follow with his eyes and see for himself. To make sure that Tim focused on the relevant spots, Andy touched the two square top faces of the cubes and then returned his gaze to see if Tim had understood now. We saw earlier in this chapter that following this exchange, Tim agreed to make a square base for his cone top. The two students had established common ground.

We see here that the artifacts assisted the two boys in communicating. In this, as in many situations, communication clearly had a material basis. It is not *just* socially constructed and a matter of talk but is fundamentally intertwined in the coordination of talk, hand movements, and shifting gaze. The coordination work was done against the material background of the two cubes and the cone top. These artifacts therefore functioned like the weft that allowed the boys to weave their emerging actions – the weft – into a network of common understanding. In other words, the artifacts were networked by means of shifting gazes and hand movements and achieved a convergence with the co-occurring utterances which thereby reduced the interpretive flexibility associated with artifacts and utterances.

My second example also clearly shows how artifacts support students' talk because discourse and material world are deeply integrated. Students' communication cannot be understood separate from the artifacts which were thus part of the conversational activities. With the following excerpt, we return to an episode in Chapter 7 during which Tim instructed Renata how to stabilize a pentagon. Figure 3 re-presents the pentagon that served as background to Tim's gestures. (Line numbers are identical to those in the initial re-presentation in Chapter 7, p. 161.)

33   It only takes two sticks to make a giant triangle.
34   Like you see.
35   [((points to the pentagon he had glued into his notebook))
36   I got the shape and then I got two pieces of spaghetti.
37   [((Traces along A_1, A_2))
38   So it's a giant triangle.
39   [((Traces along B))

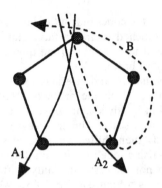

Fig. 3. The pentagon which served as topic and background for the instructions Tim gave to Renata during the stabilization task. The utterance "two pieces of spaghetti . . . so it's a giant triangle" and the hand movement over the pentagon along the lines A_1 and A_2 allowed the establishment of common ground between the two students through semantic convergence.

Renata had wondered how to stabilize and make rigid a pentagonal shape. Because she was not successfully stabilizing her pentagon in her first trials, Tim offered to explain what to do. In his attempts prior to the above transcript, Tim had tried without success to explain with words and gestures in the air how to stabilize a pentagon ("You just take, you only have to put one on the top" and "If you make a triangle, if you make a giant triangle, it keeps it stabler". As if he realized that hearing his utterance was not sufficient to establish common ground with Renata, Tim then encouraged the girl to also employ her vision, "Like you see" (line 34) and pointed to his own pentagon reiterating that he "got the shape" first. Then, he uttered "I got the two pieces" (line 36) while simultaneously tracing along the two diagonal spaghetti pieces (line 37). He reiterated that what he had done, and what he now pointed out to her, would in fact yield the giant triangle he – and the teacher before that – had talked about (line 38). The giant triangle was accompanied by another gesture that outlined where Renata should look to see a triangle. As we previously saw, this explanation was sufficient to allow Renata to successfully stabilize her own pentagon.

Here again, hand movements, gazes, and utterances were involved in establishing common ground. The pentagon including the two diagonal pieces of spaghetti, was available for visual inspection, and Tim's gestures that outlined the "two pieces of spaghetti" and the "giant triangle" encouraged a convergence of the interpretive flexibility of any one of these elements. Thus, the artifacts assisted in design discourses because they afforded a convergence of utterances, gestures, and experienced world.

So far, we saw that talk is not merely social, but has an important material basis; designing is not something that occurs in students' heads, but it has material and social aspects. If we accept these observational and theoretical descriptions, we have to draw important consequences for an issue that has long been claimed as an essential component of science instruction: Problem solving. In the following section I elaborate a different conceptualization of problem solving.

8.4. TOWARD A NEW CONCEPTION OF PROBLEM SOLVING

The approach to cognition developed in this chapter changes how problem framing and solution can be conceptualized. Rather than being an implementation of algorithms, problem framing and solution finding must be viewed as creative acts of networking resources and practices of various kinds; dealing with the interpretive flexibility of situations, people, and artifacts; and negotiating with human and non-human actors to achieve (flexible) goals. Problem framing and solution finding are more like enacting creative dances or engaging in bricolage of resources and practices, materials and human agents, into socio-technical networks.

8.4.1. *Case Studies of Problem Solving*

Here, I illustrate my perspective on problem solving in two case studies. The first case study returns to a crucial point in the history of the tower constructed by Andy, Simon, and Tim. The three have to decide what to do with Tim's cone top and whether or not it can be fit to the already existing tower. It is used to illustrate the notion of reframing, the ontology of problems and solutions, the different levels of complexity of students' problem frames, and the process of negotiating. The activities of the "Engineering for Children: Structures" unit provided opportunities for negotiating the nature of problems and solutions, and future courses of actions. My observations suggest that problems, solutions, and courses of actions are not fixed entities but have a flexible ontology and different complexity. During conversations, each of these entities was set into a convergent relationship with the whole situation which gave rise to their negotiated character.

Case 1: *Designing the Tower*
We return to Andy, Simon, and Tim at that moment where they needed to make decisions about the future design of their tower (Figure 1i). In the following episode, problem and solution frames were proposed and rejected, leading to new courses of action and therefore design. We pick up the tower construction when Tim proposed the top-most module of the tower with a tetrahedral shape ("cone top").

```
01 Tim:    OK, here is the cone top (2.2)
02              [((Holds up cone top))
03 Andy:   Don't make a triangle on the bottom
04         (1.7)
05         that's gonna be hard to put on.
```

06 Tim: I don't put it here
07 Simon: [Make a pyramid, make a pyramid, Andy
08 Andy: A pyramid. Look all these are squares.
09   [((Moves hand above top face of
10 Simon: Now.
11   (1.1)
12 Tim: Yeah, well then, we can put just a few supports like that and
             put it on.
13   [(((Gestures supports across
             top face))
14   (1.2)

In this conversation, the three students repeatedly reframed problems leading
to different possibilities: First, reuse an existing item or, second, make
changes to their present frames. Tim offered his "cone top" as a solution to
an earlier problem (the missing top), but Andy reframed the problem so
that Tim's tetrahedral cone top was represented as an inappropriate solution
(lines 03 and 05). The cone top's triangular base did not fit the square ceiling
of the current top. Rather than accepting Andy's frame, Tim suggested that
their real problem was a missing support, beams across the upper square
face to hold up his cone top (line 12). Such a solution would have required
merely two cross beams which could be fixed with four pins ("a few supports"). Simon also proposed the construction of a pyramid to replace Tim's
tetrahedron.

   However, Simon and Andy did not accept Tim's suggestion as a viable
solution and reframed the problem twice.

15 Andy: No
16 Simon: [Not really, that is too hard, Tim
17 Andy: It's too hard.
18 Tim: No it isn't.
19 Simon: Yes it is.
20 Tim: You only need
21 Andy: [It won't look good then.
22 Tim: You don't *know*.
23 Andy: No, it won't.
24 Simon: It won't work.
25 Tim: What, you want me to cut it down?
26   (1.7)
27 Simon: That would work.
28   (3.2)
29 Andy: Just cut down the bottom [one

First, Andy and Simon suggested that Tim's solution was too difficult to
implement (lines 16 and 17). Then, because Tim disagreed, Andy reframed

the problem as one of aesthetics, "It won't look good then" (line 21). Interestingly enough, Tim suggested that, in the absence of any precedence, they could not know whether or not the tower would look good. Andy and Simon steadfastly reiterated that Tim's solution would not work. It is unclear why Tim accepted the framing of the problem proposed by Simon and Andy, because for most of this project, he had taken the lead. Yet, Tim proposed a course of action which would eliminate the problem Andy framed originally, but which did not include a new solution to the problem how to construct the top (line 25). Andy agreed that Tim needed to take his cone top apart, but his utterance appeared to suggest that Tim should only cut the bottom part.

Simon reiterated a solution he had proposed earlier (line 07), a pyramid which implied a square base corresponding to the existing square bases of all the other modules (lines 30 and 32).

29 Andy: Just cut down the bottom one
30 Simon: [Just make a pyramid.
31 Andy: Just cut down the bottom and don't worry about (?)
32 Simon: [All you have to do, just make
            a pyramid.
33 Andy: The top, the top, the bottom needs to be square.
34   [((Points to the bottom of Tim's cone))
35 Tim: Square?
36 Andy: Look, all these are squares.
37   [((Touches tops of two "cubes"))
38   (1.8)

Andy and Simon continued to barrage Tim, sometimes talking at the same time (e.g., lines 29/30 and 31/32). In an episode analyzed earlier, Andy attempted to help Tim in understanding that the triangular base of his cone top could not make a good interface with the square top faces of the currently existing modules lines 33–38. After Andy pointed out that the bottom of Tim's cone top needed to be square, Tim questioned "Square?" Andy read this as if Tim had not understood their argument about the differently shaped faces that were to be attached and used to form the interface between two modules. However, after this exchange, Tim proposed a course of action that was consistent with Andy and Simon's frames.

In the following section of the episode, the three boys achieved a resolution of their problem and developed plans for new courses of action. Tim suggested that a minor change in his cone top would resolve the problem of the mismatched shapes, but constructed a new problem at the same time. The new square base of his top would be mismatched in size to the existing square top (line 42).

40 Tim: You just need one thing there. ((Cuts open base of cone.))

41   (8.4)
42   It's gonna be a small square.
43   (2.8)
44 Andy: So we make a lot of cubes and make them all smaller.
45   (1.5)
46 Tim: OK, you guys get started and I'm gonna make this.
47 Andy: And I'm making cubes. ((Picks up construction materials))
48   (3.7)
49 Simon: Make it 3 layers high, not
50 Tim: Let's make it sort of like the Empire State Building.

Andy immediately proposed a course of action that would eschew this problem (line 44). His solution consisted of making "a lot of cubes" in decreasing
size which implied a tower from progressively smaller cubes which are part
of many children's toys. All three proposed a next step. Tim would continue
on the cone top, Andy was to make cubes, and Simon suggested to make
three layers of cubes to go between the smallest currently existing cube and
Tim's cone (lines 46–49). Tim gave this solution a shape by suggesting its
resemblance to an existing structure, the Empire State Building (line 50).

Case 2: Designing a Gondola
Arlene and Chris had decided to construct twin towers connected by means
of a gondola. Just before completing the second tower, Arlene proposed to
construct the gondola that would somehow link the two towers. Again, we
can observe an instance of how these students flexibly constituted problems
and resolutions.

51 Arlene: I am gonna make the gondola. How are you gonna make the
                gondola?
52 Chris: Straws.
53 Arlene: Little straws.
54 Chris: You have to cut the pieces and put them like a wall.

During this exchange, Arlene framed the problem of how the gondola should
be constructed. In this exchange, they developed the plan to use little straws
to be assembled like the wall of their towers, that is, with cross-braces that
hold them together (line 54). They began to make short pieces of straw and
join them by sticking pins sideways through them. However, they soon found
that this was tedious so that Chris proposed using cardboard, "Wouldn't it
be easier to use cardboard?" But Arlene suggested that she wanted to
continue with their initial solution. Chris again found that the piece she
worked on was not very stable and that she would need masking tape to
keep the straws together to form a wall.

    Five minutes later, Chris suggested again to go and search for cardboard
as a different solution to their problem of making a gondola.

55 Chris: I am going to ask if we can use cardboard.
56 Arlene: I thought we weren't allowed to.

Chris proposed cardboard (line 55). In the context of the problems they faced in the gondola construction, this has to be considered as a proposal for a solution. Arlene, however, framed this solution as problematic as it was against the rules (line 56). Chris left and, two minutes later, returned with cardboard. Melinda who worked next to them immediately commented that they were "not supposed to use that" but Chris uttered that she had asked. With the cardboard, their approach changed. Chris now proposed and then proceeded to attach each short piece of straw individually to the cardboard by means of a pin. When they found that the wall was not very stable, they added masking tape which, once the cardboard roof was added, provided them with a firm box-like structure.

The problem of the gondola, initially to be constructed from the straw, changed in the course of their activities. First, Arlene and Chris framed the problem as, How do we make a gondola from straws? In the next step, their problem was how to get around the constraint that they were to use only straws. Chris needed to negotiate for more materials. Once they had the cardboard, the construction of the gondola walls changed – although it did not need to. However, the cardboard afforded new solutions; here Chris began to attach each straw individually by means of a pin rather than constructing first a series of walls to be joined later as they had done for both of their towers.

8.4.1.1. *Flexible Constitution of Problems*

These episodes illustrate the ability of Grade 4–5 children to frame and reframe problems, to negotiate solutions, and to achieve consensus with respect to future courses of actions. Specific outcomes of these conversations were virtually impossible to predict because the conversations were in many respects as contingent as the constructions which emerged from the students' actions. Individual differences could not be used to predict the functioning of the group, the dynamics of their interactions, the specific solutions proposed, or the courses of actions which the group would take. Many observers of these conversations were struck by the emergent phenomena they presented. The resulting artifacts were products of groups, emerging from the synergy of individuals rather than the opposition of different viewpoints often associated with rational discourse. As the conversations evolved, problems were framed and disappeared, solutions were proposed and rejected, and courses of actions emerged contingent upon the current state of the construction, possible future states to be achieved, and the meanings associated with different outcomes. In this, students' activities were marked by a flexibility to interpret their current situation resulting in different problem frames and solution options. Such flexibility in framing and reframing situations has been recognized as an important element in the creativity of engineers. Rather than accepting existing problem frames, creative engineers reframe their situations until they find frames which allow them to employ past experiences and solutions, and thus resolve problems (Gal, 1996; Schön, 1983). The present data illustrate that this ability for restructuring fields of experience is not only a characteristic of "creative" engineers but also evident in children's activity.[14]

Children's lack of enculturation in the practices of engineering, which entails an appropriation of cultural resources and practices, probably afforded them more flexibility in framing problems and solutions. We saw the opposite happening in the already described case of Dennis and Sandy, who saw in the glue gun their only tool for making joints. As a result, the two boys did not initially succeed in joining spaghetti to form more durable bridge-building elements. Their disciplinary frame of a glue gun culture provided a rigidity to their approach that became an obstacle to successful resolution of the situation. Only after they began to make building elements by bundling spaghetti could they continue with their cultural frames which included hot gluing.

My past research showed that students' competence in framing problems increases with their familiarity of the context of inquiry (Roth, 1994a; Roth & Bowen, 1993; Roth & Roychoudhury, 1993b). Hence, the longer students have experiences in a given domain, the more they structure their interactions with the environment (perceiving, manipulating, talking about); that is, they learn about the context of inquiry and from their interactions with it. What these studies had left unanswered was how students learned when they interacted with their settings in contingent ways. Here, I identified one process by which students learn in settings characterized by ill-defined problems. There were two steps in this learning process. First, students developed a solution as a contingent response to a problem they had framed. Then, when a similar problem was framed in a new context, previously negotiated solutions could be adapted. From this, problem frames and corresponding solutions could be reused. However, I could not predict how quickly students framed new problems in terms of old ones so that they would shift from developing contingent to reapplying earlier solutions.

8.4.1.2. *Ontology of Problems and Solutions*

In the traditional problem solving literature, the notion of "problem" is treated as if it referred to something as unproblematically as "keyboard"

---

[14] This probably means that such competencies are fundamental to our human condition. However, schools, with their need to make "learning" rationally accountable at all levels, undermine creativity for the sake of easy evaluation and grading. In this sense, schools are in the business of creating "obedient bodies" (Foucault, 1975) that fit into the factories of modernity rather than creating independent and flexible beings. It may be that these students were so flexible because schooling has not yet been able to make them entirely submissive.

refers to the specific object used to generate this sentence. A close examination of student conversations revealed that problems did not exist *a priori* but were constructed as impasses to the dynamic of agents-acting-in-settings. Situations that give rise to an impasse (problem) for some were unproblematic for others. For example, Andy indicated that it was too difficult to fit the triangular base of the "cone top" to the square ceiling of the lower part of the tower. He related the triangle to the square and suggested that the course of action required to connect the two was too difficult (line 05). Tim on the other hand did not see a problem in this contrast. He suggested a course of action that would resolve the issue: the two pieces could be connected by means of an intermediate layer of straws. He suggested mounting straws to form cross beams as an unproblematic course of action (line 12). Andy's problem did not exist for Tim. However, Andy and Simon constructed Tim's proposition as a problem in a double sense. First they considered it "too hard"; subsequently, Andy suggested that the tower "won't look good then". In a similar way, Arlene's problem of constructing the base for the gondola became a non-existing problem once Chris had acquired the cardboard. Acquiring cardboard was here constructed as a smaller impasse than trying to construct the gondola entirely from straws.

We can see that problems did not exist independent of the individuals but as relations between people, setting, and the courses of actions they presumed necessary to change the setting from its current state into some desired future state. Likewise, solutions (final products) and courses of actions which bridged the situation from its current problematic state to the solution state needed to be constructed before they existed and before they could be matched with a problem. From Tim's perspective, because there was no problem, adding two cross beams was simply a matter of course. For Simon and Andy, the proposed course of action did not constitute a feasible solution.

Traditional problem solving research might consider problems in the present context more narrowly and request students to mount the cone top onto the cube. The "correct" solution (known beforehand to the problem solving researcher, teacher, etc.) would be to search for a way so that the tetrahedral cone top would end up on top of the square. Problem solvers' attempts would then be judged against this correct solution. If students did not succeed in reaching the standard, they would be deemed deficient and in need of remediation. However, in everyday out-of-school situations people frequently change the problems with creativity, uncanniness, and innovation. In a similar way, students in this class dealt with problematic situations in creative, uncanny, and innovative ways. Rather than trying to fit the triangle and the square, the three began to consider a change of the cone top into a square pyramid. Once they had "cut down" the tetrahedron, they added another straw which yielded the trunk of a pyramid, and attached it directly to the existing tower without producing a square base. In this way, Andy, Simon, and Tim flexibly negotiated their problem with the setting and adapted their cone top so that it could be attached to the tower. Chris and Arlene also negotiated with the setting, this time a teacher, to make a new course of action possible and therefore resolve an obstacle to the gondola construction. Some readers may consider such situations as failures because students abandoned the original problem. However, out-of-school problem framing and solution finding has just this quality of flexibility: individuals-act-in-settings, construct problems, formulate courses of actions to resolve the problems they framed, exercise alternatives to the solving of a particular problem, or abandon formal approaches to some other kind of solution (Lave, 1988). The outcomes are not failures, but merely different options.

Based on my reading of the entire data corpus, I began to question the traditional ontology of "problems" and "solutions". First, problems in the present context (as those in out-of-school life) were constructed in relation to, and by interacting with, the setting. The appropriateness of a solution was determined by individuals in control both of problem frames and courses of action by which resolutions were achieved. Because objects, events, and the language used to describe them were inherently subject to interpretive flexibility, the same setting which was problematic to some is unproblematic for others. At the same time, problematic situations were reframed so that they changed into unproblematic ones. Thus, problems did not exist as such, prior to students' experience, but were dynamically constructed, reconstructed, resolved, and abandoned. Second, problems of the nature discussed here were constituted as part of lived situations rather than by means of narrowly interpreted texts (i.e., "word problems"). Students' relation to the whole situation allowed them to construct problems as meaningful. Third, "solutions" may pre-date or co-exist with "problems" without being brought together. But, such a conceptualization (pre-dating, co-existence) can be established only a posteriori when problems and solutions have been matched, for problems come to being only when framed, that is, when a particular situation is constructed as problematic, a snag or breakdown, impeding with the achievement of some goal. In a similar way, a solution is not a solution unless framed as such. So while we may say that problems and solutions co-exist, they also have to be framed as pairs.

The issue of snags or breakdowns that impede the achievement of personally relevant goals is important, for it is constitutive of the flexibility of everyday problem solving.[15] Because the everyday pursuit of goals is concernful, personally relevant, people can deal with snags and breakdowns in resourceful ways: what counts is to arrive at the goal, even if this means that goals have to be reframed in the course of action. In schools, however, the

---

[15] Under communist regimes, many people did not personally identify with problems in the economy. As a result, the kind of problem framing and solution finding processes lacked flexibility. Unresolved problems in the economy ultimately led to the fall of many communist regimes. On the other hand, one can observe great flexibility in situations where people have personal interests in the resolution of breakdowns.

goals cannot be reframed and among the potentially many solution paths, only one is legitimated by the teacher and upheld as the standard against which all solutions are compared. Furthermore, drawing on resources such as calling on another person are proscribed in schools whereas it is normal to draw on others in everyday out-of-school situations.

8.4.2. *Micro-, Meso-, and Macro-Problems*

In design environments, problems are of different complexity. Because learning to frame problems and resolutions depends on students' familiarity with the context of their activity, we can expect that it will take longer to resolve less frequently occurring complex problems. In the present classroom community, students framed problems at different levels of complexity which had different consequences for their plans of action. (Readers should keep in mind that the levels cannot clearly be distinguished; different levels are characterized for heuristic purposes only with the understanding that they constitute markers of a continuity of complexities.) The overarching macro-problems students framed were an imaginary object or creature, a tower that had some special, student-selected function or capacity (earthquake-proof, supports an object, features a moving elevator, etc.), a bridge that could span 30 cm and support as much weight as possible, and a mega-structure that provided space for several children. While developing their designs (macro-problems), students framed other problems to which they had to find solutions. For example, a construction may have been considered structurally weak, modular pieces of a tower may have been considered as not fitting, and the construction of gondolas from straws may have been considered too tedious. All of these frames were meso-problems because they themselves entailed more complex but yet undetermined sets of actions. As students pursued their projected solutions for a meso-problem, they framed new problems at a more local level (micro-problems) such as producing braces, joints, or longer beams which then led to solving the stability problem.

Within the rather loose frames set by the teachers, individual students developed their ideas. In groups, they had to come to an agreement about the features of their joint design artifacts. In some instances, students did not commit to one plan but let the joint project emerge to incorporate features of all members' designs. Thus, the tower built by Andy, Simon, and Tim was never planned in the traditional sense of the word. That is, the three did not draw up a plan which could serve as a rigid guide for their future courses of actions. Just what the final tower would be like was at first left open and became more specific as the tower emerged from local contingencies. About mid-way through the project, conceptualizations of their tower as the "Empire State Building", "Eiffel Tower", "CNN Tower", and the "Tower of London" projected possible final states of the constructions. In other cases, students drew a design before they began – such as Arlene and Chris who had prepared a sketch of the gondola. While the two

approaches (leaving details of the plan unspecified and making a drawing of
the joint project) could have led to different types of implementations,
considerable similarities were observed when the students actually began to
construct. In both cases (building with and without architectural design),
there were hosts of problems which students framed during the trajectory
from the initial conception of the design problem to its final solution.

Most of the artifacts' trajectories were continuous; new problems were
generated in view of a project's current status and possible final outcomes.
These problems were at levels different from the macro-problem, because
they arose from the interactions of the students among themselves and with
the materials. Meso-problems were those which entailed more complex plans
of actions and their implementation still gave rise to more local and more
readily solved problems. Thus, Andy, Simon, and Tim's problem to connect
the "cone top" to the existing structure was a meso-problem. Likewise,
increasing the stability of their tower which showed some structural weakness
was a meso-problem. In each case, the problem frame entailed a complex
set of actions each of which could give rise to more local problems. During
the above conversation, the three students considered the solution to include
"cutting down the bottom" of the tetrahedron (lines 25–29), changing it into
a pyramid (lines 07–08, 30, and 32), and then producing a number of cubes
of decreasing size to mediate between the current square top and the square
base of the future pyramid (lines 40–50). Each of these projected steps
entailed themselves situations which could (and did) become problems. For
example, the three straws for the current "cone top" were easily joined by
sticking one pin through all three of them. The same type of solution was
impossible when Tim tried to join four straws to make the tip. There were
also problems with some of the individual joints which the three had to repair
repeatedly. Both of these latter problems were at a micro-level.

The macro-level design problem was thus an emerging complex within an
open horizon. Problems and goals outside psychological laboratories are
themselves open so that they can be flexibly reconstituted to adjust to new
developments, new interpretations, and new knowledge. As students worked
to "concretize" their not yet fully articulated ideas of an earthquake-proof
tower into material form, hosts of meso- and micro-level problems emerged
which could entail changes to the overall plan. In the above conversations,
the three boys had resolved their problem of matching different shapes and
the two girls had overcome their gondola design problem. According to their
common plan, Tim was to reconstruct the cone top into a pyramid (with a
square base), and Andy and Simon would make tapered cubes to be used
as intermediary layers between the current top cube and the new square-
pyramidic cone top. Sometime later, Tim had transformed his cone top into
a pyramid lacking a base. At this point, Tim and Andy decided to attach
this pyramidal trunk directly to the current square ceiling without constructing either a base or intermediate cubes. Here, their previous plan changed
in response to a local contingency. The cubes were not yet ready, and Andy

and Tim saw a possibility to short-cut their earlier plan which gave a new and yet unplanned shape to the upper part of their tower (Figures 1a–c). They articulated a new solution for their tower. Rather than thinking of their project as "Empire State Building", they now conceptualized their construction as the "Eiffel Tower". Thus, the problems students constructed while pursuing their overall design (the earthquake-proof tower) were unforeseeable and could not be predicted because they were the outcomes of situated interactions. From this dynamic situation, students-acting-in-their-settings, emerged new problem frames at meso- and micro-levels.

When they can reuse their problem frames, students often move beyond contingent development of new solution strategies and begin to apply previously successful courses of actions. Being able to develop contingent solutions is part of a trajectory of participation – i.e., learning – by which newcomers to a practice become competent old-timers, that is, students move along trajectories of increasing participation as I described them in Chapters 6 and 7. For heuristic purposes, such trajectories can be divided into four parts. Students frequently begin these trajectories by observing and making tentative experiences; these are followed by significant teacher or peer assistance and subsequently move into the contingent emergence of the practice before the trajectories lead to competent practice. The three boys moved from a contingent solution (to the problem of fitting two modules) to one which they could use as a resource in the future. In a similar way, Arlene and Chris constructed walls of the gondola based on the same principle as they had designed the walls of their tower (uprights held together by diagonal braces). A necessary condition for the development of such competent practices are extended experiences in some domain. Because students solved few macro problems, more meso-, and many micro-level problems, it may be expected that they learned more about framing micro-problems, a level where they developed a greater familiarity. I observed just this. Students built increasingly stronger joints (micro-problems); it took longer to be familiar with meso-level problems such as stability, so that students needed more time to become competent (4–6 weeks). I know little about macro-problems because students had only one attempt at each at towers, bridges, and mega-structures.

8.4.3. Negotiating Problems and Solutions

Many authors in the constructivist tradition suggest that students learn as they attempt to come to shared meanings by negotiating. But this gloss leaves it open just what happens in the process; given the uncertain outcomes of these processes it is unclear how students learn by negotiating. For example, in the above conversation, Tim proposed his cone top which Andy constructed as a problem (lines 03–05). Sometime later, the three had agreed on a course of action to which each of them contributed (lines 41–50). The conversation between those two points achieved this agreement on a problematic issue. Such conversations in which problematic issues are resolved by and within a group of people are often glossed as [negotiations]. To understand what [negotiation] involves, the notion is topicalized in the present analysis.

During students' conversations, various aspects of the situation were foregrounded by the same or different people and then connected with each other. For a shared problem, solution, or course of action to emerge, several individuals had to agree on a common set of relationships between the (arti)facts, events, and discourse to which they attended. Which of the associations constituted by these relationships survived and became shared depended on the strength of the individual relationships and the social and physical settings. These aspects of students' negotiations were also exemplified in the above transcript. Here, Andy contrasted the triangle of Tim's cone top with the square ceiling of the current top. He established this contrast as problematic in emphasizing the difficulty of connecting the cubical and tetrahedral modules. Tim, in turn, established a different relation. Rather than connecting the triangle to the square, he proposed a mediating layer intermediate to the two existing geometric forms ("a few supports like that"). Both of these juxtapositions were material. Andy and Simon judged the work to mediate the latter juxtaposition as "too hard", while Tim thought that "it wasn't". Here, there was a juxtaposition of Tim's versus Andy and Simon's descriptions. The course of action to be taken by the group then depended on finding a material situation (problem and its resolution) such that it received the required support from all agents in the situation. What made it impossible to predict the course of these projects was that I could not know in principle answers to the questions, "Which aspects of the situation students will juxtapose to constitute as problems and respective resolutions?", "What counts as problem, relevant expertise and equipment, and appropriate solution?", and "Which of these juxtapositions garners enough support to determine the required course of action?" I did not observe additive processes in which knowledge was generated and integrated, but conglomerates of disorderly problem-solution conversations, in which different kinds of knowledge were contraposed and checked, and where the outcomes could be considered a function of the persuasive abilities of the participants.

The [negotiations] in this above episode can be understood in terms of a constraint minimization problem (Roth & Duit, 1998). Here, Andy, Simon, and Tim generated propositions designating problems, solutions, constraints, and courses of actions. These propositions, which can be thought of as nodes in a constraint network, included "Here is the cone top", "Don't make a triangle", "Make a pyramid", or "Put a few supports". These propositions were tagged by individual preferences, "It's too hard", "No it isn't [hard]", or "it won't look good"; these tags determine the strength of pull on the nodes. Other tags were more implicit such as a repetition of previous statements (e.g., lines 24 or 27) which usually signal agreement (Roth, 1995b).

These tags weighted the propositions, and therefore the strength of the links, more or less favorably. A final decision emerged not as the result of a deductive conclusion from all the available options, but as that course of action with the least constraints, negative tags, or stress in the network.[16] The pyramid received positive tags by Simon and Andy (lines 07, 08, 30, and 32); cutting down the current cone top was suggested by Tim and flagged positively by Andy and Simon (lines 27 and 29), and making intermediate cubic levels was flagged positively by all three students (lines 44–50). Once the constraint network was satisfied, a resolution could take place. From the pool of propositions, flags, and preferences emerged, unpredictably, the consensus for making a pyramidal top, connected to the existing tower by a few intermediary layers of "cubes". What made it difficult to predict the outcome of the negotiations was the fact that in addition to the publicly stated tags, individual preferences added unstated tags weighting the propositions differently, not only between individuals but also within individuals across time. Different and differently perceived tags give rise to different stresses that ripple through the network and lead to different end states of the network, that is, solutions.[17]

Such interdependence of students who interact with each other, the environment, and the problems, made any prediction of specific outcomes of negotiations impossible because of the simultaneous presence of several ontologically flexible elements of comparable weight, including several individuals, materials available in the setting, and current state of the project. This was not trivial as any changes in one element rippled through the network to affect all others. Such networks cannot be understood by studying each node or link in isolation, because they are more than the sum of the effects of each node or link. The characteristic features of integrated networks (social and physical setting, problem frames) depended on the constraints in the entire network and are not reducible to individual components. Thus, even in those cases where students prepared a design for their construction, it was open for renegotiation between students themselves and with the setting, and unforeseen contingencies led students to make design changes.

More specifically, the students' general plans were flexible enough to allow actions to emerge as a consequence of current individual understandings of what the final outcome should be, the artifacts' current states, and the evolving conversations (as a group achievement). That is, conversations and artifacts were residues of the dynamic situation given by holistic agents-acting-in-settings. This led to a double function of negotiations. First, nego-

---

[16] In multi-dimensional scaling, one goodness of fit index is the stress value. The smaller the stress associated with a solution, the better the fit between model and data.

[17] My model of negotiation as constraint-satisfaction network shares similarities with the connectionist models of cognition (e.g., Elman, 1993). Bereiter (1991) used a network of frisbies linked by elastics to illustrate how connectionist models settle into solutions by adjusting to the stresses and strains in the network as a whole. We used such a model of negotiation for the outcome of student-centered discussions in physics (Roth & Duit, 1998).

tiations projected into the future possible solutions to the overall problem which the students tried to address, an earthquake-proof tower or gondola. In this function, conversations went beyond the actual state of affairs and prepared fields of possible actions. Second, conversations were in important ways anchored to the emerging artifacts.[18]

8.5. DESIGNING AS CONTEXT FOR LEARNING

Here, at the end of this chapter, I want to reiterate my claim that designing constitutes an ideal learning environment distinctly different from traditional ones. The major educational goal in engineering design is that students can develop two important kinds of knowledge necessary for making increasingly intelligent choices and decisions: (a) deep familiarity within a specific domain and (b) strategies for bringing structure to complex and ill-defined (that is, unstructured) problem settings. Chapter 4 provides some of the evidence that students achieved such knowledge as outcomes of their learning. In the process, student-produced artifacts have important functions in that they serve as and support the structuring of the learning environment and problems themselves; they become tools for designing by indissolubly integrating thinking and acting. At the same time, the emerging artifacts constitute a focus and backdrop for students' discursive activities of talking, pointing, and gesturing, that allow them to make sense of each other's utterances and to negotiate shared meanings in the face of ambiguity. Emergent design artifacts embody these negotiations and thus contribute to the convergence of meanings in designer collectivities. In the progression of design processes and with the convergence of meanings, the interpretive flexibility of plans, ideas, and the artifacts themselves are increasingly constrained.

By designing in learning environments such as the community described here, students learn that tools, artifacts, materials, and spoken and written text are interpretively flexible rather than embodying and embedding specific meanings and applications. Students develop a remarkable competence to recognize, and capitalize on, the ontological ambiguities of their settings, a knowledge dimension important in engineering design. Interpretive flexibility is a crucial attribute of learning environments intended to foster student

---

[18] Critics may claim that this indeterminacy of students' plans to achieve a solution to their problem (building an earthquake-proof tower) and the associated negotiations were mere plunder. Here I want to reiterate that children's activities should not be compared to that of architects, engineers, or systems designers during routine activity, but to the design phases in which these engineers construct initial plans or scale models (e.g., Schön, 1983; Sørensen & Levold, 1992; Suchman & Trigg, 1993). Professional activity is characterized by attributes that resemble those I have described here. Furthermore, the fate of some products of professional engineering – the famous collapse of the Tacoma Narrows Bridge or Norwegian engineers who failed attempt to get a second machine to work though their first was still running perfectly (Sørensen & Levold, 1992) – makes me question the clear-cut rationality some associate with engineering design.

innovation, creativity, and negotiation, all of which characterize successful and adaptive technological change. I have additional evidence for these claims from earlier experiences as teacher and researcher. Many of the students who participated in my physics classes where students framed their research problems, developed experimental designs and constructed experimental set ups also participated in engineering competitions. Year after year, these students came in at the top because, in many instances, they flexibly modified their designs and problem framing on the spot and by reinterpreting the given situation. Being able to thrive and capitalize on interpretive flexibility allowed them, like professional engineers, to frame and reframe problems so that they became suitable for resolution, to redefine needs.

In design environments, students learn to exploit interpretive flexibility. Interpretive flexibility is important for understanding the negotiation of meaning, and for constructing design languages, which, though not necessarily those of engineers, are important and productive means for communicating and designing within the elementary students' design community. Students experience first-hand the "social" construction and maintenance of their own design language game without the concomitant emergence of an "anything goes" attitude that teachers often associate with constructivist teaching. In the process of learning through design, teachers, engineers, films, or a science museum become important resources that allow students to link their own design languages to canonical design discourses (cf., Chapters 4 and 7).

The view of elementary students' designing developed here acknowledges the situated nature of cognition by accounting for the contributions of multiple actors in the construction of design artifacts. Thus, although individual actors claim their authorship, the children's artifacts have to be seen as heterogeneous objects whose origins cannot be traced to specific individuals and times. As in the world of adult designing, children's designing is not a linear process of applying individual ideas, the sum total of which will lead to some end product, the design artifact. Rather, designing changes aspects of the setting; and changing the setting changes cognition in fundamental and irreversible ways. An appropriate metaphor for this process of change is evolution. Here, we find not simply an organism of interest adapting to a stable environment, but a complex system in which all the parts continually adapt to each other. Because of the complexity of the system, the trajectories of individual organisms over longer time scales are unpredictable. In this classroom, for example, as soon as children introduced glue guns in increasing ways, the task and artifacts were changed significantly. New tools afford new actions (e.g., building new types of joints) and are reconstructed by agents to fit their goals. The availability of new tools, joints, or engineering techniques for stabilizing structures changes designing itself. In a similar way, new interpretations of design artifacts, tools, interdictions, and materials change design environments and provide new affordances and constraints.

Design learning environments are consistent with the findings of situated cognition. Here, students learn as they are challenged by designs of considerable complexity. Rather than learning an odd collection of decontextualized principles, they contextualize their knowledge in sets of experiences which make learning more meaningful. In this, design learning environment share similarities with the goal-based scenario approach that specifies designing as one of four categories for organizing learning environments (Schank, 1994). Specifically, my students engaged in the explication of the design aspects of their tasks. Questions such as "What was your major problem and how did you solve it?", "What is it about the glue that's really good compared to pins and tape?", and "Which part would go first if you were testing it?" asked children to critically reflect on the decisions they made during the design process.

There is a tension between the notion of design as open-ended activity and the fact that the top-level goals – such as building towers, bridges, and mega-structures – are set by teachers. At a second level, in collective designing, the goals of individual students are subject to group processes in which members negotiate one collective goal. In schools, the specification of top level goals by the teacher has the purpose of specifying pedagogical goals to be achieved through the design activity. To work at all, however, top level goals need to be accessible to learners. For students in this study, the goals included the provision of opportunities for exploring and experiencing critical engineering designing issues, managing complex open-ended design situations, and learning to collectively design, make sense, negotiate, plan, and execute group projects. The learning outcomes documented that the curriculum achieved these goals. After some top level goal has been established by the teacher, students define their own goals and subgoals and subsequently complete the tasks on their own behalf. Gitte and Tammy specified the construction of creatures, towers, bridges, and mega-structures. The specifications for each of these projects was then determined by the students: groups chose to make earthquake-proof towers, wind mills, cranes, towers connected by gondolas, or strong arms (towers sticking out from wall).

Important aspects of design environments are the possibilities emerging from collective designing where group goals arise from negotiations between individual members. That is, for example, children interested in trucks are not in the position to convert the curriculum into a "truck curriculum" but have to negotiate their goals with teachers and peers. However, students can learn to flexibly interpret the overall goal so that they find ways to include their personal goals in the collective goals. Thus, the children in the present study included trucks, other "micro machines" (various toy machines), and dolls as decorative pieces to towers and bridges. In another study, we found more evidence to confirm this claim (McGinn, et al., 1995). For example, one student designed a machine in which his favorite Power Rangers were used as second class levers. Again, his personal goals found legitimate expression within the teachers' top level goals. Furthermore, pursuing one's goals constrained by those of others contributes to learning. Collective designing enriches the learning experience by providing opportunities to concretize and reflect on ideas, and by building arguments which are essentially different than those produced by individuals. Thus, although individuals submit to collective decision making, they gain from opportunities specific to collective activities. Children learn to talk and write engineering design, fundamental to engaging in the communicative aspects that constitute designing. That is, collective designing affords possibilities for learning that arise from the dialectic of flexibility and constraint and the dialectic of cultural production and reproduction of designing communities' resources and practices.

9. NETWORKING INDIVIDUALS AND GROUPS

Throughout this book, we saw that, alongside the material elements, there are essential social aspects to knowing and learning. Both knowing and learning are expressed in and through (changing) participation in everyday settings. Communities are produced as individuals interact with others and enact common practices. Most interactions are within small groups so that communities (their rules of conduct, patterns of behavior, social phenomena) are produced in face-to-face encounters between individuals. That is, students act in some way not *because* of a teacher's power; rather, in the interactions between teacher and students, power is produced. In the same way, "rules" are produced in action. Whether a student "followed" or "broke" a stated rule can only be established after the fact. As we saw in the previous chapter, "Don't use cardboard rolls!" led in some situations to the inclusion of cardboard rolls in a tower, whereas in other instances, cardboard rolls were not included. Communication and culture are therefore produced and maintained continuously through the collective activities of individual persons. Bernard, the teacher of Moussac had realized this and accepted that learning arises from the collective activity of the class. Communities are produced through the networking of people, and the concomitant circulation of stories, artifacts, materials, and practices that are relevant to members.

This chapter is devoted to my observations of the production of the classroom community by small groups. While these processes of cultural production are invisible between individuals, such as inventors and scientists who work in physical isolation,[1] they are readily observable when several individuals work together. This is equally true for homogeneous teams of scientists, engineers, and for medical doctors or heterogeneous teams such as the parents, principal, teachers, and school counselor who make school placement decisions. In all these situations, participants in collective work are likely to vary in their competencies so that they accomplish their tasks by engaging each other in dialogues that allow them to draw on and negotiate their differences. To provide a satisfactory account of individual human performance in collaboration requires on-line studies of groups in natural settings rather than studies of individual cognitive properties. Such studies investigate how students organize their collective activities, negotiate shared norms, and resolve disagreements.

To date, we know little about the structural organization of student-student

---

[1] In a detailed analysis, Whalley (1991) showed that even those artifacts invented and designed in the homes of secluded individuals bear important and significant marks of social influence. Completely "independent" inventing is a myth.

collaborations (Rafal, 1996); and the existing research often refers to pre-school children (Orsolini, 1993) and out-of-school situations (Eckert, 1990). However, these studies provide useful suggestions as to key features of students' organization of their collective activities. Important to the present study, the negotiation of shared norms affirms and requires a sense of community; this sense evolves through the development of shared accounts of common experiences. Disagreements and conflicts are important aspects in the establishment and maintenance of community standards and norms. Consensus may be negotiated by finding a position that includes opposing positions, or by refining one or more positions until disagreements disappear. Research also indicates that the structure of arguments among students differs depending on the presence or absence of the teacher (Roth & Roych-oudhury, 1993a).

Small groups are sites where the individual and culture are produced. Here, individual understandings meet other conceptions, occasioning nego-tiations of understandings until become common sense. At the same time, it is in small groups that culture itself is continuously constructed and reaf-firmed. In every encounter between two or more individuals, the apparently objective reality of social facts is a

... locally, endogenously produced, naturally organised, reflexively accountable, ongoing, prac-tical achievement, being everywhere, always, only, exactly and entirely, members' work, with no time out, and with no possibility of evasion, hiding out, passing, postponement, or buy-outs. (Garfinkel, 1991, p. 11)

Both the production of culture and the contribution of individuals to this production can therefore be documented by studying interactions. Rather than using concepts such as "power" and "attention deficit disorder" as resources for explaining teacher-student and student-student interactions, these concepts have to become topics of research. Thus, the small group is that site where culture not only affects the individual, but it is also continu-ously constructed.

This chapter is concerned with the networking of individuals into groups and into the community. That is, my fundamental question is, "How is the classroom community held together in the course of students' co-participation in activity?" To better understand the structural organization of group inter-actions during student-centered open design, I set out to answer questions such as "How do elementary students structure their collective activity?", "How do elementary students negotiate?", "How do groups interact with other groups?", "How do teachers' interactions with groups contribute to students' collective activity?", and "How do students manage different ac-tivity types in the course of a project?"

To answer these questions, I conducted two intensive case studies involving two groups of students during different design projects: Stan and Tim while designing their bridge and Chris and Arlene while they designed their twin tower connected by a moving gondola.[2] I chose to study Stan and Tim because, *a priori*, the auspices for their collective work would not appear favorable. Both had difficulties in most areas of the curriculum, though Stan's academic "problems" were much more serious than Tim's. Tim usually worked well with others but had previously shown off-task behavior. Stan was diagnosed as suffering from attention deficit hyperactive disorder and had been prescribed medication to control his behavior.[3] As indicated above, Stan's classmates generally refused to team up with him and tended to ignore his attempts to engage in design-related conversations. Even teacher mediation did not help to find him a partner so that he had to work on his own during the first seven weeks of the unit. Based on this information and my initial observations, I selected Stan and Tim as examples of a student group with academic, social, and task management problems.

Arlene and Chris appeared to constitute the polar opposite to Stan and Tim. The two girls were very similar and had been friends inside and outside school for a long time. Arlene and Chris were very independent, managed their time well, and always submitted neat and well-organized work. Both were keen and enthusiastic, enjoyed learning and experimenting with new ideas, and possessed good social and communication skills. Academically, they were above average in all subjects. Like all other Grade 5 students, Arlene and Chris never changed partners and worked together throughout the unit. Tammy emphasized the "social skills" of Arlene and Chris, both of whom described their collaboration as "No major problems". During the initial classroom observations and analyses, I was struck, however, by what appeared to be a rather unbalanced approach to group work. One of my research team members captured this imbalance in the metaphor of "the surgeon and the operating nurse". Chris often made one-word utterances ("pin!", "straw!", "tape!") which observers and Arlene ("Yes ma'am")

---

[2] I analyzed videotapes of five student groups observed over periods of 1–10 hours (two students were observed in the context of different groups). In addition to teacher interactions with these five groups, I also observed one teacher for two hours interacting with eight groups not among these five. My answers to the questions about small-group interactions are based on all data, but are presented in two illustrative case studies. These answers provide important clues to understanding interactions in small groups as the crucial link between individual students and classroom communities. A complementary set of analyses was provided elsewhere (Roth, 1997a). There, the microanalyses focus on lessons not re-presented in the figures below.

[3] Mehan (1993) provides an interesting analysis of the ways students come to receive labels, such as LD (learning disabled) or ADHD (attention deficit hyperactive disorder): the process shows all signs of a social construction in which texts provided by certain individuals with "authority" (school psychologist, school nurse) weigh much more than the voice of a parent. Drugging children seems to be a somewhat unethical quick-fix serving teachers, schools, and pharmaceutical industries rather than the students; but Western societies' ideology to take the individual as unit of analysis for problems seems to legitimize such short-cuts. Integrating such children into communities without drawing on drugs would require to engage with them and their problems and thereby to take up a social responsibility.

heard as commands for materials. Arlene handed Chris the things thus demanded.

My central interest was to document how students made sense of, and organized, their work in small groups. Thus, the unit of analysis had to account for students' momentaneous sense-making activities; these were not always related to the project on which a pair of students presently worked. For example, I filmed Tim and Stan who worked close to Clare and Shelly so that the latter's conversation could be overheard. I observed how the two independent conversations merged into one during which students discussed a problem together or during which Tim taught the girls how to operate a glue gun. At other times, only one student in a pair worked on the project while the other attended to project-unrelated work. The unit of analysis had to reflect this communal, irreducible, distributed, and dynamic nature of work in natural settings. Thus, the unit of analysis had to be (a) useful when applied to the work of the group, (b) flexible and adaptive to changing social and physical settings, and (c) encompass multiple points of views important to the observed students' activity. To establish a frame of reference for the changing units, we parsed the videotapes into 5-second intervals; each interval was then categorized according to the partners' project-related, discursive and practical activity.

As a heuristic, I used six categories for partners' activity to describe varying combinations of student-student-task orientations in dyads:

- Both students worked on aspects of their project independently (*parallel*).
- Both students worked on the same aspect of the project or talked about it (*together*).
- One student worked on the project, the other engaged in project-unrelated activity or watched (*Arlene, Chris, Stan,* or *Tim*).
- Both students engaged in project-unrelated activity (*neither*).
- Students interacted with a teacher (*w/teacher*) on aspects of their project.

I established this heuristic so that I could look for similarities and differences across time within and between groups; but I do not make the claim that these are objective units of analysis or the only possible units of analysis. This heuristic interacts with the length of the analytic segments. Because of transition times between modes of work and the irremediable ambiguity of discursive and practical actions, the characterization of students' work on the basis of the present heuristic becomes sometimes difficult. To make sure that segments were categorized consistently, I used two forms of checks. First, another research team member and I analyzed sections of the tapes independently; subsequently we negotiated the few remaining disagreements. Second, I mapped randomly chosen parts of transcripts/ethnographic description onto the graphical representations of group work and later checked my mappings against the videotapes.[4]

---

[4] The coding was so consistent that two years after first producing it and analyzing the tapes, I could produce, in an independent coding, virtually identical activity graphs.

9.1. NETWORKING WITHIN AND ACROSS GROUPS

Contrasting prior expectations according to which Stan and Tim's cooperation should have been disastrous, teachers and observers agreed that their
joint project turned out to be harmonious and successful. During the lesson
prior to their collaboration, they, as all students, had been asked to draw
bridges as starting points for their designs. However, for their collective
design, students were not required to follow either team member's drawing.
At the end of the bridge project, both teachers and students recognized Tim
and Stan's trussed bridge as a collaborative effort. Stan summarized, "I had
fun with Tim. I had ideas and he had ideas. And so we took our ideas and
made a really good and interesting bridge". During the bridge project, they
had "helped a lot" and "didn't stop working". Both felt so good about their
collective project that they decided to join forces on a subsequent project,
the mega-structure.

What did this joint work, which all participants described as collaborative,
look like? Typically, the students shifted frequently from one activity type
to another. Sometimes students worked together on the same piece or
planned how to proceed; sometimes they worked independently (parallel)
on different aspects of their project; and sometimes one team member engaged in activities unrelated to the project (but, when helping others, nevertheless contributed in important ways to maintaining the classroom community). Figure 1 re-presents the modes of work for Stan and Tim over a
period of about 40 minutes. One can see that Stan and Tim did not work
on the bridge all of the time. Rather, the two shifted back and forth between
the different types of individual and collective activities within their group
and with other groups or the teacher. Figure 1 can be seen as a state diagram
of a weaver's shuttle that works across the warp constituted by the lines of
activity types; the shifting between the activity types constitutes the weft that
constructs the network and holds groups and community together. This was
no different for Chris and Arlene (Figure 2). Here, too, we can see a
continuous shift between different modes of working that ultimately network
the individual into the group and community.

This observation seems significant to me, for it is so different from traditional schooling – especially in European countries – where children are
forced to sit and listen to the teacher and are accorded few or no interactions
with other students. Here, however, I observed in many instances a
functioning community in which resources and practices freely circulated.
These activity graphs highlight how the back and forth weaving of individuals
from attending to some task to attending to other group members, to attending to specific peers, teachers, and generalized others ("everyone else is
doing it").

In Figure 3, I summarize the information that appears in Figures 1 and 2
and contrast it with that of complementary lessons (the activity graphs of
which were provided in Roth [1997a]). Stan and Tim simultaneously engaged

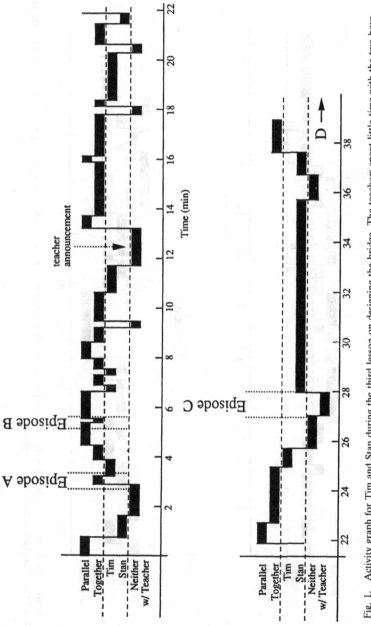

Fig. 1. Activity graph for Tim and Stan during the third lesson on designing the bridge. The teachers spent little time with the two boys.

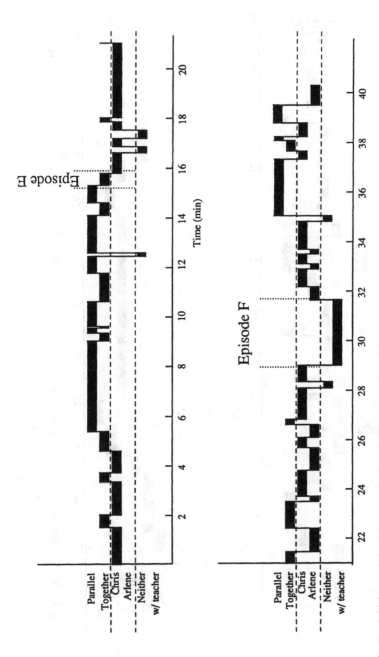

Fig. 2. Acitivity graph for Arlene and Chris during the second day of their tower project when both worked on the project at the same time, the two girls worked more frequently in parallel than together.

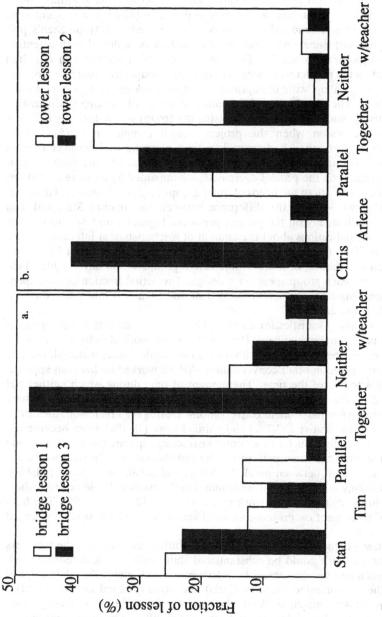

Fig. 3.   Fraction of lesson spent by four students during two lessons. For Stan and Tim, the lessons compared were the first and third on bridge design ($\chi^2 = 14.48$, $p < 0.05$). For Chris and Arlene, the lessons compared were the first and second on the tower design ($\chi^2 = 12.47$, $p < 0.05$). When Stan is paired with Chris, and Tim with Arlene, there is a significant difference between the two distributions ($\chi^2 = 28.39$, $p < 0.0001$).

in project-related discursive or practical activity for about 50% of the first lesson, most of this in a collaborative mode. Stan spent substantially more time, 25.8% of the lesson, on the project than Tim (12.2%), who spent most of his off-project time helping others. This focused activity on Stan's part was quite surprising given that he was said to be afflicted with attention deficit hyperactive disorder. The time during which neither Tim nor Stan worked on the project was spent helping other groups, listening to announcements, or dealing with disruptions by other students and visitors to the classroom. The distribution of the different modes of working (individually, collectively, and with individuals outside the group) was statistically different during the lesson when the project neared completion $(\chi^2(5) = 14.48, p < 0.05)$. Here, they had spent less time planning their project than in parallel work to finish the bridge. The total amount of time both simultaneously spent on the project decreased, accompanied by an increased attention of one or both to the needs of other groups (see also Figure 1). However, over the two lessons, the difference between the amount Stan and Tim worked individually on the project persisted. Figures 1 and 3a also provide interesting indications about the amount of teacher-student interactions with Stan and Tim. Given that there were two teachers and a total of 14 groups, an even distribution of teacher time across groups would have yielded 14% of the time each group spent on a design. The actual fraction of time spent in teacher-student interactions with Tim and Stan was much less on both days (6.5 and 2.4%).

The summary statistics for Chris and Arlene also indicate rather dramatic differences in individual work (Figure 3b). Chris worked individually on the tower project between 30 and 40% of the time while Arlene watched, listened and participated in other conversations. Arlene worked on her own approximately 4 to 6% of the time. The amount of time during which neither girl worked on the project was negligible (2.3%), and about 4% of their time was spent in teacher-student conversations. During the first lesson, the differences in collaborative (37.7%) and parallel work (18.4%) arose because the two girls spending much of the time constructing separate pieces. The second lesson showed the same pattern for the distribution of individual work, but there was a shift between parallel (16.8%) and collaborative work (30.4%). It was mainly this shift which contributed to the statistically detectable differences in the distribution of work modes $(\chi^2(5) = 12.47, p < 0.05)$. The fraction of time spent on project-unrelated activity or with the teacher changed marginally.

To test whether the observed qualitative differences between the two boys and the two girls could be substantiated statistically, I calculated the work type fractions averaged across both lessons for each group. For the comparison, the task-oriented (Stan vs Chris) and other-oriented individuals (Tim vs Arlene) were matched. While the within-group differences across periods were only slightly reliable, the between-group differences were considerable $(\chi^2(5) = 28.39, p < 0.0001)$. Although both groups spent about 50% of their time working simultaneously on the project, most of Stan and Tim's work was truly collaborative (in the everyday sense of the word) while Arlene and Chris worked together as much as they worked in parallel. The differences between project- and other-oriented individuals were more pronounced between the two girls, but both boys spent more time interacting with others on non-project related matters.

Figure 3 provides a coarse-grained summary analysis of four students' activity relative to their design tasks. One may be tempted to say that one or both students in each group spent a considerable amount of their time "off-task" and in activities not related to their own project. In this formulation, the statement carries negative connotations and value judgments. However, such a perspective captures only part of the situation and may be the least important. In their own evaluation, Stan and Tim spent a lot of time helping others. Further support for their own assessment comes from my observations, for example, in the description of their crucial role in the trajectories of Clare's and Shelly's competencies in glue-gun related practices. Both boys contributed to a considerable extent in the networking of the community and production of knowledge at the level of the community. A second assumption underlying a negative evaluation of the observed off-task time is that learning arises only from actually manipulating one's own materials. This assumption, too, does not bear out in my observations. Time and again, students who wandered around the classroom observing others at their work later returned to their own construction sites with plans for new courses of actions to problems they had earlier constructed.

9.2. CASE STUDIES OF NETWORKING

9.2.1. *Networking Within Groups*

How do students manage different activity types in the course of a project? Typically, students shifted frequently from one activity type to another (Figures 1 and 2). Sometimes students worked together on the same piece, or planned how to proceed; sometimes they worked independently (parallel) on different aspects of their project; and sometimes one team member engaged in activities unrelated to the project (but, when helping others, nevertheless contributed in important ways to maintaining the classroom community). This was no difference for Stan and Tim or Chris and Arlene. The following episodes illustrate such shifts between activity types and point out some characteristics of the interactional styles.

The following episode derives from Stan and Tim's bridge project. In Episode A (Figure 1), Tim and Stan first shifted from the "neither" mode to working together on the same thing, and ended with Tim gluing some pieces while Stan watched. This episode shows how Tim and Stan structured their interaction; from this interaction later emerged the choice of a material and the decision about a course of action. (The time added to the transcripts

in this chapter allows cross-referencing with the associated figures and
provides an indication of the overall flow and beat of events.)

```
01 00:02:33    Stan:    It's melted.
02                      (3.0)
03 00:02:37    Tim:     I can smell it.
04                      (2.1)
05                      The glue's melting. Okay, now get the piece of
06 00:02:40             wood that we're gonna use (.) for the base.
07                      (3.2)
08 00:02:47    Stan:    Here is the base.
09                      [((Picks up bridge artifact. Pulls flat piece of card-
10                      board from underneath))
11             Tim:     That's the base?
12 00:02:50    Stan:    That's the base.
13             Tim:     You just want it on the side, so you can,
                        that on the whole.
14             Stan:    [You already
15 00:02:57    Tim:     Like this.
16             Stan:    Yeah.
17                      (0.9)
18 00:02:59    Tim:     Ok, I'm gonna have to glue.
19 00:03:05             ((Tim starts gluing. Stan walks around him,
                        watches.))
```

At the beginning of this episode, both students waited around. Their glue
gun was not hot enough to be used. After Tim noted that the glue in the
gun was melting, he asked Stan to hand him the piece of wood they were to
use as base for the bridge (lines 05–06). Stan, however, picked up and
handed him a piece of cardboard (lines 08–10). Astonished, Tim asked
whether this was in fact the base. After Stan affirmed this, Tim provided
instructions for the course of action to be taken to glue the bridge onto the
base (lines 13 and 15).

In this episode, the decision about the material and the specific piece for
the base arose, unpredictably, from the interaction. Although they had earlier talked about gluing the bridge to a wooden support, Stan's presentation
of the cardboard and affirmation that this was the base became normative.
That is, the factual choice he made in this situation determined the course
of his future actions and design. The cardboard was there, whereas the
wooden base Tim asked for was not immediately available. Without further
discussion or question, the existing piece of cardboard became the material
of choice. This way of decision making was probably facilitated by the
flexibility both had shown earlier in interpreting materials; that is, they made
do with the things at hand in the way tinkerers and bricoleurs get their work
done. Here, there appeared to be enough common ground about what

requirements needed to be met by a piece of material to complete the desired job. Without hesitation, Tim then proposed the necessary course of action by specifying the position in which Stan had to hold the base so that the existing bridge structure could be glued to it.

Here, the transition between different modes of work emerged without the necessity for further discussion. Even when faced with a novelty, such as the material for the base, the two students shared enough common ground to make decisions as a matter of course and taking into account the contingency of their situation.

The following episode was excerpted from Arlene's and Chris' work on the tower (Figure 2, Episode E). As we can see from Figure 2, the episode represents a transition from the parallel mode to the together mode and back to the parallel mode of work. Here, the shift was clearly indicated by the shift in Arlene's body towards Chris, thereby making the connection which later was interrupted when she returned, resulting in the parallel mode.

20	00:15:19	Chris:	Arlene?
21			(1.8)
22			Hm::: ((Studies her artifact))
23			(1.0)
24	00:15:22		What happened?
25			[((Arlene turns head to face Chris.))
26			(1.3)
27	00:15:24	Arlene:	I killed you, that is what happened.
28			(4.7) ((Puts down pen, turns whole body towards
29			Chris' artifact.))
30			See, hold it like this.
31			[((Takes artifact and shifts joints to
32			straighten it; one joint comes undone.))
33	00:15:35	Chris:	*No*, Arly, *no*, look at that.
34			[((Points to joint that had come un-
			done.))
35			(1.6)
36	00:15:38	Arlene:	Sorry.
37			(2.2)
38	00:15:40		Maybe it is, a::m
39			(5.0)
40			flat.
41		Chris:	A flat one.
42	00:15:48	Arlene:	Maybe a (?) flap.
43	00:15:50		[((Arlene shifts body back to her own arti-
44			fact, but then looks up to listen to another con-
45			versation without continuing her own task.))

At the beginning of this episode, Chris noted that one of the joints in their second tower was shifted. In the course of the conversation, the two try together to resolve the matter of the problematic joint. However, as Arlene tried to straighten the joint, it came apart altogether (lines 30–32). In the end, they arrived at repairing the joint, yet without being able to straighten it. Both returned to their previous tasks and into the parallel mode of work. The way students made such transitions is further elaborated by the sequence of images from the videotape that show the bodily orientations of the students (Figure 4). Initially, both worked on separate pieces (Figure 4a). Chris called Arlene's attention who then turned to face her (Figure 4b, lines 24–25). Arlene then shifted her entire body and began to manipulate the artifact, but the joint came undone (Figure 4c, Lines 28–32). Chris also began to put her hands to the artifact so that, for a moment, the girls handled it simultaneously. Chris' utterance "No, Arly, no, look at that" and Arlene's response, "Sorry", suggest that the blame for the undone joint was laid on, and accepted by, Arlene. Arlene then shifted her body back, looked at the piece she worked on before, and the shift from the together mode to the parallel mode was accomplished.

The transcript and photographs make the networking and weaving metaphors strongly suggestive. Arlene's shifting body and attention allowed the work mode to shift so that the girls networked, resulting in the start of their exchange. (It is in such "exchanges" that they circulate stories, resources and practices.) The video off-prints (Figure 4) and the diagram of the shifting work modes (Figure 2) express the same shifting pattern that characterized the work of students in the community. As in the sequence of photographs that showed Andy and Tim (Chapter 8), the present sequence is centered around an artifact. Students' utterances make sense because they are over and about something that is available to both at the same time and – by default and until further notice – in the same way. The utterances are about the artifact in that students were communicating something that they wanted the other to attend to or notice. When the girls said, "Hold it like this" or "Look at that", they encouraged the other to hold the artifact in a particular way that straightens the problematic joint or to notice something about the tower. At the same time, these utterances were about the artifact, for the indexical terms "this" and "that" only make sense if there is something in the situation, available to both, to which these refer. "No, Arly, no, look at that" was meaningful to Arlene because the utterance is spoken against a background constituted by the second tower with its undone joint: The utterance was "over" the artifact.

The importance of the students' body orientation for networking and connecting was further evidenced in episodes such as the following (Figure 1, Episode B). Here, Tim suggested how they could enhance their bridge by mounting straws in a particular way indicated by holding the straws at the place he wanted them (Figure 5a, lines 46–48). At this time, Stan focused on the straws he was working on.

a. Arlene, What happened?

b. I killed you, that is what happened.

c. See, hold it like this.

d. *No*, Arly, *no*, look at that.

e. Maybe a (?) flap.

Fig. 4. Arlene and Chris shift from working in parallel to working together. The artifact is at the center and provides a supporting structure for the networking of the individuals to a collective. Arlene's body orientation to Chris and the artifact further supports the networking of the individuals.

46 00:04:59	Tim:	Stan, why don't we connect some straws like this?
47		[((holds straw dia-
48		gonally in the top part of his bridge.))
49	Stan:	[((looks up,
50		then back at his work and continues))
51		(1.1)
52 00:05:03		No, we are not allowed to use straws.
53		(7.2)
54 00:05:11		We are not allowed to use straws.
55		[((Turns to Tim and the artifact.))
56		(2.5)
57 00:05:15	Tim:	Go get some cardboard (.) so we can go like that.
58		[((gestures dia-
59		gonal between piers.))
60		I go get (?).
61 00:05:20		[((Stan walks away; Tim walks away to approach a teacher))

Stan briefly looked up and, after a short delay during which he had continued working on his piece, suggested "No, we are not allowed to use straws" (Figure 5b, lines 49–52). Tim did not say anything. After a long pause, Stan looked up again, walked towards Tim and reiterated his statement in a determined way (Figure 5c, lines 54–55). After a brief pause, Tim suggested that Stan should go and get some cardboard; his gesture indicated that he meant to use it between the piers (Figure 5d, lines 57–59). Both walked away to search for the additional materials.

In this episode, the two shifted from parallel work to a brief exchange during which they resolved a potential conflict between the course of action proposed by Tim and the teacher-set rule that specified no additional materials which was evoked by Stan. At the end of the episode, both engaged in independent searches for the construction material required to continue the design in the way Tim had suggested (line 57). It was strikingly clear that Tim did not react to Stan's first evocation of the rule. Stan, only looked up briefly, and by the time he invoked the rule, was already oriented again to his own task. This orientation was visible in the movement of his head and upper body. When Stan invoked the rule for the second time, he already approached Tim. His entire body was oriented and close to that of his class mate. This time, Tim attended to Stan's invocation of the rule and proposed an alternate course of action. Thus, in the first instance when the connection between the two was not clearly established, the rule was not circulated. During the second instance, Stan clearly sought to establish the contact with Tim. This time, Tim reacted in a way so that, *a posteriori*, the rule could be said to have been followed.

In Chapters 5 through 7, we saw that knowledge was built at the community level by means of circulating of resources and practices. In the present

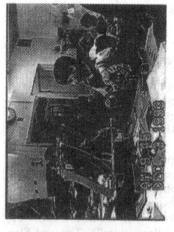

a. "Stan, why don't we connect some straws like this?"

c. "We are not allowed to use straws."

b. "No, we are not allowed to use straws."

d. "Go get some cardboard so we can go like that. I go get (?)."

Fig. 5. In his first attempt, Stan could not convince Tim that "they were not allowed to use straws." To make networking possible, Stan actually needed to shift his body, walk up to Tim, and reiterate the interdiction.

examples, we see microprocesses of networking which afford the circulation of resources – such as the rule "No additional materials" – and practices to occur. Physical closeness and body orientation are important aspects by means of which individual actors make possible groups, collective activity, and joint decision making. The video off-prints in this chapter provide a stark contrast to that in Chapter 5 which illustrated how body positioning contributed to the marginalization of Stan. While body positioning is not the entire story to networking into communities, the evidence provided here illustrates that it contributes to this phenomenon.

9.2.2. Networking between Groups

In many science classrooms, students are required to stay in their seats or assigned laboratory places, and teachers attempt to minimize the exchange of information between individuals and groups, by, among other strategies, labeling interactions as "cheating" and punishing pupils for such behavior. In the present classroom, interactions with other students were not only allowed, but made up for an important aspect of the classroom culture. When individual students or whole groups were not working on their projects, they were often helping other students or getting inspirations for problems and solutions in their own project. The videotapes revealed this was true for Tim and Stan. Furthermore, the two characterized their teamwork as helpers of others ("We wouldn't stop working and helped a lot"). Tim and Stan spent much of their off-project time helping others or interacting with them in some other, project-related ways. Thus, other students' projects cannot be understood without the contributions made by Stan and Tim. In Chapter 6, we already saw how the co-participation of Clare and Shelly with Tim, and to a smaller extent Stan, permitted the two girls to traverse a trajectory of increasing competence in the glue gun practice.

The following episode from Day 3 of the bridge project shows an additional aspect of their learning. (This episode was not part of the sequence represented in Figure 1, but was part of the overall analyses such as that represented in Figure 3 and associated statistics.) Immediately preceding the transcript, Clare was searching for Tim around the classroom. When she finally approached him, she appeared to be in despair about the glue guns which did not function. Tim looked up from his work and suggested that they use his own.

62 01:044:50	Clare:	Tim, our glue gun is not working.
63	Tim:	Use mine.
64 00:44:55	Clare:	No, it's not working either.
65		(1.1)
66		No. ((Tim and Clare turn around and move
67		towards the construction desk.))
68		(4.2)

69 00:45:02		See this!
70	Tim:	The circuit blew.
71	Shelly:	What's the matter with our circuit?
72 00:45:08	Tim:	This circuit blew, nothing's working.
73 00:45:12	Clare:	I'm gonna take it over there.
74		[((Points to other power bar.))
75	MR:	[Didn't you turn it back on?
76 00:45:15	Clare:	We tried to, we tried everything
77		(1.6)
78		We are going to take it to the other side.
79	Shelly:	[Well, our
80		circuit at home (.) never blew.

Tim did not even look up when he responded to Clare (line 63). However, when she insisted that his glue gun was not working either, Tim turned around, then walked with her to the construction table. Clare showed him the glue gun (line 69) and Tim responded by suggesting that the circuit had been interrupted. He reiterated and explained his statement by indicating that no glue gun ("nothing") at the table worked. Clare suggested that she would take her glue gun to the other outlet (lines 73 and 78). It is not clear whether she had actually tried to reset the fuse as I asked her. Given that she had tried "everything", she engaged in the course of action earlier proposed.

As in the previous exchanges, connecting with members from other groups required closeness and orientation.[5] Tim's first response, "Use mine", uttered without lifting his head could result in warding another person off. In it's brevity and lack of orientation to the other, Tim's actions might suggest that he did not want to be interrupted in his own work. Only after he began orienting himself in the direction of Clare could she be certain of his engagement. It was at this point that Clare and Shelly could learn more about operating a glue gun – because it requires electricity, one needs to make sure that the electrical circuit is working properly.

Such interactions between the two boys and other students or groups were frequent as can be judged from the amount of time either or both students spent away from their own project (Figure 1); this time increased as the boys neared the end of their project. In this way, Tim and Stan were important contributors to the formation of the classroom community. Their willingness

---

[5] In a subsequent study, we confirmed this observation (Roth & McGinn, 1996). We found that each group worked in something like a sensitive area which we conceptualized as a stage. To receive attention from a group, others – students or teachers – had to enter this sensitive area, walk on stage, so to speak. At this time, it was possible to establish a connection and therefore, to circulate resources and practices.

to abandon their own project for a while to help another group was an important aspect in the functioning of the classroom. It is possible to interpret Tim and Stan's activities away from their own project (Figures 1 and 3a) as evidence for their off-project (worse, off-task) behavior. However, talking and interacting with other students actually contributed in important ways at another level of learning; the construction of a knowledge-building community. That both teachers did not interact more with Stan and Tim can be interpreted as a recognition of their resourcefulness and ability to deal with the project successfully on their own, and that helping others did not distract them significantly. As it happened, Tim and Stan also were among the first to successfully complete their project.

These interactions among groups, then, contributed in important ways to the fact that projects incorporated contributions whose originators could not be localized. The constructions of all groups incorporated social aspects in important ways leading to moments where groups might use a technique, artifact, or design feature "because everyone else was doing it".

9.2.3. Teachers' Contribution to Network Construction

9.2.3.1. Instituting Constraints

An important aspect of classrooms are the interactions between students and teachers. In these interactions, teachers both constrained and scaffolded students' activities. The present episode illustrates how teachers attempted to constrain student activities, frequently with the intent to make specific learning experiences occur, such as learning how to use braces for stabilizing structures. During the two periods represented in Figure 3a, three of the four interactions were initiated by Tim and Stan for purposes of negotiating the release of some constraint on building materials (the initial design specifications stated specific building and fastening materials and their amounts). Gitte's interactions with students were guided by the goal of having children work under the same constraints as those under which engineers have to work. Whereas this goal was in agreement with the design philosophy of the curriculum and with my own interests in authentic learning environments, Gitte wanted to implement this goal irrespective of the children's needs. In the same way, Tammy's interactions with the students led to the circulation of rules and constraints, but also to negotiations in which rules and constraints were differently interpreted and enacted.

In the following episode (Figure 1, Episode C), Tammy's impact on the testing procedure was considerable as the boys left the meeting to cut the requested holes in the place indicated by Tammy. Here, the conversation was again initiated by Stan and Tim who were in the process of searching for the wooden blocks that others in the class, such as Jeff and John, had already used for testing the strength and stability of their own bridges. Prior to Stan's question, both boys had walked about in the classroom for over a minute to show their bridge to peers.

81	00:26:51	Stan:	Can we have the blocks for our?
82		Tammy:	Which blocks? Which kind of blocks?
83	00:26:57	Tim:	The blocks for testing.
84			(6.5) ((Tammy looks at the bridge.))
85	00:27:05	Tammy:	Interesting. You added this.
86			[((Gestures along X-shaped braces.))
87		Tim:	Yes, I that that this
88		Tammy:	[And where do you want, where is
89	00:27:09		the thirty centimeters. Where are they?
90	00:27:14	Tim:	The thirty centimeter are here.
91			[((Gestures the between outer
92			and middle pier.))
93	00:27:17	Tammy:	Ah! If you want to test, you have to test here.
94			[((Points to
95			center of half-deck.))
96			You'll have to make holes with the Exacto knife
97	00:27:21		which have the form of a square, here and here.
98			[((Points to
99			support in the middle of each bridge-half.))
100	00:27:28		So that you can suspend a bucket underneath.
101		Stan:	We are going to put the bucket over here.
102			[((Gestures at skew-
103			ers above bridge deck.))
104	00:27:35	Tammy:	No, no, the bucket will be suspended around the
105			bridge here, and we are going to take the string
106			of the bucket here and put it here.
107			[((Points to each center of bridge deck.))
108	00:27:50		((Boys go back to their construction site.))

In this episode, Tammy reiterated the conditions under which the bridge was to be tested. When Stan proposed an alternative, Tammy insisted and pointed directly to the place where the string supporting the load had to be fastened. Later, Stan actually cut square holes into the base through which the two boys fed the string to suspend it from the bridge deck. However, not all tests were conducted such that Tammy's prescription could be said to be a description. We saw that her prescription of the tower design task for Ron and Peter was changed by them ad hoc without further interactions and negotiations. In the conversation between Stan and Tim earlier in this chapter, Tim did not use straws to build braces that would have strengthened the bridge after Stan had reiterated the teachers' rule about additional material.

But a little later, Tim suggested to "just use" skewers (though he later asked one of the adults if he could use them).

9.2.3.2. *Scaffolding the Construction of Accounts of Collective Activity*

Many of the teacher-initiated student-teacher interactions became occasions for constructing accounts of past actions or plans for future design changes. Whereas it is easily apparent to recognize teachers' contributions to design changes (the teacher's part is often referred to as "scaffolding"), the collaboration of a teacher in constructing accounts of past events was less obvious. Such accounts were both helpful for teachers to get a sense of the process by means of which individual projects evolved and the interactional and design problems groups faced. Based on these accounts and the current state of the project against which accounts were always compared, teachers decided to intervene to help students reflect on their work and consider other possibilities or, when they felt an intervention was unnecessary, to move on to another group. The following analysis of Episode F (Figure 2) elaborates the point that these accounts are not simply co-constructed by the children, but that the teacher contributed in important ways. More importantly, although an observer of the videotape may have concluded that Chris did most of the work, the account emphasizes the joint responsibility taken by both Arlene and Chris.

109 00:29:49	Gitte:	This is very clever (.) how did you come up with the idea?
110	Chris:	Well, I started to build the frame a::nd
111	Arlene:	Then we did the::se to make it stronger.
112		[((Takes structure by the uprights.))
113	Chris:	No, put the triangles.
114		[((Takes structure from Arlene's
115		hands, shows small, end sides of structure.))
116	Arlene:	Yeah.
117		(1.8)
118 00:30:01	Gitte:	Good old triangles, he:
119	Chris:	Yeah
120		(1.0)
121 00:30:03	Arlene:	And then we did the::se and then made the glides.
122		[((Glides finger along brace.))
123	Chris:	[and we put this in as
124 00:30:08	Arlene:	People walk along there and then up on the ceil-
125		ing, you know, then we still have to make the other one.
126 00:30:15		And then we move.

```
127                                      [((Rocks hands back and forth))
128                    (1.3)
129                    And this is the gondola.
```

In this episode, Chris and Arlene constructed an account of their work to date and re/constructed their design plan. This account was collectively produced by the two girls who changed turns to add to and often complete the other's sentence (lines 110/111 and 123/124). At other times, they talked simultaneously (lines 121/123) or repeated parts of the other's utterance. This fit of their respective utterances is an indication that there was a lot of common ground between the two girls in regards to the project and future design plans.

The interactions with the teacher contributed to the construction of the project as a collective achievement. It was during those moments that they accounted for their work, and frequently independent of what had happened, both girls took responsibility for developing designs. The present observations underscored the symmetric nature of their relationship and the intersubjectivity (sharing a common situation definition and knowing about it) which existed between them. This symmetry was also apparent in their account which emphasized the shared responsibility for the project's current state and the overall goal. In their generalized account, both emphasized the joint nature by using the plural pronoun "we" for their own and the other's contribution. Although Arlene frequently assisted Chris when asked (see the transitions in the activity graph from "Chris" to "Together", or from "Parallel" to "Together" [Figure 2]). This extensive use of "we" was not only observed during their construction of an account, but was the exclusive pronoun during much of their planning effort such that a typical sequence of utterances went like "I don't think we are allowed to do the bottom thing", "So how are we gonna? What do *we* do?", "*We* attach these and these", "*We* should tack them probably", and "*We've* got another one on the end".

9.3. NETWORKING AND THE EMERGENCE OF CULTURE, POWER, AND NORMS

The present chapter constituted an attempt to open "Pandora's box of horrors" on collaboration. I wanted to understand the role of individuals, groups, and classroom community in collective activities in an elementary science classroom. That is, the present study topicalized "collaboration" by providing answers to questions such as "How do elementary students structure their collective activity?", "How do elementary students negotiate?", "How do groups interact with other groups?", "How do teachers' interactions with groups contribute to students' collective activity?", and "How do students manage different activity types in the course of a project?" The foregoing analyses showed that "collaboration" can describe quite distinct patterns of inter/actions and may include joint conversational and material action, parallel and individual work within the group, and interactions with other groups and teachers. For this reason, the notion of "collective work" may be more appropriate then cooperation or collaboration. Furthermore, the labels "on-task" and "off-task" appear too simplistic to describe the complex interactional structures between members of this classroom community.

In the context of two different groups and projects, characteristic and prevalent interactional patterns (within and across groups) have been documented and analyzed. Thus, despite overall structural differences between groups, there were similarities across groups in the interactional patterns that could be observed. The overall differences between the two groups were quite apparent from the summary graphs and the distributions across time (Figure 3). Arlene and Chris' project developed continuously, rhythmically punctuated by their occasional attention to others or to interaction with a teacher. Stan and Tim, on the other hand, frequently interacted with others in various forms, helping them in the execution of material actions or providing suggestions for dealing with problematic issues. Another important difference lay in the inordinate amount of individual work done by Chris while Arlene either watched or attended to activities elsewhere in the classroom. On the other hand, the time during which Stan and Tim individually or collectively did not attend to their bridge project was spent by helping others. Much of the time during which the two boys worked simultaneously on the project was spent working on the artifact at the same time, one partner supporting the other's activities; but especially during the second period analyzed, the two girls worked much of the time in parallel while talking not at all or about project-unrelated topics. However, much like Rafal (1996), I do not take the differences in the frequency of talk turns as an indication of the significance of a group member's contribution to a collective project. What is counted as contribution is an empirical matter that has to be solved anew each and every time. It takes microanalyses of collective task situations to unravel the dynamics and interactive qualities of interactions.

Whereas the two projects (both as products and processes) were at least partly well described by the label "collaborative" in its everyday use, there were some striking aspects in each of them. First, I was struck by the rather conflict-free relationship between the two boys. Tim had been a leading figure in his earlier group projects. Stan was rejected by his classmates and could not find a partner. A priori, there was a potential for a dysfunctional group. Quite surprisingly, however, the most fitting adjectives were harmonious and productive. On the other hand, despite their matching abilities, the girls' activities were, at least on the surface, characterized by Chris' apparently dominant role and the more or less supporting function Arlene seemed to play. The importance of Arlene's contributions to the overall success was evidenced only through microanalysis of those moments when important decisions about design, plans of action, and choice of materials were made.

An important common aspect of both groups, and with few exceptions, of all groups in this class, was the preference for agreement during planning and execution of projects and during the production of accounts of the activities and projects (during group presentations to the class and interactions with the teachers). There were no observable differences in this preference between groups constituted of girls or boys. In this, my observations differed from those made in some scientific laboratories (Woolgar, 1990) or physics classes of older students (Roth, 1994b). Scientists appear to have a preference for disagreement (either overtly or in polite form) which they use productively to achieve solutions on the basis of which they can proceed. Among high school physics students, the boys showed a preference for disagreement which was expressed in their attempts to "shoot down" discourse contributions of their classmates, their preference for argumentation, and their use of voting procedures to overcome differences impossible to negotiate. The girls, on the other hand, had shown a preference for agreement. In the present study, preferences for agreement reflected an interactional style at the classroom level, and also characterized the interactions between students and Tammy, their regular classroom teacher. Decisions regarding classroom norms and even many decisions with respect to structuring classroom time were taken jointly by students and teacher. During such decision-making processes, Tammy reiterated the preference for agreement with statements such as "Either positive comments or no comments at all". On the reverse side of the coin, this norm may have contributed to the already-mentioned abundance of positive comments and the lack of critical reflections and questions. These might have provided for opportunities to contribute to the improvement of individual projects and to the processes of appropriating the language games of engineering design.

Social order does not simply affect individuals and the interactions between them. Rather, the present data provide evidence for the contention that social order is produced and accounted for in and as ordinary everyday activity. The relationship of power and submission between Tammy and Gitte on the one hand, and Stan and Tim on the other did not simply exist *a priori*; it was constructed in the interaction. But power was constructed differently when the two boys converted the cardboard pad despite the interdiction implicit in Gitte's refusal. The fact that the teachers were present in some situations while absent in others does not undermine the claim: in other groups, and at other times in Stan and Tim's interactions, "I think we are not allowed to" was constructed such that students did not extend the existing amount and type of materials. Thus, a teacher-set rule was interpreted so as to lead to significantly different sequences of actions. Power and control of a teacher was reconstructed in direct interaction, but its effect on events did not extend the immediate presence of the teacher. In the same way, the failure to reconstitute a previously established norm may lead to redefinitions as to the applicability of a norm, and thus to changes in possible or forbidden actions. The teachers' failure to reconstruct in each and every encounter the need to provide a convincing rationale for changing materials, tasks, and constraints opened new possibilities for students' actions. Thus, as Stan and Tim's cutting up of the pad illustrated, students interpreted this lack as tacit permission to make these changes without providing a rationale at all. Because of the highly interactive nature of this classroom which allowed knowledge to be circulated rapidly throughout the community (cf. Chapters 5 through 7), the factual nature of the student-originated changes in norms became normative in itself. This led to changes in the nature of the tasks as children accomplished them. Thus, students and teachers, rather than being cultural dopes, were engaged in activities during the course of which phenomena of social order were continuously produced.

In the same way, Stan and Tim produced their relationship as a local and endogenous phenomenon, which contributed to the fact that I could not predict the observed absence of conflicts; nor could I predict the asymmetric division of labor between Arlene and Chris, because they constructed their "collaboration" locally and endogenously. In both cases, there were not simply different individuals who interacted with their intact and unchanging psychological attributes. Rather, being produced locally and endogenously, collective activities constitute social phenomena, which are *sui generis* and cannot be reduced to irregular and psychologically motivated individual activities (Lee, 1991).

In Chapter 8, I described the cognitive flexibility of elementary students which allowed them to draw benefits from, and to productively use, learning environments based on ill- or poorly-defined problems. The data and analyses in the present chapter attest to an interactional flexibility of elementary students in the same environment. This flexibility manifested itself during the distribution of the task (division of labor) and in the willingness to change a position or point of view to achieve consensus, allowing groups to get on with their projects. Contrasting rigorously structured task distributions in cooperative learning models, students in this study structured their interactions themselves. The distribution of tasks arose spontaneously within each particular setting taking account of extenuating circumstances. Sometimes a student indicated to take on one task, followed by the partner's announcement to complete another; one of the students may have assigned the tasks for both; or a student asked for help so that parallel work changed to joint work; one student's statement about a state of affairs may have been followed by the partners' statement to bring out the implicitly desired change; or a student may have simply stated to do something without a prior statement by a peer.

I observed a similar flexibility during decision-making processes, even when students had differing point of views. In Stan, I observed a quite striking and *a priori* unlikely example of such flexibility. According to the information provided by Tammy and according to my own observations prior to this "Engineering for Children: Structures" unit, Stan was not accepted by his peers who thought that he was "impossible to work with", often off-task, and too inflexible when it came to making decisions. Surprisingly enough, when he actually got to work with Tim – who himself had shown off-task behavior and some inflexibility in previous activities – a partnership emerged in which divergent points of view were negotiated and resolved in a harmonious manner.

Implicit in the previous accounts of group interactions in an elementary classroom is the complexity of the process by means of which students' artifacts came about. Hence Stan and Tim's bridge or Arlene and Chris' twin towers are poor representations of the multitude of decisions, conversational and practical actions necessary to achieve these artifacts as products. As we saw in Chapter 8, much of the design process, decision making, problem solving, and solution finding was deleted once the material outcome alone was considered. Similarly, much of the work as part of a collective was not re-presented in the artifact. It is clear, that the projects were to a significant extent shaped by the members of a group. But the present data also illustrated that there were non-negligible influences and constraining effects which arose from the interactions of the groups with peers from other groups and teachers, and from contingencies of the setting. Stan and Tim's project unfolded leaving a bridge as the residue of a series of conversational and material activities including the constraints provided by Gitte's refusal to supply further cardboard strips, Tim and Stan's conversion of a pad into a source of new building materials, or Tim's problems with cutting building elements and Stan's expertise on the same task. Clare and Shelly's bridge (with its extremely long piers) could not be understood apart from the length of the supplied skewers, Tim's help in learning to use the glue gun, Stan's refusal to provide scissors, Clare's decision to use the skewers in their full length rather than to get another pair of scissors, or Clare and Shelly's re/production of an interdiction. Thus, the children's buildings unfolded from the local and endogenously produced constraints which showed elements of small group interactions, a teacher's interdiction, the nature of materials and the form in which they were supplied, the cognitive make up of individuals, or the culture, to name but a few. "Social" constitutes an unsatisfactory adjective to describe "construction" because it backgrounds important material and cognitive aspects of the situation. The present analysis, rather than focusing on any of these aspects individually, brought them into focus at once. This procedure renders visible the multiplicity of the shaping factors on a project which contrasts in a striking way the rather bland gloss of "social construction".

PART IV

CONCLUSIONS

10. DESIGNING KNOWLEDGE-BUILDING COMMUNITIES

Throughout this book, I emphasized the role of networking to bring about communities in which knowledge-building and learning are central concerns. The resulting networks are reflexively constituted with the circulation of resources and practices. In this chapter, I want to reflect about the conditions that need to exist for knowledge-building communities to emerge.

This study documented the transformation of a collectivity in terms of its resources (facts, objects) and practices and the differences between them. These differences are most readily apparent in the relative speed of the transformation and the extent to which individual resources and practices are incorporated and transformed. Much of traditional science teaching is concerned with students' acquisition of facts. The data provided in Chapters 5 through 7 showed that the circulation of facts and artifacts is the simplest of the processes by means of which a community is transformed and, there-fore, by means of which it learns. Objects, materials, and simple artifacts such as the Canadian flag were rapidly circulated, and therefore "shared" within the community. Furthermore, Chapter 7 illustrated that children ra-pidly became triangle-sayers, though without making the associated practices a central part of their design. This circulation also has to be considered as that of a fact. It took much longer and was frequently much more difficult to circulate material and discursive practices. These took much longer to become ready-to-hand to a larger membership in the community. We saw that some of these transformations were student-centered; that is, the circu-lation of discursive and material practices was entirely driven by children's interests and goals. On the other hand, for some discursive and material practices – those surrounding "triangles" and "catastrophic failures" – both teachers expended much effort and energy to enroll individuals into a net-work that would continue to circulate these practices – and thus transform the classroom community as a whole.

The view of children's designing described here provides new perspectives on at least three aspects of knowing. First, learning through design is inte-grally related to those aspects of the setting which student-actors constructed as important. Designing is not simply a psychological, social-psychological, or sociological phenomenon as radical and/or social constructivist theorists want to have it, but is, in important ways, networked with the material world in and for which students design. These material aspects do not only lie in the artifact nature of students' products in engineering design, but much of students' individual and collective activity – designing and sense making – are constituted by the manipulation of materials, tools, and artifacts. Second, the elements of a setting – discourse, materials, tools, artifacts, rules, community standards, etc. – do not exist in any absolute sense but have a flexible ontology such that their meanings are determined and elaborated in the context of their use. Third, the artifacts created by students are not just ends in themselves, but are important structuring resources in children's designing: (a) as integral aspects of students' cognitive activity during design and (b) as tools that facilitate negotiating, constructing collective meaning, thinking, and planning in groups.

The research presented here has important implications for classroom teaching. The study provides evidence that children develop complex ways of talking science and engineering issues *without* direct instruction, recitation, or drill and practice activities. Rather, children learn to talk and write science and engineering design while engaging in activities which are personally relevant to them. In such situations, the teacher's task, as that of Bernard in the village of Moussac, lies in developing and maintaining a functioning community in the pursuit of meaningful goals. This community, its resources and practices, and the goals, have to exist for the children; that is, the children have to construct them as meaningful aspects of their setting. Therefore, there exist several important tasks for research including:

- The identification of contexts which allow the emergence of self-sustaining, knowledge-building classroom communities.
- The identification of conditions which afford or constrain this emergence.
- The conduct of longitudinal studies which describe and theorize changing language games as they arise from specific conditions in the setting.

The findings reported in Chapters 8 and 9 have important implications for classroom teaching because they question some central assumptions of traditional education particularly about the nature of designing, problem framing, and solution finding. My central claim is that the kind of activities teachers and curriculum experts design will radically change once we take children's perspectives on the world: They see, interact with, and understand the world differently than adults do. Once we take this position we recognize its incompatibility with, and the simple-mindedness of, "discovery" learning and many approaches to "hands-on" activities.

I also provided descriptions of the micro-structure of networking people into small groups and larger collectivities. We saw that "collaboration" can describe quite distinct patterns of (inter)actions and may include joint discursive and material action, parallel and individual work within the group, and interactions with other groups and teachers. For this reason, the notion of "collective work" may be more appropriate then cooperation or collaboration. Furthermore, the labels "on-task" and "off-task" appear too simplistic to describe the complex interactional structures between members of this classroom community. Furthermore, I do not take differences in the frequency of talk turns or the amount of time spent directly manipulating materials as indicators of the significance of a group member's contribution to a collective project. In the following four sections, I draw implications from these findings for the design of curriculum and subsequently provide an example of how I used the recommendations to design and enact a curriculum about simple machines in a Grade 6–7 class.

10.1. DESIGNING FOR THE CIRCULATION OF RESOURCES AND PRACTICES

One of the central features curriculum designers and teachers must address is what conditions they need to set up, and how they can foster, the free circulation of resources and practices throughout the community. The teacher of Moussac had his answer to the question: Set up a community, and then keep it going, always accepting new members – the younger ones, who just have reached schooling age – and loosing members – the older ones who continue their schooling in a different type of school. The key is, as he pointed out, to get the community going. It can then support the normal change overs – much in the same way that a team on a ship's navigation bridge is rejuvenated as newcomers enter the team, then traverse a trajectory of competence, only to leave once their career cycle requires so – which constitutes a trajectory of increasing competence in navigation. The integrity of the team as such is not changed. What then are the conditions we, as science teachers, need to set up to support a community that continuously networks members and setting, and circulates resources and practices?

The contemporary move towards student-centered classroom organization requires knowledge of those conditions that favor the collective transformation of resources and practices. The present study suggests four such conditions. Students learn when:
- they interact with others by moving about;
- they interact with others by meeting them in areas of high pupil density;
- they interact over artifacts of their own design so that resources and practices are highly desirable in the pursuit of goals; and
- resources and practices are well promoted (by the teacher or other students) while acknowledging the primacy of students' currently available resources and practices.

These dimensions are discussed in the following paragraphs. However, a claim is not made that any of these conditions is necessary or sufficient for the appropriation of knowledge at the level of the collectivity.

First, a critical component in the appropriation of competence, resources, and practices is local access to the embodied knowledge of practitioners (even in the era of jet planes and electronic media). That is, conditions need to encourage students and teachers to co-participate in material and discursive activities. Consequently, innovation depends on the infrastructure of the community, an infrastructure that provides the link from individual to community. In the present study, moving about the classroom, talking to other students, or seeing different structures became an important aspect of the classroom culture. Students moved freely about the classroom to pick up new materials, returned to their seats to pick up something from their desks, or crossed the classroom for a glue gun job. This movement brought them in contact with new ideas, problems and solutions, or material and discursive practices. When they returned to their work stations, they often constructed solutions to problems that had previously seemed intractable. Students did not necessarily intend to copy specific solutions but sought inspiration for new perspectives on their own problems.

Second, high density areas are likely places where exchanges (the basis for transformation of the collective) occur. I observed a lot of common resources and practices around the Grade 5 table, material storage areas, and electrical outlets. Moments of high pupil density occurred when the whole class came together to talk about, celebrate, and praise specific projects. In all of these situations, students had opportunities to engage with the processes and products of other students' work. However, these encounters with new ideas or techniques were not necessarily conscious in such a way that students attended to them (e.g., even the investigators present had not attended to Brigitta's public mention of the Canadian flag on Day 1 of the tower construction; we only noted this a day later during transcription). But these meeting places and times provided opportunities for interacting with others. Here, students were asked to praise and/or question peers about their work that encouraged their active engagement with this work. Similarly, when the teachers asked students to construct a recollection of the difficult moments in their collaboration or difficult problems in their work, they encouraged the presenters to reflect on their learning.

Third, resources and practices appear to be readily appropriated if they are desirable, when students recognize a need for them, and when they fit into a system of interlocking meaningful practices. Thus, the Canadian flag was readily adopted by the classroom culture when a considerable number of students found them attractive and desirable. In this case, much of the collective knowledge transformation occurred without the teacher. This is consistent with my advertising metaphor. If a product is such that it meets the needs of many, it will spread readily. In a similar way, tools (and with it material practices) and discourses diffuse readily throughout scientific and technological communities if they are desirable or meet specific needs. In educational terms, desirability and "intrinsic motivation" are closely associated. Thus, learning settings that foster high levels of intrinsic motivation are likely to be more successful than others. When desirability/need is low, promotion (a term from the advertising metaphor) of resources and practices can help but will not guarantee successful circulation.

Fourth, the collective transformation of the triangles as an instance of an engineering practice appears to be closely related to the amount of promotion the concept receives by teachers. Both teachers used a variety of methods to promote the concept of triangles including those of reading aloud student reflections, pointing out and praising triangles in existing student work, and "drawing out the concept". Consistently, from one lesson to the next, in whole-class and small-group sessions, teachers promoted the notion of tri-angles as ideal geometric configuration to achieve structural stability. In this way, over considerable time and to varying degree, the triangle became part of the classroom as resource and practice. If, on the other hand, the level of promotion was too low, a concept had little chance of becoming adopted in the discourse of the community. In the case of concepts such as "catastrophic failure", the teachers' efforts were more sporadic. The notions were used by only a small number of students. Thus, whether a discourse practice was adopted by the class depended on the degree of promotion it received on the part of the teachers. I have shown in the triangle case that this kind of teaching harbors the danger that children appropriate resources only rather than the associated practices.

10.2. ARTIFACTS AND THE NETWORKING OF COMMUNTIES

The work reported in this book should have implications for another aspect of classroom teaching because it questions some central assumptions of tradi-tional education:

- The myth that students' activities and artifacts are the result and sum total of individual cognition that can be entirely attributed to specific human actors (students are evaluated on "individual" work or "contributions" to a joint project);
- the myth that artifacts, tools, teacher-set constraints, trends, or rules un-equivocally embody specific meanings; and
- the myth that the manipulation and production of artifacts are ends in themselves in order to motivate students and get their "hands on science".

These myths, de facto, allow educators to uncouple learning outcomes from learning processes. While I acknowledge that learning outcomes such as those documented in this study are important aspects of schooling, the main purpose of this book is to show that learning processes (and therefore their products) are in fundamental ways linked to the social and material settings of students' work. This view has important implications for thinking about and using artifact-related activities in science teaching.

First, my studies showed that students' activities include multiple actors who interact in non-additive fashion. The conversations, ideas, or artifacts resulting from this activity are situated and emergent and therefore heteroge-neous assemblages. Individual aspects cannot be isolated and attributed to individual students which is a common teacher practice for marking group activity. It follows that the activities described here do not easily lend them-selves to traditional evaluation practices that prize individual prowess over functioning as part of a collective. That is, the artifacts – the towers or bridges – by themselves have limited value as indicators for assessing "designing". As in this study, multiple process indicators need to be included to show how children design, negotiate, communicate, and use tools and materials. Video assessment such as that proposed by Collins *et al.* (1993) is an important alternative for evaluating student performance in such contexts. But video assessment will not work if the ultimate purpose is to assign a number or letter grade.

Second, artifacts and tools are commonly regarded as repositories of cultural knowledge which merely have to be decoded appropriately by their users to bring about learning. Furthermore, teacher-set constraints, interdictions, and rules are conferred absolute ontologies, that is, they have single right meanings, namely the respective teacher's meanings. Students who fail to do this decoding, use these tools and artifacts in "inappropriate" ways, or arrive at different interpretations are penalized with low grades. If, as I suggested here, constraints, interdictions, and rules are constructed in every situation, it makes little sense to confer to them any absolute character; rather, one has to recognize the accomplishment each time they are applied/observed. If artifacts, tools, constraints, interdictions, and rules are interpretively flexible, teachers can no longer penalize students for "not getting it", but have to establish an environment in which interpretations acceptable to the classroom community can be negotiated. In the same vein, teachers can no longer expect students to "discover" canonical forms of knowledge (patterns, scientific laws, rules, application of tools). One way of introducing students to specific, culturally devised forms of knowledge is through participation in culturally organized activities in which this knowledge plays a role; that is, activities where students experience these canonical forms of knowledge used by someone who already has a certain degree of competence. This form of appropriation is captured in learning environments that are built on the metaphors of cognitive apprenticeship, legitimate peripheral participation, or practicum. In all three learning environments, newcomers to a practice co-participate in authentic activities of the domain with more experienced old-timers.

Third, the construction of artifacts is far more important to learning than simply to motivate students. Materials, tools, and artifacts serve in important ways as structuring resources to design and make sense of the learning environment and as backdrop against and with which students can construct individual understandings and negotiate shared meanings. Design activities in which students collectively construct artifacts by networking a wide range of resources and practices should thus become central aspects of learning rather than appendices with mere motivational purpose. "Hands-on" alone cannot be the focus of instruction, nor its replacement "hands-on, minds-on". Rather, educators need to realize the possibilities for learning through design: Designing and constructing artifacts produces a good deal of problem solving in ill-structured settings, allows students to construct an experience-based design-related discourse, and facilitates interactions and sharing of knowledge in the classroom.

10.3. DESIGNING AND ASSESSING COLLECTIVE LEARNING EXPERIENCES

The microanalyses reported in Chapter 9 are important to classroom teaching in at least two ways because they inform the planning and implementation of collaborative classroom environments and the assessment of individual and group student activity. Contrary to practices in the cooperative learning literature (e.g., Johnson et al., 1985; Johnson et al., 1985), my collaborating teachers did not structure students' group activities but let them choose partners on their own. As in previous studies, I found that such groups (with the exception of those in which one particular girl was involved) are very productive in interactional and cognitive terms. I did not find the kind of conflictual and often dysfunctional patterns of interaction when students are forced to work with others to achieve groupings based on some criterion (homogeneous or heterogeneous grouping according to ability, race, gender, etc.) which others observed (e.g., Eichinger et al., 1991). However, the existence of functional groups could not be understood apart from the teacher's (Tammy) efforts since the beginning of the school year to establish, together with students, a set of classroom norms that guided the interactions of collaborating partners. "Norms" in themselves are not sufficient, for the understanding of rules and norms depends as much on the extent to which community members follow them as on the extent to which they break them. Rules and norms are not external to student behavior, but are locally and endogenously produced achievements which may need to be facilitated. It appears that the establishment and maintenance of non-conflictual groupings can provide "safe" learning environments which have a function similar to parking lots for beginning drivers.

Most current school assessment practices focus on outcomes, that is, artifacts left as a trail of student activity. The present study showed that students' activities were underdetermined by the artifacts (tower, bridge) however impressive these artifacts appeared. The processes by means of which students accomplish their tasks, the decisions they make, the material problems which they construct and resolve, the rocky aspects of their relationships which they learn to deal with, the interactions and help they provide to other students, all have disappeared, become invisible in, and unrepresented by, the actual piece handed in for evaluation. Nevertheless, these processes and events are, for the students, part of their experience; from their perspective, any evaluation is also an expression of these achievements inaccessible to the evaluator of the final product. It is clear that evaluation has to be based on more than the "hard" evidence provided by artifacts, and has to include at least some of the process aspects of students' work. Simply observing a group and trying to assess how much each student is contributing may not do justice to each student's contribution to a project. Apparent inactivity may hide the impact a students' contributions had on shaping a design, constructing and resolving problems, and making other crucial decisions.

Video provides a very different view of students' social competencies because it can record their ongoing activities and explanations in rich detail. Some readers may want to argue that such assessment practices increase the amount of work that has to be done and will become unmanageable. Here again, I have to ask what it is that we are trying to achieve in science education. If we are truly interested in learning and education (rather than schooling and the construction of obedient bodies) then our assessment methods need to reflect the complexities of learning and knowing. The dimensions which should be assessed include students' competencies to explain ideas and answer questions that challenge their understanding, their competencies to listen to others, their competencies to collaborate, and their integration of eyes, hands, voices, and minds in the performance of tasks and experiments. As this entire book shows, video-based performance assessments would likely yield different evaluations of student work than if assessments were based on the students' artifacts alone. Video-based assessments, for example, question the kinds of assessments made by a variety of school personnel about Stan. Video-based assessments would have particularly questioned previous assumptions (by peers and teachers) about his "inability" to function in group situations and the inadequacy of much of his work. My video materials also provided a view of Arlene's contributions that exceeded by far the gloss "a student with good social skills". I have begun to assist teachers in using video for their assessment purposes. They have found this very helpful in establishing anecdotal reports.[1]

10.4. DESIGNING FOR AUTHENTIC PROBLEM SOLVING

Authentic problems are those that people frame themselves and which they attempt to resolve because of intrinsic goals and needs. These problems have little to do with traditional "problems" that are more like puzzles to which their creators withhold the solution. My observations support the contention that students in their elementary years can learn to exploit interpretive flexibility to frame and solve problems in very creative ways. In fact, they might find this easier than traditionally-trained engineers because children designers still lack the dominant disciplinary constraints that frequently limit the activities of their professional counterparts. School science instruction should provide settings in which students can develop their competence to deal with open-ended and ill-defined situations while at the same time providing opportunities in which children designers can draw on and appropriate canonical resources and practices. Such learning environments could

---

[1] Unfortunately for students in British Columbia, politicians have reversed the attempt to make assessment in elementary schools more holistic (through anecdotal reports), and have reintroduced grades. Thus, whereas Tammy was still able to write complex reports about the activities and achievements of the students who participated in this study, this is no longer possible under current government policies.

curb traditionally heard student questions, "What do you want me to do?" or "How do you want me to do it?" and foster instead creative approaches to dealing with open-ended and ill-defined situations. Students who learn to flexibly frame and resolve snags and breakdowns may have an easier time adapting to the out-of-school work place (the "real" world as some refer to it) where flexibility is at a prime. Those engineers and engineering companies are successful who can reuse existing solutions and apply them to new situations or who reframe existing problems and solutions such that a match can be achieved. These engineers and their companies exploit interpretive flexibility to adapt to market situations, customer demands, changing problems and solutions.

In science and engineering, interpretive flexibility is exploited through institutional arrangements by capitalizing on the affordances of collaborations. In schools, this means that students frame and resolve problems by working in collectivities (rather than in the solitary work mode that characterizes much of traditional schooling). From heterogeneous meanings, common meanings can be negotiated which leads to new and innovative combinations of problems and solutions. A challenge for teachers is to make use of children's existing material and discursive resources and practices and to foster a classroom climate in which these can be circulated among all members of classroom communities. At the same time, teachers have to make provisions so that students can re-produce cultural knowledge which then becomes an important resource in their own design activities. Having children work in isolation or removing them completely from the class (for disciplinary reasons or to provide "special" services) is antithetical to the idea of knowledge-building communities.

Interpretive flexibility of (arti)facts, plans, and instructions is unavoidably present no matter what teachers do to communicate equivocal meanings. Rather than overdetermining children's actions with ever more detailed instructions – which turn science experiments into lessons of "how to follow instructions" – teachers could follow the example of successful instruction in tool-use. Here, teachers provide minimal initial guidance and repair troubles as they arise. At the same time, troubles allow students to explore the affordances and limitations of their materials, tools, and ways of understanding. They develop familiarity with the domain of interest which has been identified as a constitutive feature for developing experimental practices and conceptual understandings.

Instruction also has the goal of familiarizing children with the current paradigmatic views of science and technology. "Discovery" learning approaches have failed because they do not account for the inevitable interpretive flexibility of facts, artifacts, or discourse which made it unlikely that children would reinvent the agreed-upon material and discursive practices that have developed in scientific communities over the course of two millennia. Discouraged, teachers often returned to traditional teaching in the form of telling. However, this teaching method has also failed to lead to conceptual understanding or improved problem solving. This and other studies in my research program suggest that students can both learn to exploit interpretive flexibility in the form of creative framing of problems and of solutions and, through appropriate teacher constraints (in the form of modeling scientific practices and scaffolding students' initial attempts), learn paradigmatic ways of looking at and talking about the world.

10.5. FROM RESEARCH TO PRACTICE: CURRICULUM ON SIMPLE MACHINES

After completing the study reported in this book and after team-teaching a unit on simple machines in a French-immersion Grade 7 class at Mountain Elementary School, another teacher asked me to model teaching in his class in the context of a unit on simple machines. He left it up to me how to teach, but asked me to meet the curricular objectives outlined in the directives of his school board. This provided me with an opportunity to design a curriculum based on my experiences and understandings of learning in the Grade 4–5 class under the constraint that traditional objectives also needed to be met. To elaborate these recommendations for curriculum planning, I offer the following description of my curriculum design effort.[2]

I developed the unit on simple machines based on the notion of learning as an increasing generation of, and participation in, scientific material and discursive practices. My overarching goal was the development of students' discourse related to simple machines. However, I did not want just any discourse to emerge, but was interested in fostering one that had some resemblance to canonical discourse. At the same time, I knew that any founded material and discursive practice related to simple machines, if it was intelligible and plausible, had to develop from students' current practices. By choosing models of simple machines such as a pulley or pulley system and their re-presentations in diagrams projected onto the screen above the pulley, I set the stage for students to produce, maintain, and develop new discursive practices. My design of artifacts that were to become the focal point of whole-class and small-group activities was influenced by the following observation of professional designers of computer systems:

One of the hardest challenges for the designer seems to be to *create a design language game that makes sense to all participants*; the designer in the role of play-maker. In this role the

---

[2] This teaching experience was also documented in an intensive research study (McGinn *et al.*, 1995; McGinn & Roth, 1996; Roth, 1996g, 1996h, Roth & McGinn, 1996). The research study was in part designed as a "confirmatory ethnography", that is, as an ethnography in which the same researchers tested the transportability and viability of observational and theoretical descriptions generated in one context by conducting a study in another context. Here, the school remained the same, but teachers, students, grade level, and subject matter were different.

designer sets the stage by finding and supporting ways for useful cooperation between professional designers and "designing users". (Ehn & Kyng, 1991, p. 177, emphasis in the original)

Consistent with my recommendation that there need to be meeting places in
which students can begin to tell and circulate stories, discourses, ideas, etc.,
my unit included whole-class conversations on the topics of simple pulleys,
block and tackle, equal arm balances (1st class levers), second and third class
levers, inclined planes, work, and energy. To provide opportunities in which
students could intensively co-participate in talking simple machines, I asked
them to conduct small-group activities with equal arm balances, second and
third class levers, or summarize their ideas about specific simple machines
on specially designed forms. Both types of activities allowed students to
interact with materials, re-presentations, and discourse. I also wanted students to spend a lot of time in activities where they could set their own
agendas and develop and gain experience and confidence in their simple-
machine-related discourse. Therefore, I asked students to design a variety
of machines (top level goal). The first three machines were to lift loads,
move loads over a long distance, and move loads by means of a self-propelling
mechanism. In the fourth machine, students were to combine a minimum of
four processes, two of which had to be based on one of the simple machines
discussed in the unit. At the end of each design project, all student groups
presented their artifacts in whole-class sessions and directed the subsequent
question and answer session.

All activities were designed to support students' participation in talking
science and engineering design, each with different amounts of structure
provided. Because of my concerns for cultural production and cultural reproduction, I planned lessons that had both student-designed and teacher-designed artifacts as foci for the conversations; because of my concern for
active participation in conversations, I provided opportunities for large-group
and small-group conversations. Thus, the four resulting activity structures
differed in terms of the social configuration (whole class, small group) and the
origin of the central, activity-organizing artifact (teacher-designed, student-
designed). The different activity structures followed the same underlying
pattern. First, students took about three lessons to design machines – which
included the construction of working models. The subsequent three to four
lessons were split into two parts. During one part, I led whole-class conversations about a physics topic (simple machines, conceptual issue) or small
groups of students completed investigations and/or recorded their ideas on
previously discussed topics. During the second part of the lesson, students
presented their machines in whole-class sessions and then directed sense-
making conversations.

10.5.1. *Whole-Class Conversations around Teacher-Designed Artifacts*

To allow for opportunities in which I had some control over what and
how to see a machine, I designed whole-class conversations around simple

machines and associated representations which I had designed and constructed. For example, I moderated a class discussion about two different
ways of setting up a pulley. In one, the effort remained the same, in the
other, the effort was cut in half. The purpose of the lesson was for students
to develop a language game, descriptions and explanations, of all forces
acting in the system, and to measure these forces. In the case of each force,
I asked students to predict the forces, one force at a time, operating at various
points in the system. After about 10 students had made their predictions, and
after tallying the commitments of other students, I involved students seated
closely to the pulley in measuring the forces. The magnitude of each force,
along with student predictions, were recorded on an overhead transparency
displaying key features of the system. I subsequently asked students to
explain the outcome. Some of these discussions followed students' own ex-
ploratory activities with the device (balanced lever, second and third class
levers), while others, because of time constraints, were treated in whole-
class sessions only (pulley, block and tackle, inclined plane including work
and energy).

These activities were suitable for introducing canonical ways of seeing and
describing simple machines because they allowed me to make particular
aspects of a simple machine the topic of discussion. For example, I added a
Newton scale in a pulley set up and asked students to predict how much it
would read once I left the system alone. I also marked the place in the
overhead transparency that re-presented the pulley system. Thus, there were
at least three ways in which I demonstrated canonical practices. First, I
attached the pulley in a place that is relevant to scientists in showing the
mechanical advantage of the machine. Second, I had used a re-presentation
as it is used in the scientific community. Third, I introduced in this manner
the language about mechanical advantage.

In another activity, I first challenged students to beat me in a tug of war.
However, because the unit was on simple machines, I used a block and
tackle to mediate the tug of war. Although about 20 students pulled on the
other end (including one student in a wheel chair who had pulled the brakes),
they lost the competition. I subsequently challenged them to construct
explanations and alternative configurations of the pulley that would change
the outcome. Given the competitive aspect of a tug of war, it was no wonder
that the whole-class conversation also had a adversarial and argumentative
character that is fundamental to scientific discourse. Using diagrams, several
students argued their case, cheered on by their peers. Again, students engaged in typically scientific practices, argumentation and use of inscriptions
(diagrams).

On the negative side, whole class conversations with 26 pupils in the class
make for a lot of listeners and few that can talk in any one lesson. To provide
more opportunities for each student to engage in activities that focused on
canonical concerns, I designed tasks which were to be solved in groups of
two and three students.

10.5.2. *Small-Group Conversations around Teacher-Designed Artifacts*

Small group investigations and reflections organized around simple machines, written instructions, and summary sheets allowed me to highlight canonical concerns. These activities provided students with opportunities to manipulate and explore simple machines, develop familiarity, and come up with tentative explanations of the machines' functioning. I required each group to produce one agreed-upon answer. This forced students to negotiate any differences. That is, in small groups, students have opportunities to develop, experiment with, and participate in language games before going public and presenting them to the whole class. After students completed their activities, the same sketches and artifacts served as focal points in whole-class conversations which I directed.

10.5.3. *Small-Group Conversations around Student-Designed Artifacts*

In the two activity structures discussed so far, the conversations focused on issues which were by and large my canonical concerns. To provide students with opportunities to establish and pursue their own goals (under the larger umbrella of my top-level goal), students received design tasks that were subject to a small number of constraints. Students spent a considerable amount of time designing models of machines by working in groups of two or three students (e.g., design a load-moving machine; construct a model of the machine able to carry a 100-g load over a 2-m distance). Design tasks are not well-suited for reproducing canonical science and engineering but for participating in activities central to design practice (such as designing and testing artifacts, generating hypotheses, making presentations, describing systems, etcetera). Because of the generally unstructured nature of the tasks, students had opportunities to engage in specifying goals, framing troubles and breakdowns in the pursuit of these goals, and finding resolutions that allow a continuation of the main activities. Furthermore, I knew from the present Grade 4–5 study that in open-design classroom communities, new language games evolve and circulate among student members.

10.5.4. *Whole-Class Conversations around Student-Designed Artifacts*

To extend the development and evaluation of design artifacts and discourses, the final activity structure provided students with opportunities to present and talk about engineering design as exemplified in their own work. The comments, questions, and critiques by other students and the teachers provided further opportunities for arguing design. These conversations were directed by the presenting students. The community also used these sessions for public tests designed to evaluate the mechanical advantage the machine afforded to its users. I usually intervened only when conversations seemed to come to a halt, when old animosities resurfaced, or other communicative problems surfaced. The inclusion of this activity structure was based on my experience (as a teacher and researcher) that students who are required to elaborate, explain, or defend their own ideas tend to evaluate, integrate, and elaborate knowledge in new ways.[3] This activity also provided opportunities for, and encouraged, changing language games. Students were more likely to engage in these processes when the questions and critique came from peers because they are framed within the same language game. This activity structure allowed students to take greater responsibilities for their participation and feel in charge of classroom activities.

10.5.5. Making it Work

A recurrent problem in elementary schools are tight budgets allowing for little or no money to be spent on science equipment. I was therefore interested in designing a unit on simple machines that teachers could do without great expenditures.[4] Thus, my research assistants and I built individual or class sets of the artifacts for small group activities and demonstration purposes from materials that we bought in local hardware and office supply stores (including nuts, stripping, elastics, glue sticks, masking tape, paper clips, and grommets for fastening tarpaulin that could be used to make pulleys). For the small-group conversations about student-designed artifacts, we asked students to bring scrap materials and tools from home. We provided one set of simple tools including a hammer, drill, saw, glue gun, Exacto knife, and C-clamps. There were two community work places that could be accessed from several sides and therefore afforded concentrations of students necessary for the rapid circulation of resources and practices: one large table that we protected with sheets of $\frac{3}{4}$-inch plywood. A small table across the room and next to a receptacle afforded operating the glue gun. (In contrast to Gitte, I attempt to foster bricolage and tinkering, creative innovation, and reinterpretation of tools.)

For the students, teaching and learning in this science unit differed from that in their other subjects. During most of their school day, the children

---

[3] Videotaping my students while I was teaching physics at a private university preparatory school brought about a considerable change in my teaching (Roth, 1995b). Prior to that and grounded in my Piagetian training, I was mainly concerned with the individual construction of knowledge. Once I began watching the videotapes in which students collaborated on constructing concept maps or designing and conducting experiments, it became clear to me that the social organization of students constitutes an important part of knowing and learning. This was the starting point of a research agenda concerned with the investigation of my teaching practices. As a result of my research (e.g., Roth & Roychoudhury, 1993) I could help students appropriate and develop science discourse by introducing the rule that students could challenge each other to elaborate, explain, or defend any statement. A subsequent study showed that this rule had become a central aspect in students' learning (Roth, 1994b).

[4] As a result of this development effort, we produced a box containing materials, simple machines, artifacts, overheads showing diagrams of simple machines, and a curriculum guide which the teachers at Mountain Elementary could use even after I had left the site.

experienced traditional teaching with a focus on facts and right answers. Students listened to the presenting teacher, completed silent seat work that consisted of routine activities (repetitive practice exercises in mathematics, fill in the blanks in social studies), or teacher-directed recitation of facts. Their tests reflected the same factual orientation of teaching ("What do you call the distance around a polygon?" "The capital of Nova Scotia is . . . Halifax or Moncton?"). Teachers also took major responsibilities for structuring students' work (routines) and organizing their notebooks (placement of pages, location of specific entries such as title or date on each page).

Changing from these activity structures to the open lessons I conducted was the most difficult aspect for students. With time, however, many students adjusted to the format of this unit on simple machines which required greater participation than their normal school experience. Students therefore felt (especially in the beginning) that the unit presented too great a challenge. Students thought that I had unwarranted expectations more appropriate to my university students, permitted too many students to voice their ideas instead of immediately providing the right answer, did not give them enough time, did not immediately give them right answers when requested, and gave too much homework. Later during the unit, students no longer expected to be given right answers; they participated to varying degrees in whole class conversations, and many were enthusiastic about the activities during which they constructed models. One of the most audible changes in the classroom conversations was the fact that students increasingly included explanations and justifications for their answers. For example, when asked for an hypothesis about the magnitude of a tension, students initially stated a number. I continuously insisted that students provide an explanation or justification by following their laconic answers by "Because?". Over time, students increasingly included justifications on their own, "I agree (disagree) with . . . , because . . .", or "I think it's . . . because . . .".

In this chapter, I showed how the results of research can inform curriculum design and teaching practice. Furthermore, I showed that it is possible for teachers to achieve a great deal of learning during open-ended inquiry activities when the classroom is organized as a knowledge-building community around the production of artifacts.

11. EPILOGUE

11.1. PARTICIPATING IS LEARNING

In recent years, many researchers have changed their conceptualization of knowledge. Traditional perspectives conceived of knowledge as (a) information stored somewhere in long term memory and (b) more or less content-independent "thinking skills". Throughout this book, I presented a view of knowledge as practice that is increasingly being accepted as a more appropriate way of thinking about thinking. Practices are constituted by sequences of repeatable actions which arise, at their origin, from the concerned engagement of people-acting-in-settings. Thus, rather than thinking of cognition in terms of agents who re-present their settings and the objects in them, we may think in a holistic way of organic wholes, persons-acting-in-settings, as the basic unit of analysis of cognition. Agents, such as the Grade 4–5 students with whom I shared a lifeworld for some time, leave their setting mostly uninterpreted and do not re-present it. This entails that not everything that makes competent practice can be described and explicated; there is always another angle from which we can learn about knowledge as practice. Even highly trained scientists often cannot learn a new practice such as amplifying DNA using the PCR technique – the technique used in the O. J. Simpson case – by reading about it: They go and work with people who already know how to use this technique.

Learning from others, in the ongoing work of everyday activity, can be found in many places. During a visit to my bank, I was served by two customer representatives, one a specialist in loans, the other in retirement plans and investments. Their organization had paired them up so that they could teach each other; the loans officer learned about retirement plans and investments, the retirement plans and investment expert learned about loans. They taught each other in the context of the ongoing daily and ordinary activity of banking, at each other's elbows, so to speak. Their teaching-learning relationship was part of the ongoing banking activity; that is, networking and circulating resources and practices co-occur with purposeful activity. Researchers like myself, who think about knowledge in terms of practices use such vignettes to argue that the sites of knowledge are people-acting-in-settings and the communities of practice to which they belong. Learning in a community of practice means participating in ongoing, that is, "authentic" activity. Throughout this book I showed how various aspects of knowing were circulated as members in the community co-participated in activities which they constructed for themselves as desirable – in some instances, this meant that students reinterpreted or renegotiated tasks to fit their own goals.

How does this relate to teaching and learning science and other subjects in the curriculum such as history or mathematics? If our students are to learn to think historically, mathematically, and scientifically, that is, if they are to appropriate the practices of historians, mathematicians, and scientists they have to do so in the context of a community where they do what experts in these fields do. Thus, taking more of the same courses where one can get an A by cramming the night before the exam cannot be the answer. Learning a practice, like learning to ride a bike, does not come overnight. At the same time, being competent in a practice, like knowing how to ride a bike, is not something that one forgets after the exam is over. What we need instead – at primary, secondary, and university levels – are courses in which students can participate in doing history, mathematics, and science as the practitioners in these fields do: Analyzing how a particular document shaped a historical event, making conjectures and conducting proofs, designing experiments, testing alternative hypotheses, and so on.

Throughout this book I showed that by giving students some freedom over the goals, and by specifying only some overarching goals, children do learn a tremendous deal individually and as a community. Some readers may think that this works only at the elementary level and that intermediate and secondary students are not ready to appropriate practices in this way. But this is not so. A few years ago, I documented learning in an innovative biology curriculum for Grade 8 students. The task of each pair of student was to stake out a $40 \, m^2$ plot on the school campus and then to find out as much as they could about this plot, that is, the relationships between biological and physical characteristics. Students were asked to write reports and present their results to peers in the most convincing manner. That is, the criteria of successful reports were not some external standards but to a large part their rhetorical strengths in the student community. In this environment, these students developed considerable research competence, and the competence to convincingly argue why the research they did was valuable, and why their results are plausible. Their teacher acted as an advisor, modeled new scientific practices as needed, and showed how to use tools *just-in-time*, when students really needed it.

We had documented the entire 10-week unit taught in two classes on videotape (Roth & Bowen, 1995). In our estimation, all students had learned a lot. As part of the unit, our young researchers took a test in which they were asked to make inferences based on a set of data provided. Eleven of 19 pairs of students used some mathematical-scientific representation practice to solve a problem; they acted in similar ways as the scientists to whom I had also presented the problem. The other Grade 8 groups argued their case on various qualitative grounds. The competencies became quite clear to us when we gave the same test to preservice science teachers all of whom had previously obtained at least a bachelors degree in one of the sciences or in mathematics. Over the years, 46 preservice teachers did the same problem. Despite their backgrounds and experiences which by far exceeded that of the Grade 8 students, only six preservice teachers provided solutions based on mathematical-scientific grounds. The others constructed answers based on verbal descriptions.

What can we learn from this? First, Grade 8 students can develop a great deal of competence in scientific and mathematical practices when their learning environment is organized as a knowledge-building community. Second, rather than saying that the preservice teachers did not have the "skills", I prefer to say that they had not had the opportunities to participate in authentic practices of science, and therefore did not develop the competence. The upshot of this lack of competence of preservice teachers was that they said one could not teach in the way we had taught the Grade 8 students. They wanted to tell students science, rather than have them *participate* in scientific practice. They wanted to teach the way they had learned and often talked about constraints: no time to cover the curriculum, students who are not appreciative of open-ended inquiry, my supervising teacher never did it this way, and so on.

When I talk about our experiences in the organization of learning environment, I frequently get the response that my research was conducted either in a private school or in an elementary school. That is, "private school" and "elementary school" are used as discursive resources to dismiss my claims about alternative learning environments that give students more opportunities to take charge of their learning. Examples such as that in the school of Moussac are equally dismissed because of the rural nature of the school and the small number of students. Teachers and administrators argue that these examples are not transportable to the (sometimes tough) conditions of their own schools. At this point, I would like to relate the experience of members of several self-managed schools in France organized as communities. These schools are for the toughest clientele of students teachers can imagine: dropouts with little hope to be re-integrated into the normal educational system.[1] The students in these schools are approximately equivalent to junior and senior level students in North American and Australian schools.

For more than fourteen years, several high schools have existed in France that have two main mottoes: "a place where the pleasure of learning is not assassinated" and "a place where everyone participates". These high schools are organized as communities where everyone participates in all aspects of daily life. That is, these schools are self-managed: all decisions are taken by its members – students and teachers – and everybody participates in the functioning of the community including administration (there are no principals), house keeping (there are no cleaning staff), running the canteen (meal planning, shopping for groceries, cooking), and secretarial work. There is no attendance requirement and students have free choice of their courses and activities. Teachers are no longer centers of information distribution but

---

[1] The following descriptions are based on a documentary featured by an international, French-language channel as part of a series on teaching in France (Nauer *et al.*, 1996).

partners in conversations; classrooms are arranged accordingly without the habitual military-like arrangement of desks but such that members face each other. Even when they discuss their chemistry laboratory experiments, students and teachers sit facing each other much like in graduate seminars, the experimental set-up visible to all members. The teacher assists them in framing observations in a way that will later help them in formulating theoretical descriptions. There are no grades, but continuous evaluations of students' interests, goals, and progress in face-to-face encounters with their mentor-advisor.

In the experience of students and teachers, these schools are tremendously successful despite the absence of control and punishment and the tool for implementing them, grades. Past students from these schools highlight that, because they participated in all aspects of the school, they were able to develop a spirit of independence and adapt to the adult world in ways that other schools prevent. In terms of a more traditional measure of success, 30% of the students pass their national examinations and receive high school diploma which they would otherwise not have received. Universities and colleges acknowledge that the graduates from self-managed schools that they admit are more successful than normal students.

So why do so few teachers organize their classrooms as knowledge-building communities which give students the freedom to make their own choices, pursue phenomena of their own interests, and receive feedback continuously but without grades? My own hunch is, substantiated by the teachers and students in the French documentary, that the core issue is that of control. Communities of learning are by and large based on democratic ideals of learning processes and products. They are built on principles of co-participation with others as the basic mode of teaching and learning. In the present contexts of education, however, neither students' teachers nor their parents are interested in nurturing individuals who challenge authority; question traditional forms of scheduled classes that disrupt the continuity of learning; question form, content, and evaluation of assigned work; and challenge traditional modes of teacher-student relationships. With the removal of grades, teachers no longer have their most important technology of disciplining as a resource. Rather, *author*-ity (of students and teachers) derives from the fit of resources and practices with the values of the entire community, not simply with those of the teacher-wardens.

11.2. NETWORKING TEACHERS – LEARNING TO TEACH SCIENCE BY PARTICIPATING IN THE PRACTICE OF SCIENCE TEACHING

Here, I want to shift and ask readers to think about teaching as practice in the same way as I have described knowing and doing engineering design throughout this book. In the process of their work together, Gitte and Tammy learned from each other in the authentic setting of a teacher's work place (see Chapter 3) in the same way as the banking people learned by co-participating in authentic activity. Here, this activity concerned teaching as part of the normal curriculum. Tammy "picked up" a lot about the subject matter and how to ask productive, open-ended questions that facilitate children's learning without leading them. Gitte learned a lot about facilitating learning communities.

Tammy: I just improved so much in teaching kids to think for themselves by asking productive questions. I don't think three university courses could have given me what collaborative teaching gave me in these two months.

Gitte: It was the same thing for my learning. Even for someone who knows the unit theoretically; it is another thing to read it, but it is a whole different thing to do it with you . . . . This collaborative teaching experience has changed my thinking about this unit although I wrote it, tested it, and had done workshops with teachers on it for the past three years . . . . I am just dying to re-write that manual now, to me it is looking so amateurish now.

This excerpt from a conversation during which Tammy and Gitte talked about their experience of team teaching an engineering unit to a class of Grade 4–5 children expresses some of the fundamental dilemmas that divide theory and practice in teaching. There are fundamental but not sufficiently recognized differences between learning to *teach* and learning *about* teaching. Tammy felt that she had learned more during two months of team teaching than she could have learned in "three university courses". In this, her comments reflect a common complaint of preservice and inservice teachers about the irrelevance of university science methods and content courses to their everyday practice. Tammy credited Gitte for much of her learning. Gitte also learned and became a better teacher and curriculum developer. She experienced in person the discrepancy between the curriculum as a set of instructional materials and the actual teaching experience in classrooms. Gitte credited Tammy for much of her learning. What these two teachers learned from each other by co-participating in shared lessons was tremendous both on their own accounts and my independent account supported through the video-based data.

As Tammy and Gitte implemented the "Engineering for Children: Structures" (EfCS) curriculum, it became clear to both that much of what it took to make the curriculum a success was not communicated in the teachers' guide; much of what made the unit a success in this classroom was due to Gitte's implicit knowing. As the following quote illustrates, Gitte was not aware of this knowledge that appeared in practice but was not articulated:

I don't know if I know [a practice]. It is only through interacting with Tammy and the kids that that is coming to the surface, that experience that I have had in the past. 'Oh yeah, that fits here and that fits there.' I don't know that that would happen in any other way.

In this way, Gitte acknowledged that much of her knowledge about questioning was tacit. She received feedback, primarily through Tammy, that a question was productive and "that not everyone might have asked it". Gitte recognized the difficulty in communicating what she knew by means of photographs, instructions, and even stories that would make her revised version of the teachers' guide too bulky and thereby discourage teachers from reading it entirely. She suggested in this respect that she "would model that for other teachers, how to teach the unit, which I did not realize until I talked and worked with Tammy. She is saying a lot of things like, 'I would not have known to do that'". In a similar way, there were important aspects of Tammy's practice which also contributed to the successful implementation of the unit, but which had not been part of Gitte's repertoire. When Gitte occasionally pointed out such a practice, Tammy often responded, "I just do it without thinking about it" or "I think most teachers would do that".

As they worked together, Tammy and Gitte began to resemble each other in striking ways. My field notes and annotations of the transcripts reflected this sense in remarks such as, "Tammy is asking a question, but I hear Gitte" or "Gitte is doing what Tammy would have done". More so, they resembled each other in their mannerisms, a supportive stance toward some students and a more overbearing stance towards others; an individual, pensive movement of a hand to the chin followed the same motion of the hand by the other; a turn of the head or the whole body of one, reflected in the movements of the other. After about two months, the two were so attuned in their practices that they could conduct a class without previously orchestrating their roles. Gitte and Tammy performed their questioning because they "just knew to alternate like that without having spoken about it"; like improvising jazz musicians in the process of creating a beautiful piece, they "sensed" the development of the situation, "picking up" in their questions the lines of development outlined by the other. This harmony, the often tacit understanding of the collaborating partner, attested to the fact that they had learned more from each other than they could state in just so many words. The Aristotelian notion *mimesis* refers to the similarity of practices that arises from the homogeneity of conditions of lived experience; such similarities of practice are thought to arise in, and are indicative of, "schools of practice" as opposed to "schools of thought" which give rise to similarities in discourse about practice (MacKinnon, 1993). Tammy and Gitte's learning appears to have the same roots as the circulation of practices and resources in the classroom.

Much of Gitte and Tammy's knowing- and learning-in-practice was co-constructed by means of collaboratively constructed accounts of events (stories) – much in the same way as stories provided students with ways of circulating their understandings. These stories indexed Tammy's and Gitte's learning and individual and collective practice. Here, too, we can observe processes very similar to those that I described earlier in the context of children's learning. That is, Tammy and Gitte formed the nucleus of a community. Unfortunately, we could not enroll more teachers to form a larger community. Ideally, the teachers in the school would have come to form a community by working together in the same classrooms at the same time to build a stock of common experiences.

Importantly, the essence of Tammy's and Gitte's stories could only be understood by those who shared the common background, for they left unsaid that "which goes without saying" and which cannot be communicated other than by "sense" or "feel". For example, Gitte's timing of productive questioning increased with her participation in Tammy's practice. Her accounts of good timing and her own learning of it were contained in specific stories of classroom events. Just what made timing good in one instance and inappropriate in another, however, was never explicit. Both teachers could only indicate in which situation the timing was appropriate, and where it was not. Even their elaboration of background information about the children involved, the current status of the curriculum, the extent of shared knowing and experience in the class, and so forth, could not communicate to visitors just what constituted the fine line between appropriate and inappropriate actions.

There was a reflexive relationship between experience and stories. Gitte and Tammy spent much time in constructing and elaborating stories; that is, their knowing and learning was communicated and shared through the collaboratively constructed stories of their joint experience; and experience was shared through the construction of stories. Such collaborative construction of stories are analytical tools for understanding complex economic situations and for the exchange, modification, and appropriation of distributed knowledge in communities of learning; stories can capture the highly emotive and complex aspects of craft knowledge. It is through such stories that teachers' resources and practices are circulated in important ways without the need for direct and declarative pedagogical discourse. Co-participation in practice and the concurrent construction and elaboration of shared narratives, which have traditionally been important aspects in (cognitive) apprenticeship models, are an important outcome of team teaching, and may also be fruitful alternatives to current teacher preparation practices.

The collaborative articulation of experience and practice also allowed Tammy and Gitte to bring new phenomena into the domain of their discourse; that is, it allowed them to construct new dimensions of their environment. Thus, the articulation of experience and the recognition of new phenomena were mutually constitutive. Once the teachers had articulated an aspect of their practice, they sometimes linked this to other propositional knowledge. Again, there are striking similarities between the learning of the two teachers and those of the students, despite the fact that the teacher community was so small.

For Tammy and Gitte it was important that their learning and professional development was driven by their articulation not by some other process or person who made them aware. Both thought it was important that their learning was driven by their own articulation as a result of watching the videotapes or in the course of their practice. Much like the children, they wanted and needed to set goals for their own learning. Then they could bring about change, immediately and on the spot ("there", "right in that state"). Both considered unsolicited outside critique inappropriate; but they sought my help in collecting gender-related data or to serve as sounding boards. They indicated that others could not know if their practices had successful histories ("because I learned that in the past") and that outside-critique was disempowering and encouraged them to react negatively to and reject it ("'anti' reaction"). Co-participation in practice that allows collaborating teachers to bring into discourse many aspects of their practice therefore provides opportunities for professional development that do not rely on traditional top-down supervisory models. Traditional models, based on an epistemology of knowledge as object that can be transferred and applied often fail to appreciate the personalized nature of teaching and the endemic uncertainty of the linkage between teaching and learning. Co-participation in practice, by design, overcomes these shortcomings because it allows practitioners to construct common ground, experience, and interpretive frames.

Practice does not have the luxury of reflection and consideration. Making contextually appropriate pedagogical moves depends on teachers' situation definitions. Classroom events, however, are different from sports such as football or soccer where associations have experimented with instant replay options to assist referees in making "the right" decisions. The question is not whether or not a situation definition is "correct", but if it is flexible and therefore contextually appropriate. Gitte was able to learn a lot from Tammy about constructing appropriate situation definitions, without being able to state – in just so many words – what it was that she had learned. Most human knowledge of "right" discursive and material action is taken for granted and embedded in a tacit background as common sense, for even the simplest cognitive action requires an infinite amount of knowledge. When newcomers co-participate in practice with an old-timer, they acquire such knowledge much faster than if they have to reconstruct it on their own.

Much like the children, Gitte and Tammy had wrestled with rules of conduct. During their planning or debriefing, both teachers had time to deliberate conflicting rules of practice or establish the adequacy of rules for a specific context. They could weigh the benefits and short-comings of alternatives, and negotiate a resolution on which both could agree. Tammy had two rules for interacting with children, "Show consistency over time" and "Show consistency between discourse and action". In some situations, however, these rules were inappropriate and Tammy acted differently, without hesitation, inner conflict, or awareness of illogical behavior. In one situation, she had announced to the class that for their next assignment, partners would be teacher-selected. However, because the two teachers knew that "the bridges (the next activity) were one of the most difficult things" and "everything in the class has worked so well when they had chosen their own partners", they "knew just exactly" and "in one minute" that the children should select their own partners. In such a situation, Tammy could communicate her reasons to other members of the community that she was inconsistent for sound reasons.

Teaching, however, is different from planning and debriefing. The difficulty with theoretical precepts, plans, or laws was that their practical adequacy had to be established in each and every case. This process of establishing the practical adequacy of a direction was not embedded as knowledge in the precepts, plans, or laws. This knowing could not be provided, for practical adequacy depended on the setting, and the practitioners "feel" for what was right, here and now. The work of establishing adequacy was a situated accomplishment. Such establishment of adequacy is not unlike that which can be observed with the introduction of a new law into a legal system. Not only does a law's own adequacy have to be established as lawyers and judges attempt to apply it in praxis, but other laws become more or less applicable because of the former's introduction. The main difference between the legal example and teaching practice lies in the fact that teachers do not have the luxury of time to deliberate between alternate courses of action but have to act, irrevocably, in the here and now of the classroom. It is only after the fact that they know whether their actions were successful and whether they were consistent with their rules of conduct. The practical adequacy of rules is not self-evident but requires work. This adequacy is neither obvious nor self-evident, but has to be constructed because all discourse, artifacts, and phenomena of the world are inherently ambiguous and interpretively flexible. By participating with an experienced and competent practitioner, other practitioners and newcomers can see important principles in practical operation. They observe how a practitioner's disposition or knowing-in-action structures the setting, reacts to unpredicted and unpredictable contingencies and in the face of practical choices. In this way, newcomers can learn essential aspects of the practice even when they cannot be made explicit and communicated other than by means of a silent pedagogy.

In an important way, this view of teaching as practice is similar to the view of designing and learning of the Grade 4–5 students in the present community. This community afforded a kind of learning that is still absent from many science classrooms. However, I believe that all students can get excited about learning when they see their teachers excited about learning. This realization struck home with me one evening in the private school where I used to work. One of my students approached me and said, "Doc, I think you really like to learn yourself". I suddenly realized that my enthusiasm for learning may be the most important "secret" of my teaching.

11.3. REFLEXIVE CODA

Throughout this book, the network analogy constitutes the heart of my analysis of resources and practices at the individual and community levels, and the heterogeneous nature of the artifacts created by students. As students designed their structures, they networked heterogeneous elements to result in "impure" artifacts that could not be accounted for by descriptions of individual minds. I showed that although students ultimately took ownership, the artifacts themselves were multiply and heterogeneously connected with teachers, other students, rules, emerging community standards, tools, materials, and so on. This network analogy goes deep into the construction of this book, its central argument, and its role as an artifact that is circulated in the community of science educators. That is, my network analogy works not only for the analysis of student knowing and learning but also for the analysis of the text in which it is represented.

As any other non/scientific text, this book makes connections with other texts and their authors through citations and references. It works and reworks words and entire discourses and inserts them into new relationships with, at times, newly created discourses. The words, ideas, concepts and phrases I use, describe a population of human and non-human actors. The non-human actors from which my text draws support include tables, transcripts, drawings, diagrams, summary statistics, stories, and off-prints from videotapes. These actors are woven together into the text to form a heterogeneous network. The text also draws on the human actors, students and teachers, who populate these pages. Taken together, they define, explore, test, and stabilize each other's identities. Students, student-teacher interactions, artifacts become who they are through their relation in the network established by the narrative of the text. In this sense, this text networks the outside – various authors from the communities of science education, sociology, cognition, anthropology, philosophy, etc. – and the inside, the classroom community I observed. Inside and outside are enfolded in the textual network; and my *author*-ity provides the interface through which inside and outside are translated into each other. The text of this book, like other non/scientific texts, is a network whose description it creates.

The analogy goes further, for the book, once published and read, also becomes an artifact that is circulated through a network and thereby establishes a community. It becomes an actor that other authors can draw on, or use to create the identities of their own texts through contrast and disagreement. As a textual artifact, this book can also serve functions similar to the artifacts I described in Chapter 8 – it can assist in negotiating meanings, limiting interpretive flexibilities to construct common grounds, and providing a backdrop against which to design further knowledge-building communities.

REFERENCES

Amerine, R. & Bilmes, J. (1990). 'Following instructions'. In M. Lynch & S. Woolgar (Eds.), *Representation in scientific practice* Cambridge, MA: MIT Press, pp. 323–335.

Anderson, J. R. (1990). *The adaptive character of thought*. Hillsdale, NJ: Lawrence Erlbaum Associates.

Ashmore, M. (1989). *The reflexive thesis: Wrighting sociology of scientific knowledge*. Chicago and London: The University of Chicago Press.

Association for the Promotion and Advancement of Science Education (APASE). (1991). *Engineering for children: Structures* (A manual for teachers). Vancouver, BC: Author.

Barnes, B., Bloor, D. & Henry, J. (1996). *Scientific knowledge: A sociological analysis*. Chicago, IL: University of Chicago Press.

Bereiter, C. (1991). 'Implications of connectionism for thinking about rules', *Educational Researcher* 20(3), 10–16.

Bijker, W. E. (1993). 'Do not despair: There is life after constructivism', *Science, Technology & Human Values* 18, 113–138.

Bijker, W. E., Hughes, T. P. & Pinch, T. J. (1987). (Eds.). *The social construction of technological systems*. Cambridge, MA: MIT Press.

Bloom, J. W. (1993). *Children's understanding of machines: Meanings, conceptual organization, and discourse*. Paper presented at the annual meeting of the Canadian Society for the Study of Education, Ottawa, ON.

Bond, A. H. (1989). 'The cooperation of experts in engineering design'. In L. Gasser & M. N. Huhns (Eds.), *Distributed artificial intelligence*, Vol. 2. London: Pitman and San Mateo, CA: Morgan Kaufman Publishers, pp. 463–484.

Bourdieu, P. (1990). *The logic of practice*. Cambridge, UK: Polity Press.

Bourdieu, P. (1992). 'The practice of reflexive sociology' (The Paris workshop). In P. Bourdieu & L. J. D. Wacquant (Eds.), *An invitation to reflexive sociology*. Chicago, IL: The University of Chicago Press.

Brown, A. L. (1992). 'Design experiments: Theoretical and methodological challenges in creating complex interventions in classroom settings', *The Journal of the Learning Sciences* 2, 141–178.

Brown, J. S. & Duguid, P. (1992). 'Enacting design for the workplace'. In P. S. Adler & T. A. Winograd (Eds.), *Usability: Turning technologies into tools*. New York: Oxford University Press, pp. 164–197.

Brown, J. S., Collins, A. & Duguid, P. (1989). 'Situated cognition and the culture of learning', *Educational Researcher* 18(1), 32–42.

Callon, M. (1986). 'Some elements of a sociology of translation: Domestication of the scallops and the fishermen of St. Brieux Bay'. In J. Law (Ed.), *Power, action and belief: A new sociology of knowledge?*. London: Routledge & Kegan Paul, pp. 196–233.

Callon, M. (1987). 'Society in the making: The study of technology as a tool for sociological analysis'. In W. E. Bijker, T. P. Hughes & T. J. Pinch (Eds.), *The social construction of technological systems*. Cambridge, MA: MIT Press, pp. 83–103.

Callon, M. (1991). 'Techno-economic networks and irreversibility'. In J. Law (Ed.), *A sociology of monsters: Essays on power, technology and domination*. London and New York: Routledge, pp. 132–161.

Callon, M. (1994). 'Is science a public good?', Fifth Mullins lecture, Virginia Polytechnic Institute, 23 March 1993, *Science, Technology & Human Values* 19, 395–424.

Chapman, D. (1991). *Vision, instruction, and action*. Cambridge, MA and London, England: MIT Press.

Collins, A., Brown, J. S. & Newman, S. (1989). Cognitive apprenticeship: Teaching the crafts of reading, writing, and mathematics'. In L. Resnick (Ed.), *Knowing, learning and instruction: Essays in honor of Robert Glaser*. Hillsdale, NJ: Lawrence Erlbaum Associates, pp. 453–494.

Collins, A., Hawkins, J. & Frederiksen, J. R. (1993). 'Three different views of students: The role of technology in assessing student performance', *The Journal of the Learning Sciences* 3, 205–217.

Collins, H. M. (1982). 'Tacit knowledge and scientific networks'. In B. Barnes & D. Edge (Eds.), *Science in context: Readings in the sociology of science*. Cambridge, MA: MIT Press, pp. 44–64.

Constant, E. W., II. (1984). 'Communities and hierarchies: Structure in the practice of science and technology'. In R. Laudan (Ed.), *The nature of technological knowledge. Are models of scientific change relevant?*. Dordrecht, NL: D. Reidel Publishing, pp. 27–46.

Constant, E. W., II. (1989). 'Cause or consequence: Science, technology, and regulatory change in the oil business in Texas, 1930–1975', *Technology and Culture* 30, 426–455.

Derrida, J. (1986). *Glas*. Lincoln, NE: University of Nebraska Press.

Doyle, A. C. (1930). *The complete Sherlock Holmes*. Garden City, NY: Doubleday.

Eckert, P. (1989). *Jocks and burnouts: Social categories and identity in the high school*. New York: Teachers College Press.

Eckert, P. (1990). 'Adolescent social categories – information and science learning'. In M. Gardner, J. G. Greeno, F. Reif, A. H. Schoenfeld, A. diSessa & E. Stage (Eds.), *Toward a scientific practice of science education*. Hillsdale, NJ: Lawrence Erlbaum Associates, pp. 203–217.

Eco, U. (1984). *The name of the rose*. Boston, MA: G. K. Hall.

Edwards, D. (1993). 'But what do children really think?: Discourse analysis and conceptual content in children's talk', *Cognition and Instruction* 11, 207–225.

Edwards, D. & Potter, J. (1992). *Discursive psychology*. London: Sage.

Ehn, P. (1992). 'Scandinavian design: On participation and skill'. In P. S. Adler & T. A. Winograd (Eds.), *Usability: Turning technologies into tools*. New York: Oxford University Press, pp. 96–132.

Ehn, P. & Kyng, M. (1991). 'Cardboard computers: Mocking-it-up or hands-on the future'. In J. Greenbaum & M. Kyng (Eds.), *Design at work: Cooperative design of computer systems*. Hillsdale, NJ: Lawrence Erlbaum Associates, pp.169–195.

Eichinger, D. C., Anderson, C. W., Palincsar, A. S. & David, Y. M. (1991, April). *An illustration of the roles of content knowledge, scientific argument, and social norms in collaborative problem solving*. Paper presented at the annual meeting of the American Educational Research Association, Chicago, IL.

Elman, J. L. (1993). 'Learning and development in neural networks: the importance of starting small', *Cognition* 48, 71–99.

Erickson, F. (1982). 'Money tree, lasagna bush, salt and pepper: Social construction of topical cohesion in a conversation among Italian Americans'. In D. Tannen (Ed.), *Analyzing discourse: Text and talk*. Washington, DC: Georgetown University Press, pp. 43–70.

Erickson, F. (1986). 'Qualitative research on teaching'. In M.C. Wittrock (Ed.), *Handbook for research on teaching*, (3rd ed.). New York: Macmillan, pp. 119–161.

Faulkner, W. (1994). 'Conceptualizing knowledge used in innovation: A second look at the science-technology distinction and industrial innovation. *Science, Technology & Human Values* 19, 425–458.

Feyerabend, P. (1975/76). *Against method: Outline of an anarchistic theory of knowledge*. London: NLB; Atlantic Highlands: Humanities Press.

Foucault, M. (1975). *Surveiller et punir: Naissance de la prison* [Discipline and punish: Birth of the prison]. Paris: Gallimard.

Fujimura, J. (1987). 'Constructing doable problems in cancer research: Articulating alignment', *Social Studies of Science* 17, 257–293.

Fujimura, J. (1992). 'Crafting science: Standardized packages, boundary objects, and "translation"'. In A. Pickering (Ed.), *Science as practice and culture*. Chicago, IL: The University of Chicago Press, pp. 168–211.

Gal, S. (1996). 'Footholds for design'. In T. Winograd (Ed.), *Bringing design to software*. New York, NY: ACM Press, pp. 215–227.

Garfinkel, H. (1967). *Studies in ethnomethodology*. Englewood Cliffs, NJ: Prentice-Hall.

Garfinkel, H. (1991). 'Respecification: evidence for locally produced naturally accountable phenomena of order*, logic, reason, meaning, method, etc. in an as of the essential haecceity of immortal ordinary society', (I) – 'An announcement of studies'. In G. Button (Ed.), *Ethnomethodology and the human sciences*. Cambridge: Cambridge University Press, pp. 10–19.

Garfinkel, H. & Sacks, H. (1986). 'On formal structures of practical action'. In H. Garfinkel (Ed.), *Ethnomethodological studies of work*. London: Routledge & Kegan Paul, pp. 160–193.

Gooding, D. (1990). *Experiment and the making of meaning: Human agency in scientific observation and experiment*. Dordrecht: Kluwer Academic Publishers.

Gooding, D. (1992). 'Putting agency back into experiment'. In A. Pickering (Ed.), *Science as practice and culture*. Chicago, IL: The University of Chicago Press, pp. 65–112.

Greeno, J. G. (1991). 'Number sense as situated knowing in a conceptual domain', *Journal for Research in Mathematics Teaching* 22, 170–218.

Guba, E. & Lincoln, Y. (1989). *Fourth generation evaluation*. Beverly Hills, CA: Sage.

Hanson, N. R. (1965). *Patterns of discovery: An inquiry into the conceptual foundations of science*. Cambridge: Cambridge University Press.

Harel, I. (1991). *Children designers: Interdisciplinary constructions for learning and knowing mathematics in a computer-rich school*. Norwood, NJ: Ablex.

Harel, I. & Papert, S. (Eds.). (1991). *Constructionism: Research reports and essays, 1985–1990*. Norwood, NJ: Ablex.

Harlen, W. (1985). *Primary science: Taking the plunge*. London, UK: Heineman.

Heidegger, M. (1959). *Gelassenheit* [Discourse on thinking]. Pfullingen: Verlag Günther Neske.

Heidegger, M. (1971). *Poetry, language, thought* (Trans. A. Hofstadter). New York, NY: Harper & Row.

Heidegger, M. (1977). *Sein und zeit* [Being and time]. Tübingen, Germany: Max Niemeyer.

Henderson, K. (1991). 'Flexible sketches and inflexible data bases: Visual communication, conscription devices, and boundary objects in design engineering. *Science, Technology & Human Values* 16, 448–473.

Hutchins, E. (1995). *Cognition in the wild*. Cambridge, MA: MIT Press.

Jayyusi, L. (1991). 'Values and moral judgement: communicative practice as a moral order'. In G. Button (Ed.), *Ethnomethodology and the human sciences*. Cambridge: Cambridge University Press, pp. 227–251.

Johnson, R. T., Brooker, C., Stutzman, J., Hultman, D. & Johnson, D. W. (1985). 'The effects of controversy, concurrence seeking, and individualistic learning on achievement and attitude change', *Journal of Research in Science Teaching* 22, 197–205.

Johnson, R. T., Johnson, D. W., Scott, L. E. & Ramolae, B. A. (1985). 'Effects of single-sex and mixed-sex cooperative interaction on science achievement and attitudes and cross-handicap and cross-sex relationships', *Journal of Research in Science Teaching* 22, 207–220.

Jordan, B. (1989). 'Cosmopolitical obstetrics: Some insights from the training of traditional midwives', *Social Science in Medicine* 28, 925–944.

Jordan, B. & Henderson, A. (1995). 'Interaction analysis: Foundations and practice', *The Journal of the Learning Sciences* 4, 39–103.

Jordan, K. & Lynch, M. (1993). 'The mainstreaming of a molecular biological tool: A case study of a new technique'. In G. Button (Ed.), *Technology in working order: Studies of work, interaction, and technology*. London and New York: Routledge, pp. 162–178.

Kafai, Y. B. (1994). *Minds in play: Computer game design as a context for children's learning*. Hillsdale, NJ: Lawrence Erbaum Associates.

Knorr-Cetina, K. D. & Amann, K. (1990). 'Image dissection in natural scientific inquiry. Science', *Technology & Human Values* **15**, 259–283.

Kuhn, T. S. (1970). *The structure of scientific revolutions* (2nd ed.). Chicago: The University of Chicago Press.

Latour, B. (1987). *Science in action: How to follow scientists and engineers through society.* Milton Keynes: Open University Press.

Latour, B. (1988a). 'Mixing humans and non-humans together: The sociology of a door-closer', *Social Problems* **35**, 298–310.

Latour, B. (1988b). *The pasteurization of France.* Cambridge: Harvard University Press.

Latour, B. (1992). *Aramis ou l'amour des techniques* [Aramis or the love of technology]. Paris: Éditions la Découverte.

Latour, B. (1993). *We have never been modern.* Cambridge, MA: Harvard University Press.

Lave, J. (1988). *Cognition in practice: Mind, mathematics and culture in everyday life.* Cambridge: Cambridge University Press.

Lave, J. (1993). 'The practice of learning'. In S. Chaiklin & J. Lave (Eds.), *Understanding practice: Perspectives on activity and context.* Cambridge: Cambridge University Press, pp. 3–32.

Lave, J. & Wenger, E. (1991). *Situated learning: Legitimate peripheral participation.* Cambridge: Cambridge University Press.

Law, J. (1987). 'On the social explanation of technical change: The case of the Portuguese maritime expansion', *Technology and Culture* **28**, 227–252.

Law, J. (1994). *Organizing modernity.* Oxford, UK: Blackwell.

Law, J. & Callon, M. (1988). 'Engineering and sociology in a military aircraft project: A network analysis of technological change', *Social Problems* **35**, 284–297.

Lee, J. (1991). 'Language and culture: the linguistic analysis of culture'. In G. Button (Ed.), *Ethnomethodology and the human sciences.* Cambridge: Cambridge University Press., pp. 196–226

Lemke, J. L. (1990). *Talking science: Language, learning and values.* Norwood, NJ: Ablex Publishing.

Lemke, J. L. (1995, April). *Emergent agendas in collaborative activity.* Paper presented at the annual meeting of the American Educational Research Association, San Francisco, CA.

Lieberman, A. (1992). 'Commentary: Pushing up from below: Changing schools and universities', *Teachers College Record* **93**, 717–724.

Livingston, E. (1987). *Making sense of ethnomethodology.* London: Routledge & Kegan Paul.

Luff, P. & Heath, C. (1993). 'System use and social organization: Observations on human-computer interaction in an architectural practice'. In G. Button (Ed.), *Technology in working order: Studies of work, interaction, and technology.* London and New York: Routledge, pp. 184–210.

Lynch, M. (1985). *Art and artifact in laboratory science: A study of shop work and shop talk in a laboratory.* London: Routledge and Kegan Paul.

Lynch, M. (1991). 'Method: measurement – ordinary and scientific measurement as ethnomethodological phenomena'. In G. Button (Ed.), *Ethnomethodology and the human sciences.* Cambridge: Cambridge University Press, pp. 77–108.

Lynch, M. & Bogen, D. (1996). *The spectacle of history: Speech, text, and memory at the Iran-contra hearings.* Durham: Duke University Press.

Lynch, M., Livingston, E. & Garfinkel, H. (1983). 'Temporal order in laboratory work'. In K. D. Knorr-Cetina & M. Mulkay (Eds.), *Science observed: Perspectives on the social study of science.* London: Sage Publications, pp. 205–238.

MacKinnon, A. (1993). 'Examining practice to address policy problems in teacher education', *Journal of Educational Policy* **8**, 257–270.

McArthur, D., Stasz, C. & Zmuidzinas, M. (1990). 'Tutoring techniques in algebra', *Cognition and Instruction* **7**, 197–244.

McGinn, M. K. (1995). *Teachers mathematical practices inside and outside their classrooms.* Unpublished Masters thesis, Simon Fraser University.

McGinn, M. K. & Roth, W.-M. (1996). 'Assessing students understandings about levers: better test instruments are not enough', *International Journal of Science Education*.

McGinn, M. K., Roth, W.-M., Boutonné, S. & Woszczyna, C. (1995). 'The transformation of individual and collective knowledge in elementary science classrooms that are organized as knowledge-building communities', *Research in Science Education* 25, 163–189.

Mehan, H. (1993). 'Beneath the skin and between the ears: A case study in the politics of representation'. In S. Chaiklin & J. Lave (Eds.), *Understanding practice: Perspectives on activity and context*. Cambridge, England: Cambridge University Press, pp. 241–268.

Middleton, D. & Edwards, D. (1990). (Eds.). *Collective remembering*. London: Sage.

Nauer, B., Airaud, C., Melaye, S. & Millet, S. (1996). 'Lycées de rêves?' [High schools of dreams?] (V. Vedel, director)'. In P. Nahon & B. Benyamin (Producers), *Envoyé special*. Paris: France 2. (Rebroadcast September 11, 1996, TV5.)

Newman, D., Griffin, P. & Cole, M. (1989). *The construction zone: Working for cognitive change in school*. Cambridge: Cambridge University Press.

Orr, J. E. (1990). 'Sharing knowledge, celebrating identity: Community memory in a service culture'. In D. Middleton & D. Edwards (Eds.), *Collective remembering*. London: Sage, pp. 169–189.

Orsolini, M. (1993). '"Dwarfs do not shoot": An analysis of children's justifications', *Cognition and Instruction* 11, 281–297.

Packer, K. & Webster, A. (1996). 'Patenting culture in science: Reinventing the scientific wheel of credibility. Science', *Technology & Human Values* 21, 427–453.

Pea, R. D. (1993). 'Learning scientific concepts through material and social activities: Conversational analysis meets conceptual change', *Educational Psychologist* 28 , 265–277.

Pickering, A. (1995). *The mangle of practice: Time, agency, and science*. Chicago, IL: The University of Chicago Press.

Pinch, T. J. & Bijker, W. E. (1987). 'The social construction of facts and artifacts: Or how the sociology of science and the sociology of technology might benefit each other'. In W. E. Bijker, T. P. Hughes & T. J. Pinch (Eds.), *The social construction of technological systems*. Cambridge, MA: MIT Press, pp. 17–50.

Pollner, M. (1987). *Mundane reason: Reality in everyday and sociological discourse*. Cambridge: Cambridge University Press.

Poole, D. (1994). 'Routine testing practices and the linguistic construction of knowledge', *Cognition and Instruction* 12, 125–150.

Potter, J. & Wetherell, M. (1987). *Discourse and social psychology: Beyond attitudes and behaviour*. London: Sage Publications.

Quine, W. V. (1987). *Quiddities: An intermittently philosophical dictionary*. Cambridge, MA: The Belknap Press of Harvard University Press.

Rafal, C. T. (1996). 'From co-construction to takeovers: Science talk in a group of four girls', *The Journal of the Learning Sciences* 5, 279–293.

Rawlings, A. (1994). 'The AIDS virus dispute: Awarding priority for the discovery of the Human Immunodeficiency Virus (HIV)', *Science, Technology & Human Values* 19, 342–360.

Roberts, P. (1994). 'The place of design in technology education'. In D. Layton (Ed.), *Innovations in science and technology education*. Paris: UNESCO, pp. 171–179.

Rogoff, B. (1990). *Apprenticeship in thinking: Cognitive development in social context*. New York: Oxford University Press.

Rorty, R. (1989). *Contingency, irony, and solidarity*. Cambridge: Cambridge University Press.

Rorty, R. (1991a). *Objectivity, relativism, and truth: Philosophical papers*, Vol. 1. Cambridge: Cambridge University Press.

Rorty, R. (1991b). *Essays on Heidegger and others: Philosophical papers*, Vol. 2. Cambridge: Cambridge University Press.

Roschelle, J. (1992). 'Learning by collaborating: Convergent conceptual change', *The Journal of the Learning Sciences* 2, 235–276.

Roth, W.-M. (1994a). 'Experimenting in a constructivist high school physics laboratory', *Journal of Research in Science Teaching* 31, 197–223.

Roth, W.-M. (1994b). 'Student views of collaborative concept mapping: An emancipatory research project', *Science Education* **78**, 1–34.

Roth, W.-M. (1995a). 'Affordances of computers in teacher-student interactions: The case of Interactive Physics™', *Journal of Research in Science Teaching* **32**, 329–347.

Roth, W.-M. (1995b). *Authentic school science: Knowing and learning in open-inquiry laboratories*. Dordrecht, Netherlands: Kluwer Academic Publishing.

Roth, W.-M. (1995c). 'From "wiggly structures" to "unshaky towers": Problem framing, solution finding, and negotiation of courses of actions during a civil engineering unit for elementary students', *Research in Science Education* **25**, 365–381.

Roth, W.-M. (1995d). 'Inventors, copycats, and everyone else: The emergence of shared (arti)-facts and concepts as defining aspects of classroom communities', *Science Education* **79**, 475–502.

Roth, W.-M. (1996a). 'Art and artifact of children's designing: A situated cognition perspective', *The Journal of the Learning Sciences* **5**, 129–166.

Roth, W.-M. (1996b). 'Engineering talk in a grade 4/5 classroom: A pre- and post-unit assessment', *International Journal of Technology and Design Education* **6**, 107–135.

Roth, W.-M. (1996c). 'Knowledge diffusion* in a Grade 4–5 classroom during a unit on civil engineering: An analysis of a classroom community in terms of its changing resources and practices', *Cognition and Instruction*, **14**, 179–220.

Roth, W.-M. (1996d, March). *Learning to teach through participation in practice*. Paper presented at the annual meeting of the National Science Teachers Association, St. Louis, MO.

Roth, W.-M. (1996e). 'Teacher questioning in an open-inquiry learning environment: Interactions of context, content, and student responses', *Journal of Research in Science Teaching* **33**, 709–736.

Roth, W.-M. (1996f). 'The co-evolution of situated language and physics knowing', *Journal of Science Education and Technology* **3**, 171–191.

Roth, W.-M. (1996g). 'Thinking with hands, eyes, and signs: Multimodal science talk in a grade 6/7 unit on simple machines', *Interactive Learning Environments* **4**, 170–187.

Roth, W.-M. (1996h). 'Unterricht über einfache Maschinen im 6. und 7. Schuljahr – geplant und analysiert aus einer sozial-konstruktivistischen Perspektive des Lernens' [Learning about simple machines in Grade 6/7: A social-constructivist perspective]. *Zeitschrift für Didaktik der Naturwissenschaften* **2**, 39–52.

Roth, W.-M. (1997a). 'Interactional structures during a grade 4–5 open-design engineering unit', *Journal of Research in Science Teaching* **34**, 273–302.

Roth, W.-M. (1997b). *Situated cognition and assessment of competence in science*. Paper presented at the 7th International Conference of the European Association of Research in Learning and Instruction, Athens, Greece.

Roth, W-M. (1998). Science teaching as knowledgeability: A case study of knowing and learning during coteaching. *Science Education* **82**.

Roth, W.-M. & Bowen, G. M. (1993). 'An investigation of problem solving in the context of a Grade 8 open-inquiry science program', *The Journal of the Learning Sciences* **3**, 165–204.

Roth, W.-M. & Bowen, G. M. (1995). 'Knowing and interacting: A study of culture, practices, and resources in a Grade 8 open-inquiry science classroom guided by a cognitive apprenticeship metaphor', *Cognition and Instruction* **13**, 73–128.

Roth, W.-M. & Duit, R. (1998). 'Knowing and learning in real time: Towards an understanding of student-centered science activities', *Cognitive Science*.

Roth, W.-M. & McGinn, M. K. (1996, July). 'Differential participation during science conversations: The interaction of display artifacts, social configuration, and physical arrangements'. In D. C. Edelson & E. A. Domeshek (Eds.), *Proceedings of ICLS 96*. Charlottesville, VA: Association for the Advancement of Computing in Education, pp. 300–307.

Roth, W.-M., McRobbie, C., Lucas, K. B. & Boutonné, S. (1997a). 'The local production of order in traditional science laboratories: A phenomenological analysis', *Learning and Instruction* **7**, 107–136.

Roth, W.-M., McRobbie, C., Lucas, K. B. & Boutonné, S. (1997b). 'Why do students fail to learn from demonstrations? A social practice perspective on learning in physics', *Journal of Research in Science Teaching* **34**, 509–533.

Roth, W.-M. & Roychoudhury, A. (1992). 'The social construction of scientific concepts or the concept map as conscription device and tool for social thinking in high school science', *Science Education* **76**, 531–557.

Roth, W.-M. & Roychoudhury, A. (1993a). 'The concept map as a tool for the collaborative construction of knowledge: A microanalysis of high school physics students', *Journal of Research in Science Teaching* **30**, 503–534.

Roth, W.-M. & Roychoudhury, A. (1993b). 'The development of science process skills in authentic contexts', *Journal of Research in Science Teaching* **30**, 127–152.

Roth, W.-M., Woszczyna, C. & Smith, G. (1996). 'Affordances and constraints of computers in science education', *Journal of Research in Science Teaching* **33**, 995–1017.

Royal College of Art (1976). *Design in general education.* Part One: *Summary of findings and recommendations.* London, Royal College of Art. (Report of an inquiry conducted by the Royal College of Art for the Secretary of State of Education and Science.)

Saxe, G. B. (1991). *Culture and cognitive development: Studies in mathematical understanding.* Hillsdale, NJ: Lawrence Erlbaum Associates.

Scardamalia, M. & Bereiter, C. (1994). 'Computer support for knowledge-building communities', *The Journal of the Learning Sciences* **3**, 265–283.

Schank, R. C. (1994). 'Goal-based scenarios: A radical look at education', *The Journal of the Learning Sciences* **3**, 429–453.

Schank, R. C., Fano, A., Bell, B. & Jona, M. (1994). 'The design of goal-based scenarios', *The Journal of the Learning Sciences* **3**, 305–345.

Schoenfeld, A. (1989). 'Ideas in the air. Speculations on small-group learning, environmental and cultural influences on cognition and epistemology', *International Journal of Educational Research* **13**, 71–88.

Schön, D. A. (1983). *The reflective practitioner: How professionals think in action.* New York: Basic Books.

Schön, D. A. (1987). *Educating the reflective practitioner.* San Francisco: Jossey-Bass.

Schön, D. A. & Bennett, J. (1996). 'Reflective conversation with materials'. In T. Winograd (Ed.), *Bringing design to software.* New York, NY: ACM Press, pp. 171–184.

Schot, J. W. (1992). 'Constructive technology assessment and technology dynamics: The case of clean technologies', *Science, Technology & Human Values* **17**, 36–56.

Sørenson, K. H. & Levold, N. (1992). 'Tacit networks, heterogeneous engineers, and embodied technology', *Science, Technology & Human Values* **17**, 13–35.

Star, S. L. (1991). 'Power, technology and the phenomenology of conventions: on being allergic to onions'. In J. Law (Ed.), *A sociology of monsters: Essays on power, technology and domination.* London and New York: Routledge, pp. 26–56.

Suchman, L. A. (1987). *Plans and situated actions: The problem of human-machine communication.* Cambridge: Cambridge University Press.

Suchman, L. A. & Trigg, R. H. (1993). 'Artificial intelligence as craftwork'. In S. Chaiklin & J. Lave (Eds.), *Understanding practice: Perspectives on activity and context.* Cambridge: Cambridge University Press, pp. 144–178.

Tobin, K. (1984). 'Effects of extended wait time on discourse characteristics and achievement in middle school grades', *Journal of Research in Science Teaching* **21**, 779–791.

Toulmin, S. (1982). 'The construal of reality: Criticism in modern and post-modern science', *Critical Inquiry* **9**, 93–111.

Traweek, S. (1988). Beamtimes and lifetimes: The world of high energy physicists. Cambridge, MA: MIT Press.

Turkle, S. & Papert, S. (1992). 'Epistemological pluralism and the revaluation of the concrete', *Journal of Mathematical Behavior* **11**, 3–33.

Walsh, J. P. & Bayma, T. (1996). 'Computer networks and scientific work', *Social Studies of Science* **26**, 661–703.

Watson, J. D. (1968). *Double helix.* New York: Atheneum Publications.

Whalley, P. (1991). 'The social practice of independent inventing'. *Science, Technology & Human Values* **16**, 208–232.

Wilenski, U. (1991). 'Abstract meditations on the concrete and concrete implications for mathematics education'. In I. Harel & S. Papert (Eds.), *Constructionism: Research reports and essays*, 1985–1990. Norwood, NJ: Ablex, pp. 193–203.

Winograd, T. (1995). 'Heidegger and the design of computer systems'. In A. Feenberg & A. Hannay (Eds.), *Technology and the polititics of knowledge*. Bloomington, IN: Indiana University Press, pp. 108–127.

Winograd, T. (Ed.). (1996). *Bringing design to software*. New York, NY: ACM Press.

Winograd, T. & Flores, F. (1987). *Understanding computers and cognition: A new foundation for design*. Norwood, NJ: Ablex.

Wittgenstein, L. (1994/1958). *Philosophical investigations* (3rd ed.). New York: Macmillan.

Woolgar, S. (1990). 'Time and documents in researcher interaction: Some ways of making out what is happening in experimental science'. In M. Lynch & S. Woolgar (Eds.), *Representation in scientific practice*. Cambridge, MA: MIT Press, pp. 123–152.

INDEX

Action
- concernful 140f, 174, 240
- situated 4, 10, 155, 179
Activity
- collective viii, x, 12, 95, 195, 229, 250f, 266, 271, 279
Actor network xiif, 14ff, 119, 129, 149ff, 172
Affordance xiii, 15, 21, 129ff, 199, 212, 215, 247, 287
Agent 11, 16, 67, 128f, 139, 141, 195, 215, 229, 233, 239, 245, 247, 294
Anthropology
- symmetric 14
Apprenticeship 12, 139, 145f, 150, 284, 390
Authenticity 13f, 46, 99, 192, 268, 284, 294

Background 10, 18, 44, 74, 83, 195, 231, 299f
Breakdown 91, 143f, 240
- snag 15, 240, 286
- trouble 58, 124, 186, 287, 291

Circulation 14, 92, 101ff, 139ff, 195
Cognition
- situated 66, 75, 199, 247
Community of practice 10ff, 101, 139, 152, 158, 294
Conceptual change 15, 173

Embodiment 126f, 147f
Emergence 9, 50, 100ff, 133, 149, 172, 179, 190, 192, 211ff, 229, 242
Epistemology 9f, 301
Ethnomethodology 10, 38, 108, 222

Familiarity 6ff, 22, 38, 73, 78, 150, 200, 217, 222, 242
Flexibility
- interpretive 76, 150, 229ff, 240, 287
- ontological 5, 188

Heterogeneity 12ff, 66, 74, 92, 143ff, 199f, 225, 247, 283ff
Horizon 17, 67, 93, 212f, 228, 242

Ideology 11f, 284
Indexical 175, 178, 182, 186, 262
Intersubjectivity 228, 271

Learning environment vii, xiii, 12ff, 18f, 54f, 62f, 246, 283f, 296

Mundaneity 6, 10, 72

Ontology xiii, 14, 102, 140, 199, 218ff, 277
Openness 212, 214

Participation
- competent 12
- legitimate peripheral 12, 145, 148, 285
- trajectory xii, 10ff, 132, 144ff, 154ff
Practices
- circulation of 129ff, 154ff
- discursive 3, 35, 73ff, 99f, 103, 152, 172ff, 193f, 288
- material 101f, 129, 138f, 148, 168ff, 180, 193f, 217, 277, 282
- social 4, 6, 126, 149
Problem
- design 32, 40, 66, 94f, 197, 216, 225, 241f, 270
- solving (solution finding) 15, 27ff, 43, 49, 116, 201, 233ff, 280
Production
- cultural xii, 11ff, 14, 40, 101, 147ff, 194f, 248f, 290

Questioning
- technique 54ff, 163f, 173f, 187ff

Reflexivity xiii, 14, 89, 197f, 215, 225, 231, 250, 280, 302
Representation 4ff, 67, 254, 275, 290
Resource
- structuring xiii, 66, 200, 280, 284

312

Resources
- circulation of   101ff, 145ff, 279f, 292
- discursive   10, 76, 93, 103, 133, 154, 287
- material   38, 101ff, 136

Rule
- following   xiii, 10, 200ff, 221ff, 271f, 301f

Scaffolding   43, 51, 54ff, 154, 193ff, 268